This book is an introduction to polymers and focuses on the synthesis, structure and properties of the individual molecules that constitute polymeric materials.

The science and technology connected with polymeric materials has grown into an immense subject. Thus there is a daunting task confronting the person who, wishing to become acquainted with such materials, attempts to master some areas of special interest. This book approaches polymeric materials from a molecular basis in the belief that there is a common core of knowledge and principles concerning polymer molecules that can be set out in an introductory work.

Subjects treated include an introductory overview of synthesis, an introduction of the concept and definition of molecular weight and its distribution, experimental methods for measuring molecular weight, a more detailed view of polymerization including kinetics and mechanism, and the three dimensional architecture of polymers as determined by conformation and stereochemistry. Since much of how polymers behave depends on the fact that the molecules can be conformationally disordered, the statistical description of this disorder is covered and then built upon in treating rubber elasticity and polymer solutions.

The text is geared for an introductory-level graduate course in polymers for students of chemistry, chemical engineering and materials science. In addition to its primary use as a textbook to accompany a formal course of lectures, the book can also be used as a self-study guide to learning the principles covered.

Titles in print in this series

THE SCIENCE OF POLYMER MOLECULES

© 1990 by Sidney Harris – *American Scientist* magazine

Polymer molecules often form random coils in solution or in melts. An important result in the statistical treatment of this disorder is that, for Gaussian chains, the mean-square end-to-end distance is six times the mean-square radius of gyration, Section 6.2.2..

THE SCIENCE OF
POLYMER MOLECULES

An introduction concerning the synthesis, structure
and properties of the individual molecules that
constitute polymeric materials

RICHARD H. BOYD

*Distinguished Professor of Materials Science and Engineering
and of Chemical Engineering, University of Utah*

PAUL J. PHILLIPS

*Professor of Materials Science and Engineering,
University of Tennessee*

CAMBRIDGE
UNIVERSITY PRESS

Published by the Press Syndicate of the University of Cambridge
The Pitt Building, Trumpington Street, Cambridge CB2 1RP
40 West 20th Street, New York, NY 10011-4211, USA
10 Stamford Road, Oakleigh, Melbourne 3166, Australia

© Cambridge University Press 1993

First published 1993

Printed in Great Britain at the University Press, Cambridge

A catalogue record for this book is available from the British Library

Library of Congress cataloguing in publication data available

ISBN 0 521 32076 3   hardback

KW

Contents

Preface

The science and technology connected with polymeric materials has
grown into an immense subject. It is not possible in any single work to
cover the field in useful detail. Thus there is a daunting task confronting
the person, who, wishing to become acquainted with such materials,
attempts to master some of the areas of his or her special interest. It is
our belief that polymeric materials are best understood from a molecular
basis and that there is a common core of knowledge and principles
concerning polymer molecules that can be set out in a single introductory
work.

We have taken the viewpoint that an introduction or textbook should
undertake to explain and develop the principles selected and not just
present results. That means, for most of the subjects, we have proceeded
from a very elementary starting point and presented in fair detail the
steps. The goal has been to arrive at a point where the reader or student
can understand the principles and profitably read the literature connected
with that subject.

A number of subjects have been selected based on answering the
questions: 'how are polymers made?', 'what do they look like?' and 'how
do they behave?' With respect to the third question we have deliberately
stayed away from properties associated directly with the aggregation of
polymer molecules in bulk materials. It is of course the interest in bulk
materials that is the basic motivation of many, if not most, of the readers
and students we hope to reach. However, as stated, it is an unfortunate
fact that a collection of these properties cannot also be included, at the
level of understanding we hope to achieve, in an introductory work of
reasonable size. We do believe that mastering the behavior of material
properties in later study is greatly facilitated by and best proceeds from
a molecular introduction of the type presented here. In fact, we do

comment liberally on where the connections lie between the molecular picture and material behavior.

Our treatment starts out with addressing the question of how polymers are made via an overview of polymerization processes. Then, in the second chapter, the subject of molecular weight is introduced. This includes how averages are defined and how the distribution of molecular lengths is expressed. The subject of branching and gelation is introduced. The experimental methods available for measuring molecular weight are covered in the next chapter. The polymerization process is returned to in the fourth chapter. Much more detail concerning mechanisms and how they affect observables in polymerization is covered. With the chemistry of polymerization in hand, the structure of the molecules thus made is then addressed in the next chapter. This necessarily invokes the three dimensional aspects of what the molecules look like. This includes the subjects of molecular conformation and the effects of stereochemistry on conformation. With the introduction of conformational behavior it becomes apparent that one of the most important aspects, if not the most important one, of polymers concerns their ability to take up many different conformations or spatial configurations. The rest of the chapters deal with subjects intimately related to the configurational disorder associated with polymers.

The sixth chapter introduces the basics of the statistical description of conformationally disordered polymer chains. One of the important advances in polymer science has been the development of statistical methods that connect averaged properties, such as average dimensions, with the realistic details of the conformational behavior of a chain. These methods are a natural outgrowth of the basic concepts developed in Chapter 6. They are sufficiently complex and mathematical in nature that they are best presented separately, in Chapter 7. Thus they can be studied or not depending on the interests of the reader. Rubber elasticity is covered next, in the eighth chapter. This would seem to be a contradiction to the statement that bulk material properties are not taken up. However, the conceptual underpinnings of rubber elasticity are based on the statistical description of polymer chains. The classical molecular theories have essentially been based on the statistics of single chain segments. Considerable attention is paid to clearly presenting these theories so that some of the confusion that has arisen around them historically is, hopefully, resolved. Finally, polymer solutions are considered.

Like many introductions or texts, this one is the outgrowth of a course of lectures. It is based on an introductory course for beginning graduate students who are interested in learning about polymers. Many of them have continued studying polymeric materials in more depth, including bulk material behavior and engineering properties. Although it has been hoped that the students would have had an undergraduate introduction to physical chemistry, and organic chemistry as well, many students from an entirely engineering background have been very successful in mastering the subjects covered. The lectures have been formally offered to students from chemistry, chemical engineering and materials science. Mechanical engineering and bioengineering students have been regularly represented also.

Although the book can be used as a textbook in a formal course of lectures it is also specifically our intention that the term 'Introduction' means that it can serve outside this venue as a self-study guide to learning the principles covered. It is our hope that it will be useful in this context as well.

Salt Lake City　　　　　　　　　　　　　　　　*R. H. Boyd*
Knoxville　　　　　　　　　　　　　　　　　*P. J. Phillips*

1
Polymerization: an overview

Polymers are large molecules made up of many atoms linked together by covalent bonds. They usually contain carbon and often other atoms such as hydrogen, oxygen, nitrogen, halogens and so forth. Thus they are typically molecules considered to be in the province of organic chemistry. Implicit in the definition of a polymer is the presumption that it was synthesized by linking together in some systematic way groups of simpler building block molecules or monomers. Although the final molecular topology need not be entirely linear, it is usually the case that the linking process results in linear segments or imparts a chain-like character to the polymer molecule.

Most of the synthetic methods for linking together the building block molecules can be placed into one of two general classifications. The first of these results when the starting monomers react in such a way that groups of them that have already joined can react with other already joined groups. The linked groups have almost the same reactivity towards further reaction and linking together as the original monomers. This general class of reactions is called *step* polymerization. In the other general method, an especially reactive center is created and that center can react only with the original monomer molecules. Upon reaction and incorporating a monomer, the reactive center is maintained and can keep reacting with monomers, linking them together, until some other process interferes. This mechanism is *chain* polymerization.

1.1 Step polymerization

Aliphatic polyesters are interesting examples of step polymerization as historically they were the first such polymerizations to be carried out and studied in a systematic way (by W. H. Carothers, see Mark and Whitby,

1940). One might start with a dibasic acid such as adipic acid and a diol such as ethylene glycol. They can react to form an ester linkage and eliminate a water molecule (**1.1**). After linking, however, carboxyl groups

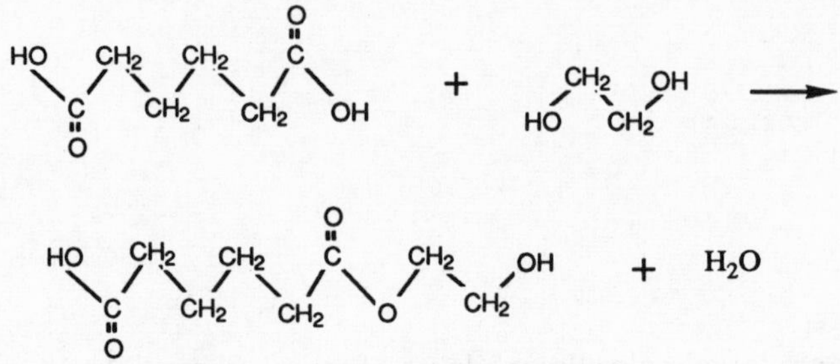

1.1

and hydroxyl groups remain for further reaction and in fact the reaction can continue until very long sequences of ester linkages are built up (**1.2**).

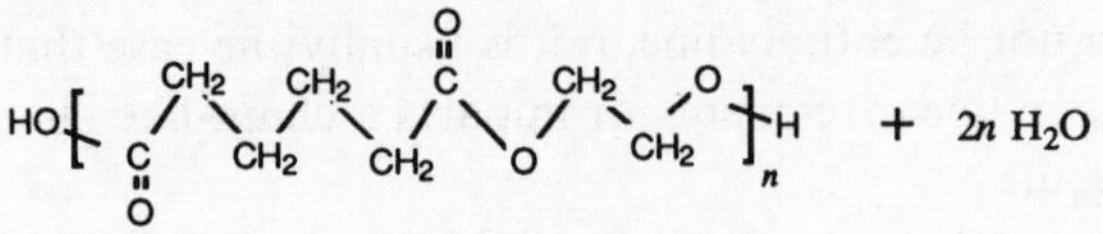

1.2

Molecular length builds up gradually through the course of polymerization (Figure 1.1). Accomplishing the building of long sequences requires that the numbers of carboxyl and hydroxyl groups be closely balanced, that the reaction proceed at a reasonable rate and that the by-product water be efficiently removed. These questions will be expanded on later in discussing molecular weight and the kinetics of polymerization (Chapters 2, 4).

The aliphatic polyesters of the type illustrated form crystalline solids and, as indicated above, they were seminal in demonstrating the viability of the concept of step polymerization and indeed of the concept that very large chain molecules held together by primary covalent bonds could be synthesized by conventional well-known chemical reactions. As a practical matter it turned out that the melting points of such polymers, only about 60 °C, were too low for significant applications. However, by replacing the diol in the reaction with a diamine, similar condensation takes place

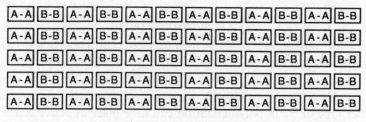

a group of unpolymerized monomers

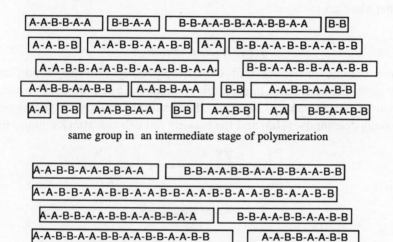

same group in an intermediate stage of polymerization

same group in a later stage of polymerization

Fig. 1.1 Molecular weight build-up in a step polymerization.

with the formation of amide linkages and elimination of water molecules (Mark and Whitby, 1940). Examples are items 6 and 7 in Appendix 1.1. Linear aliphatic polyamides were found to be crystalline solids also but the melting points were much higher, in the neighborhood of 260 °C. These materials could be drawn into strong fibers and thus immediately found important applications in this context, i.e., as 'nylon' and later as moulding resins and to some extent as films. There are now also a number of examples of polyesters as useful materials. For example, polymerization of terephthalic acid with ethylene glycol, item 10, Appendix 1.1, gives a high melting polymer useful as a fiber, as film and as a moulding resin.

1.2 Chain polymerization

As stated above, in this mechanism a reactive center is created that will react with the monomeric building blocks in such a way that the center is maintained and can continue to react with monomers. The reactive moiety can involve a free radical center, an anionic or a cationic species. Another type of chain polymerization can take place at active centers on a catalyst that might involve metal ions acting as a coordination catalyst rather than an initiator.

1.2.1 Radical initiation

Some organic molecules that contain especially unstable bonds can thermally dissociate into appreciable numbers of free radicals. Peroxides are an important class of examples. For example, tertiary butyl peroxide partially decomposes (**1.3**) and the ensuing radicals can attack a monomer

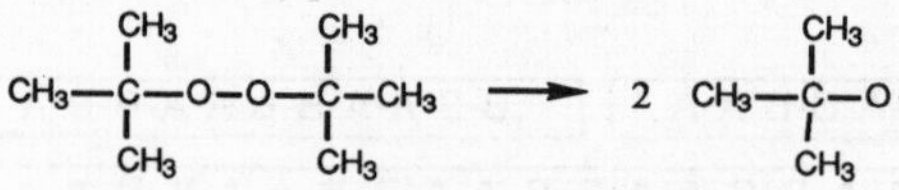

initiator decomposition

1.3

such as styrene (**1.4**). The unpaired electron is now centered on the

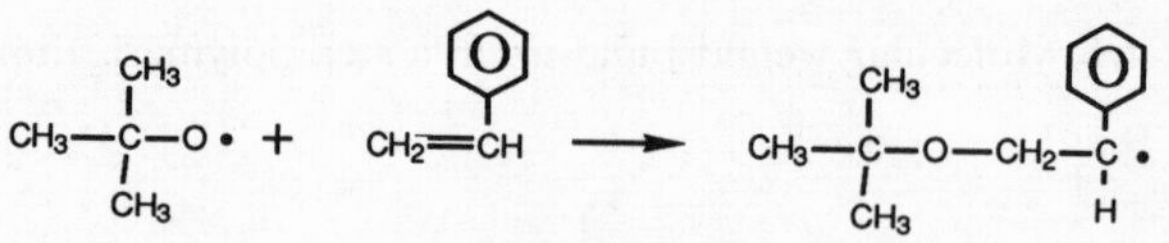

initiation

1.4

carbon containing the phenyl ring because this radical is more stable than the alternative position on the methylene carbon ($\bullet CH_2 -$) of the styrene. The free radical center can continue to add styrene units to build up long sequences (**1.5**). The unpaired electron will continue to reside on the phenyl-ring-containing carbon as it is still the more stable position. Thus the polymerization is very selective with respect to head-to-tail linking and the resulting polymer has the pendant phenyl rings on alternating chain carbon atoms.

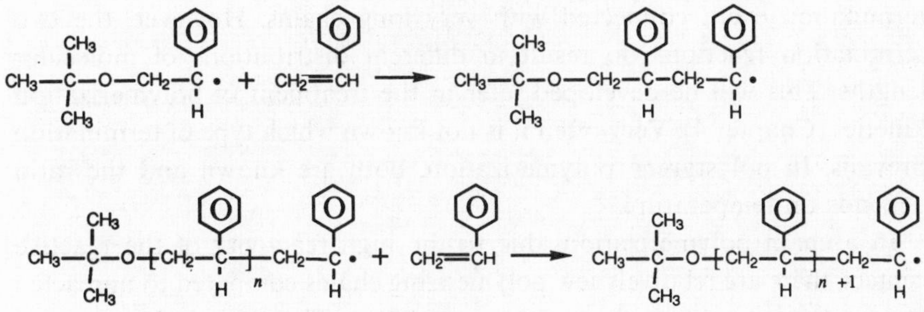

propagation

1.5

Notice that the peroxide, although it causes an otherwise unreactive system to proceed to polymerize, is chemically combined with the polymer chains and is thus consumed. It is therefore called an *initiator* rather than a catalyst. The polymerization step could continue, in principle, until all the monomers are consumed. However, an interfering reaction usually takes place. The free radical centers on two growing chains can react with each other to stop the polymerization. One way this can happen is readily apparent; two radical ends can simply combine directly to form a bond joining the two growing chains into one and stopping further polymerization. A less obvious but competitive reaction results from a radical end abstracting a hydrogen alpha to the center on another chain and forming *two* stable non-growing chains (**1.6**). The direct structural differences between the two termination reactions are minimal since there is only one

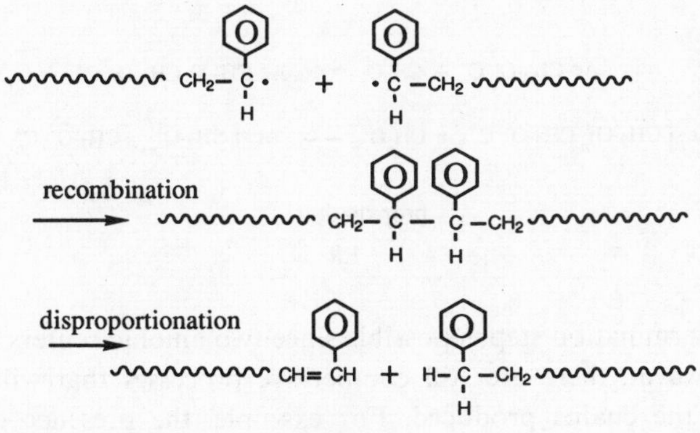

recombination

disproportionation

termination

1.6

termination event connected with very long chains. However, the two termination reactions do result in different distributions of molecular lengths. This will be developed later in the treatment of polymerization kinetics (Chapter 4). Very often it is not known which type of termination prevails. In polystyrene polymerization, both are known and the ratio depends on temperature.

In a chain polymerization, due to the high reactivity of the reactive centers, there are relatively few polymerizing chains compared to unreacted free monomer or finished terminated chains. Thus the reaction vessel throughout the polymerization contains largely only these latter entities, monomer and rather long completed chains, until all monomer is consumed (Figure 1.2). This is in contrast with step polymerization where the chain lengths increase through the course of polymerization.

1.2.2 Anionic initiation

Some monomers are capable of forming relatively stable anions and are thus susceptible to anionic initiation. For example, an acetate ion, Ac^-, introduced as a salt, C^+Ac^-, of a generic cation C^+, can attack formaldehyde (**1.7**), and the anionic center created can continue to

$$C^+ Ac^- + CH_2O \longrightarrow Ac\text{-}CH_2\text{-}O^-\ C^+$$

initiation

1.7

polymerize to form polyoxymethylene (**1.8**). In contrast to the radical case,

$$Ac\text{-}CH_2\text{-}O^-\ C^+ + CH_2O \longrightarrow Ac\text{-}CH_2\text{-}O\text{-}CH_2\text{-}O^-\ C^+$$

$$Ac\text{-}[\,CH_2\text{-}O\,]_n\text{-}CH_2\text{-}O^-\ C^+ + CH_2O \longrightarrow Ac\text{-}[\,CH_2\text{-}O\,]_{n+1}\text{-}CH_2\text{-}O^-\ C^+$$

propagation

1.8

no direct termination step is possible since two anionic centers would not react. However, there are still competitive processes that will limit the length of the chains produced. For example, the presence of a small amount of water in the system can result in reactions that stop the *physical* chain from growing but allow the reactive center to persist and keep the

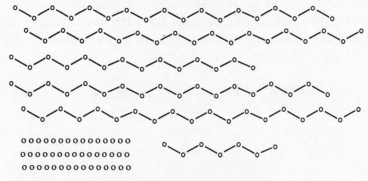

monomers before polymerization

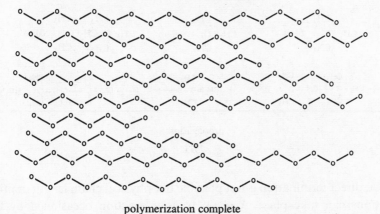

intermediate stage of polymerization

polymerization complete

Fig. 1.2 Molecular weight build-up in a chain polymerization.

kinetic chain alive and growing (see **1.9**). This phenomenon is known as *chain transfer*. In one form or another it is common in chain polymerizations, see Chapter 4.

$$\text{Ac -[CH}_2\text{-O]}_n\text{-CH}_2\text{-O}^-\ \text{C}^+ + \text{H}_2\text{O} \longrightarrow \text{Ac -[CH}_2\text{-O]}_n\text{-CH}_2\text{-OH} + \text{C}^+ \text{OH}^-$$

$$\text{C}^+ \text{OH}^- + \text{CH}_2\text{O} \longrightarrow \text{HO-CH}_2\text{-O}^-\ \text{C}^+$$

chain transfer

1.9

1.2.3 Cationic initiation

The reactive center can be cationic as well. As an example, the carbon–carbon double bond can be protonated by a very strong acid and a cationic center created. The Lewis acid BF_3 forms a very strong acid with halogen acids HX, $BF_3X^-H^+$. This acid can protonate isobutene (or 'isobutylene') to create a reactive cationic center at the tertiary carbon (**1.10**) and polymerization ensues (**1.11**). As in the case of an anionic

$$BF_3X^-H^+ \ + \ CH_2::C\overset{CH_3}{\underset{CH_3}{}} \quad \longrightarrow \quad CH_3:C^+\overset{CH_3}{\underset{CH_3}{}} \quad BF_3X^-$$

initiation

1.10

$$CH_3:C^+\overset{CH_3}{\underset{CH_3}{}} BF_3X^- \ + \ CH_2::C\overset{CH_3}{\underset{CH_3}{}} \ \longrightarrow \ CH_3:C:CH_2:C^+\overset{CH_3 \quad CH_3}{\underset{CH_3 \quad CH_3}{}} BF_3X^-$$

$$H\text{--}\!\!\left[\!CH_2\text{--}\underset{CH_3}{\overset{CH_3}{C}}\!\right]_n\!\!CH_2\text{--}\underset{CH_3}{\overset{CH_3}{C^+}} \ BF_3X^- \ + \ CH_2\text{=}\underset{CH_3}{\overset{CH_3}{C}} \ \longrightarrow \ H\text{--}\!\!\left[\!CH_2\text{--}\underset{CH_3}{\overset{CH_3}{C}}\!\right]_{n+1}\!\!CH_2\text{--}\underset{CH_3}{\overset{CH_3}{C^+}} \ BF_3X^-$$

propagation

1.11

center, direct termination is not possible but physical chain length limiting reactions can take place. A chain transfer reaction occasioned by the reversibility of the initiation reaction to regenerate the initiator is shown in **1.12**.

$$H\text{--}\!\!\left[\!CH_2\text{--}\underset{CH_3}{\overset{CH_3}{C}}\!\right]_n\!\!CH_2\text{--}\underset{CH_3}{\overset{CH_3}{C^+}} \ BF_3X^- \ \longrightarrow \ H\text{--}\!\!\left[\!CH_2\text{--}\underset{CH_3}{\overset{CH_3}{C}}\!\right]_n\!\!CH\text{=}\underset{CH_3}{\overset{CH_3}{C}} \ + \ BF_3X^- \ H^+$$

$$BF_3X^- \ H^+ \ + CH_2\text{=}\underset{CH_3}{\overset{CH_3}{C}} \ \longrightarrow \ CH_3\text{-}\underset{CH_3}{\overset{CH_3}{C^+}} \ BF_3X^-$$

chain transfer via reversal of initiation reaction

1.12

1.2.4 Coordination catalysis

The preceding three types of chain polymerization all involve an initiator that creates a reactive center by attacking a monomer and becoming bonded chemically to it. There is a very important class of polymerizations where a true catalyst acts to create the active center. The most important of these involves coordination to a transition metal ion. The catalyst is usually heterogeneous in nature. Historically, important examples are the Ziegler–Natta catalysts. They usually consist of an organometallic compound in concert with the transition metal compound. The combination of triethyl aluminum, $(C_2H_5)_3Al$ ('TEA'), and titanium tetrachloride, $TiCl_4$, is a well-known example.

In an initiated polymerization, usually carried out homogeneously, the initiator itself plays little direct role after the initiation step. The addition of a new monomer in the propagation step depends on the structure and nature of the reactive center at the end of the growing chain and the monomer itself. In contrast, in the catalyzed polymerization the end of the growing chain, the entering monomer and the catalyst are all three bonded or coordinated together in the transition state. This three-party event gives rise to relatively severe and strong steric interactions or restrictions on the way in which the arriving monomer approaches the growing end of the chain. This is to be compared with the less restricted two-party environment of the monomer approaching the reactive center in an initiator induced polymerization. In turn, this restrictive environment gives rise to a phenomenon that tends to be less noticeable in initiated polymerizations. In the initiated polymerizations discussed above, attention was drawn to the strong tendency toward head-to-tail ordering in polymerizations at double bonds. That results from the reactive center having greater stability at one of the two possible positions than at the other. There is another type of ordering possible, in addition to head-to-tail. If the incoming monomer has different *substituents* about the polymerizing bonds there is a choice of orientation of the substituents. For example, in the free radical initiated polymerization of styrene the placement of the phenyl rings versus the hydrogen substituent on the same chain carbon tends to be random (Figure 1.3). However, in the Ziegler–Natta polymerization of propene (or 'propylene') the placement of the substituent methyl groups is very regular in the direction of the substituent methyl group in each monomer as it adds to the chain. Such polymerizations are very *stereoregular* (Figure 1.4). The consequences of this will be taken up in more detail in Chapter 5.

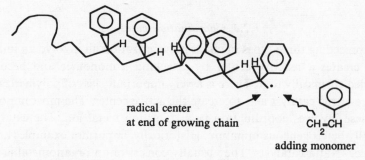

Fig. 1.3 Non-stereospecific radical polymerization of styrene.

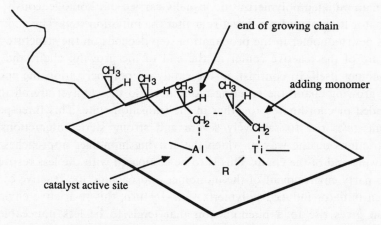

Fig. 1.4 Stereospecific Ziegler–Natta catalyzed polymerization of propene.

1.3 Further reading

The history of the development of the concept of synthesis of high molecular weight chain molecules is quite interesting and in no small part because considerable controversy was involved. Two good accounts are Flory (1953) and Morawetz (1985).

General books on synthesis are Lenz (1967), Odian (1991) and Stevens (1990). The last work has extensive references to other books and reviews and also has a good compendium of continuing serial publications covering the subject. The multivolume series, *Comprehensive Polymer Science*, edited by Allen and Bevington (1989) deals largely with chemistry and synthesis.

Appendix A1.1 Some common polymers

1. 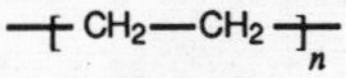 $-\!\!\left[\!CH_2\!-\!CH_2\right]_n$

polyethylene

2. 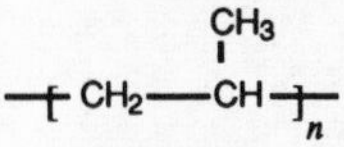 $-\!\!\left[\!CH_2\!-\!\underset{\underset{CH_3}{|}}{CH}\right]_n$

polypropylene

3. 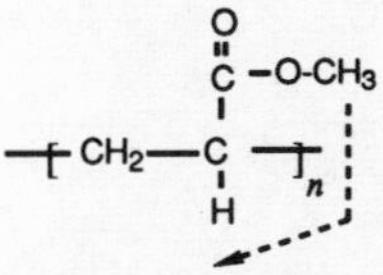 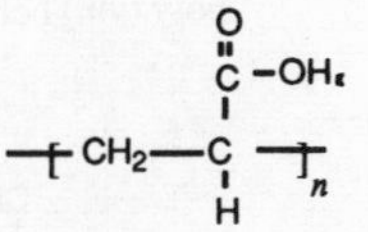 $-\!\!\left[\!CH_2\!-\!\underset{\underset{H}{|}}{\overset{\overset{O}{\overset{||}{C}-O-CH_3}}{C}}\right]_n$ $-\!\!\left[\!CH_2\!-\!\underset{\underset{H}{|}}{\overset{\overset{O}{\overset{||}{C}-OH}}{C}}\right]_n$

poly (methyl acrylate)   poly (acrylic acid)

4. 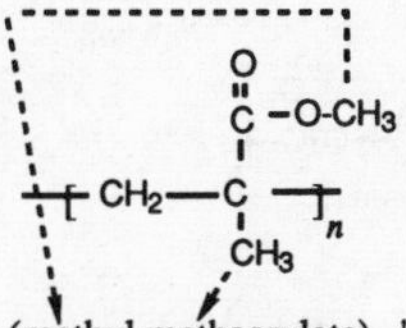 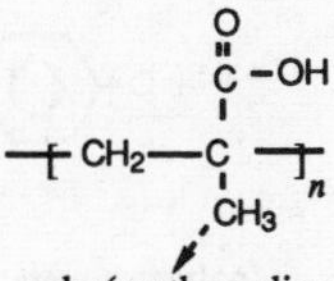 $-\!\!\left[\!CH_2\!-\!\underset{\underset{CH_3}{|}}{\overset{\overset{O}{\overset{||}{C}-O-CH_3}}{C}}\right]_n$ $-\!\!\left[\!CH_2\!-\!\underset{\underset{CH_3}{|}}{\overset{\overset{O}{\overset{||}{C}-OH}}{C}}\right]_n$

poly (methyl methacrylate) 'PMMA'   poly (methacrylic acid)

5. 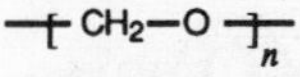 $-\!\!\left[\!CH_2\!-\!O\right]_n$

polyoxymethylene

6. 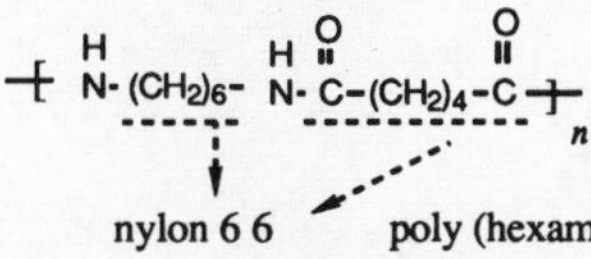 $-\!\!\left[\!\underset{}{\overset{H}{N}}\!\cdot(CH_2)_6\!-\!\underset{}{\overset{H}{N}}\!\cdot\overset{\overset{O}{||}}{C}\!-\!(CH_2)_4\!-\!\overset{\overset{O}{||}}{C}\right]_n$

nylon 6 6   poly (hexamethylene adipamide)

7. 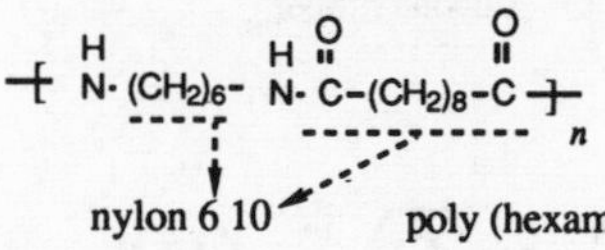 $-\!\!\left[\!\underset{}{\overset{H}{N}}\!\cdot(CH_2)_6\!-\!\underset{}{\overset{H}{N}}\!\cdot\overset{\overset{O}{||}}{C}\!-\!(CH_2)_8\!-\!\overset{\overset{O}{||}}{C}\right]_n$

nylon 6 10   poly (hexamethylene sebacamide)

8. 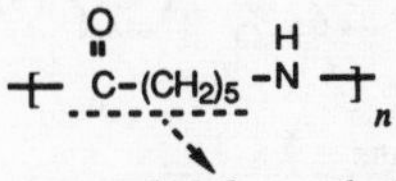 $-\!\!\left[\!\overset{\overset{O}{||}}{C}\!-\!(CH_2)_5\!-\!\overset{H}{N}\right]_n$

nylon 6   poly (caprolactam)

9. 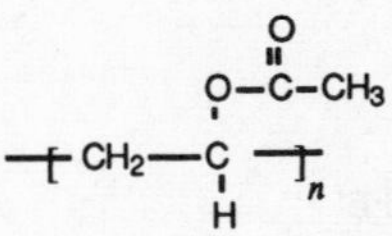 $-\!\!\left[\!CH_2\!-\!\underset{\underset{H}{|}}{\overset{\overset{O-\overset{\overset{O}{||}}{C}-CH_3}{|}}{C}}\right]_n$

poly (vinyl acetate) 'PVAc'

10.

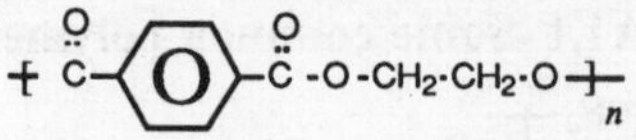

poly (ethylene terephthalate) 'PET'

11.

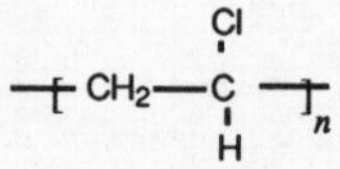

poly (vinyl chloride) 'PVC'

12.

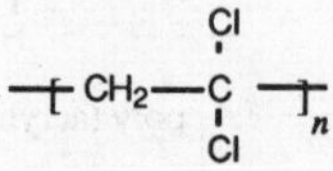

poly (vinylidene chloride)

13.

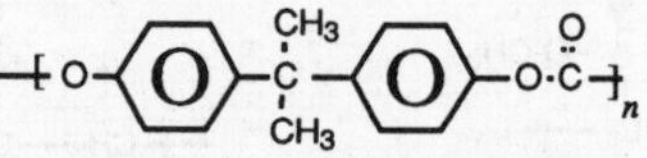

'polycarbonate' $\left(-O-\overset{O}{\overset{\|}{C}}-O- \text{ linkage} \right)$

14.

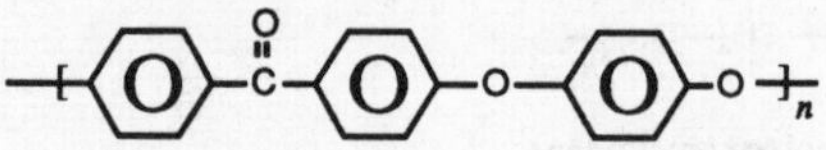

'poly ether ether ketone' , 'PEEK'

15.

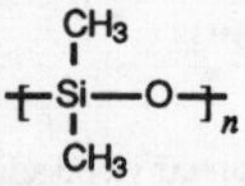

poly (dimethylsiloxane) 'silicone rubber'

16.

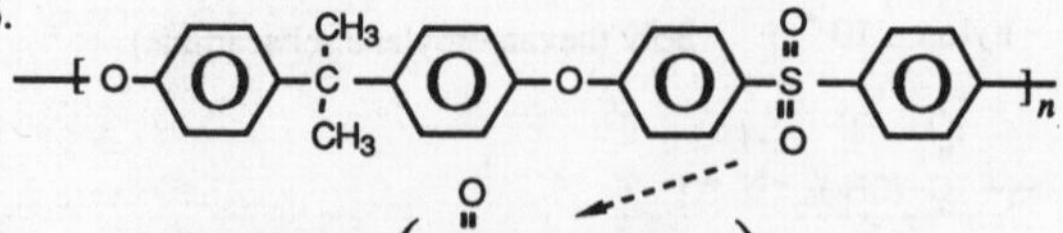

'polysulfone' $\left(-\overset{O}{\underset{O}{\overset{\|}{\underset{\|}{S}}}}- \text{ linkage } \right)$

17.

$$+CF_2-CF_2+_n$$

poly (tetrafluoroethylene) 'PTFE'

18.

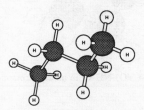

poly (p-phenylene terephthalamide)   'Kevlar ® '

19.

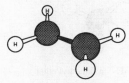

a 'polyimide' (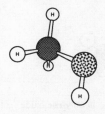 linkage)

Appendix A1.2 Some common functional groups occurring in polymers

tetrahedral sp^3 carbons in CH$_2$ and CH$_3$ groups in butane

planar sp^2 carbons in ethylene, CH$_2$=CH$_2$

hydroxyl group, OH, in methanol

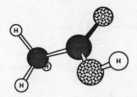

planar carboxyl group (-C=O)-OH in acetic acid

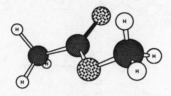

planar ester group, (-C=O)-O-C, in methyl acetate

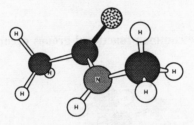

planar amide group, (-C=O)-NH-C, in methyl acetamide.

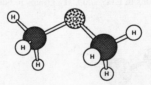

ether linkage -O- in dimethyl ether

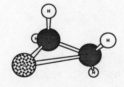

epoxy group, -CH-CH-, in ethylene oxide
                   \ /
                    O

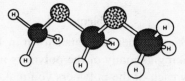

acetal linkage, **-O-CH₂-O-**, in dimethoxymethane

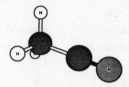

nitrile or cyano group, **-C≡N**, in acetonitrile

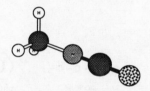

isocyanate group, **-N=C=O** , in methyl isocyanate

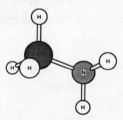

amine group, **-NH₂**, in methyl amine

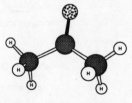

ketone group, **C-(C=O)-C**, in acetone

Problems

1.1　Using monomers and other reagents shown in **1.13**, write chemical reactions for making as many of the polymers in Appendix A1.1 as possible. Classify each of the polymers you make as made by either *step* or *chain* polymerization.

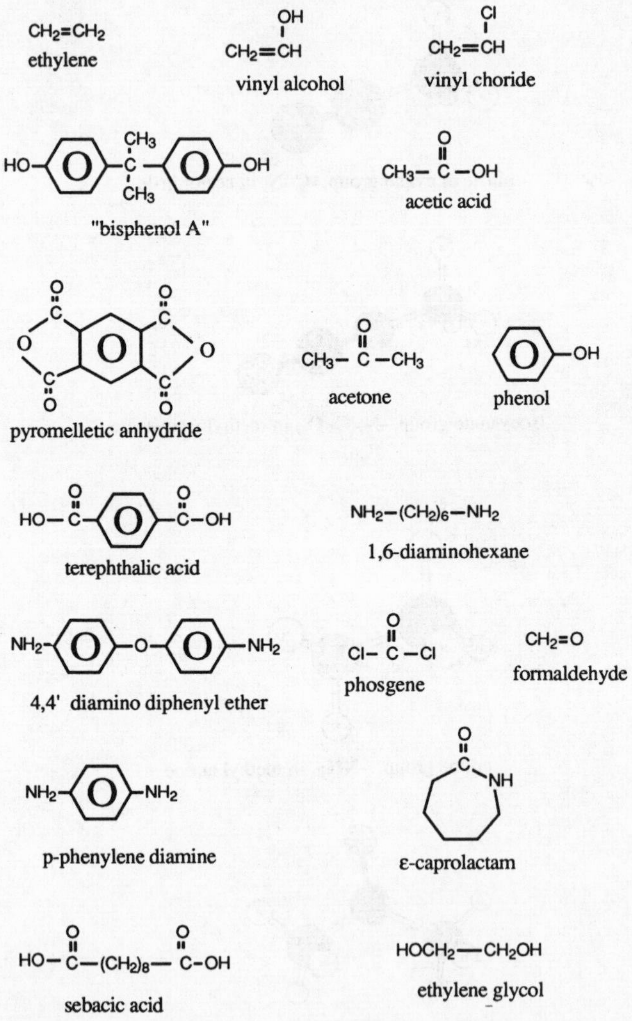

$CH_2{=}CH_2$
ethylene

$CH_2{=}CH$ (with OH)
vinyl alcohol

$CH_2{=}CH$ (with Cl)
vinyl choride

"bisphenol A" (HO–C₆H₄–C(CH₃)₂–C₆H₄–OH)

$CH_3{-}\overset{O}{\underset{}{C}}{-}OH$
acetic acid

pyromelletic anhydride

$CH_3{-}\overset{O}{\underset{}{C}}{-}CH_3$
acetone

phenol (C₆H₅–OH)

$HO{-}\overset{O}{\underset{}{C}}{-}C_6H_4{-}\overset{O}{\underset{}{C}}{-}OH$
terephthalic acid

$NH_2{-}(CH_2)_6{-}NH_2$
1,6-diaminohexane

$NH_2{-}C_6H_4{-}O{-}C_6H_4{-}NH_2$
4,4' diamino diphenyl ether

$Cl{-}\overset{O}{\underset{}{C}}{-}Cl$
phosgene

$CH_2{=}O$
formaldehyde

$NH_2{-}C_6H_4{-}NH_2$
p-phenylene diamine

ε-caprolactam

$HO{-}\overset{O}{\underset{}{C}}{-}(CH_2)_8{-}\overset{O}{\underset{}{C}}{-}OH$
sebacic acid

$HOCH_2{-}CH_2OH$
ethylene glycol

1.13

1.2　As is shown in **1.14** formaldehyde can react (**1**) with phenol in either the *para* or *ortho* positions marked to form methylolphenols.

1.14

Then further reaction (**2**) of the methylolphenols with themselves or with phenol at either *para* or *ortho* positions leads to condensation to form polybenzyl ethers. The latter are thermally unstable and upon heating ('curing') decompose (**3**) leaving a methylene bridge and release formaldehyde for further reaction.

(*a*) By continuing these reactions, using all the ortho and para positions, draw a final structure that is an indefinite three dimensional network.

(*b*) How many moles of formaldehyde per mole of phenol are required to form the most dense network?

2

Molecular weight and molecular weight distribution

Since the achievement of the physical properties characteristic of high polymers depends critically on molecular weight, it is most important to be able to define and measure this quantity. In this chapter definitions are taken up and basic concepts illustrated, largely by the example of step polymerization. The companion conclusions with respect to chain polymerization are necessarily deferred until the treatment of polymerization kinetics is undertaken (in Chapter 4). The subject of branching and crosslinking is also introduced. In the next chapter (Chapter 3) the various methods for measuring molecular weight are considered.

Before proceeding, it is worth giving a brief overview of how certain key material properties depend on molecular weight. In doing so, no effects of molecular weight distribution are considered, only some broad generalizations that are based on average molecular weight are taken up. These properties naturally group themselves into three categories and the effect of molecular weight on each of them is different. The first category is what might be considered typical thermophysical properties such as melting point, heat capacity and, especially important, elastic properties. All of these have the characteristic that they 'saturate' with molecular weight. That is, they become independent of molecular weight as the latter increases. Figure 2.1 shows the melting points of alkane or paraffin crystals as chain length increases toward their high molecular weight counterpart, polyethylene. Within a homologous series where the chain length is long enough for solids to prevail at room temperature, the elastic constants of the crystals versus chain length do not vary greatly. A conceptual plot for the Young's modulus (the reciprocal of the elastic tensile compliance along one of the crystal axes) is given in Figure 2.2.

The second category is 'ultimate properties'. Prominent here would be mechanical strengths, such as breaking stress or yield stress. Strength is

18

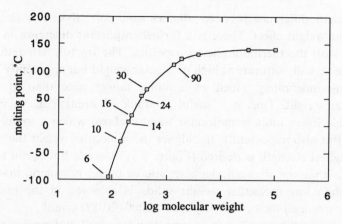

Fig. 2.1 Melting points of alkanes versus molecular weight. Chain lengths of paraffins expressed as number of carbon atoms are marked. Others are polyethylene fractions.

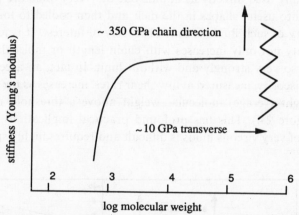

Fig. 2.2 Idealized plot of stiffness versus molecular weight. Values for Young's modulus approximately characteristic of polyethylene along the chain axis and transverse to it are shown.

very sensitive to molecular weight also. Instead of a progressive increase approaching a saturation limit, the strength is essentially zero at low molecular weights, i.e., a liquid might be dealt with. Even if a solid is encountered, the strength may be effectively zero. Considering the alkanes again, they are highly crystalline substances. Yet the very low yield stress of an alkane is quite evident. The material is easily deformed permanently by smearing with the finger tips. Since the chemical structures are identical, the strength difference between alkanes and polyethylene is clearly a molecular weight effect. There is a further important difference in comparison with the thermophysical properties. The fracture strength does not quite actually saturate at high molecular weight but apparently keeps increasing indefinitely, albeit at a much slower pace than at lower molecular weight. Thus it is useful to think of strength as having an acceptable lower limit in molecular weight, above which a useful solid results. But also importantly, the higher the molecular weight the better, if the highest strength is desired (Figure 2.3). A crude but useful rule of thumb is that the strength characteristic of useful polymeric materials rather than low molecular weight solids is achieved in the range of (number average) molecular weight of 10 000–20 000 g/mol.

Polymeric solids usually can exist as liquids as well. Polymeric crystals, as has been seen, have melting points and non-crystalline, amorphous polymeric glasses become fluid above a glass transition temperature. The third category of molecular weight behavior deals with liquids. Polymers are not usually used directly as liquids but they very often are processed or formed into useful shapes in the melt and then cooled to form solids. The viscosity of such fluids is thus obviously of interest. It is found that melt viscosity not only increases with chain length or molecular weight but it does so very strongly and without limit. In fact, an accepted rule is that the viscosity, measured at low shear rates, increases as the 3.4 *power* of the (weight-average) molecular weight above a threshold molecular weight (Figure 2.4). This has profound practical implications since the processing of very viscous fluids is difficult and requires highly specialized

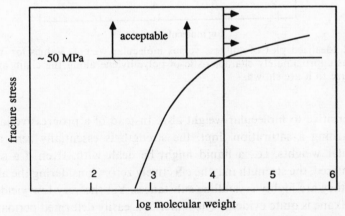

Fig. 2.3 Idealized plot of fracture stress versus molecular weight.

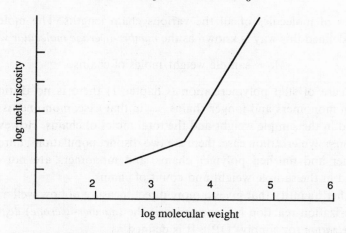

Fig. 2.4 Idealized plot of melt viscosity versus molecular weight.

equipment. Yet, as has been seen, for achieving high strength the rule is 'the higher the molecular weight the better'. This means that, in practice, a balance may have to be struck between processability and strength by choosing a molecular weight lower than the highest achievable synthetically.

2.1 Average molecular weight

The definition of molecular weight in chemistry is considered a simple matter, being just the weight of one mole of the compound in question. For a given known compound it is easily calculated as the sum of the atomic masses of the constituent atoms in the molecular formula. The alternative empirical definition is the weight of a sample of the compound divided by the number of moles in that sample as determined in an experiment. In polymeric systems nearly pure substances, in the sense of most of the molecules being identical, are rarely dealt with. Rather, the usual result of polymerization is a mixture of molecules identical in chemical structure of the chains but differing widely in the number of monomers incorporated into the chains. Clearly an *average* molecular weight is called for.

2.1.1 Definition of number-average molecular weight

The simplest way of defining an average molecular weight is the empirical definition alluded to above: sample weight divided by the total number of moles of molecules of all the various chain lengths. The molecular weight defined this way is known as the *number-average molecular weight*,

$$M_N = \text{sample weight/moles of chains.} \tag{2.1}$$

In the case of step polymerization (Chapter 1) there is no distinction between monomers and longer chains, so, in that case, monomer is to be included in the sample weight and the total moles of chains. However, in the chain polymerization case, there are two distinct populations, unreacted monomer and finished polymer chains, and monomers are not to be included in the sample weight and count of chains.

Another quantity that gives a more direct measure of how well a given polymerization reaction has succeeded is the (*number-average*) *degree of polymerization* (or simply 'DP'). It is defined as,

$$X_N = \text{total moles of monomeric units considered to be incorporated}$$

$$\text{into polymer chains/moles of chains.} \tag{2.2}$$

The same considerations that apply to M_N with respect to step versus chain polymerization and whether monomer is to be counted also apply to the moles of monomeric units incorporated as well as to the chain count.

2.1.2 Number-average molecular weight in step polymerization

In a step polymerization of the type, for example,

$$HO—CO—(CH_2)_4—CO—OH + H_2N—(CH_2)_6—NH_2 \rightarrow$$

$$HO—CO—(CH_2)_4—CO—NH—(CH_2)_6—NH_2 + H_2O,$$

the acid and amine monomers can be represented as A—A and B—B respectively, see Figure 1.1. The various molecular lengths of polymer can be represented as

$$A—A—B—B—A—A—B—B,$$

$$A—A—B—B—A—A \text{ etc.}$$

The end-groups, such as free acid or amine, can often be detected experimentally by such techniques as titration or infrared absorption. Thus,

$$C_A, \text{ the concentration of acid end-groups}$$

and

$$C_B, \text{ the concentration of amine end-groups}$$

provide a measure of the total moles of chains present. The initial concentrations of monomers (indicated by superscript zero), provide the total moles of units present. Therefore, since each initial monomer has two end-groups and each polymer chain has two ends, from equation (2.2) it follows that,

$$X_N = \frac{(C_A^0 + C_B^0)/2}{(C_A + C_B)/2}. \tag{2.3}$$

Further, since

$$C_A^0 - C_A = C_B^0 - C_B$$

(each acid that reacts must react with an amine, and *vice versa*) and if the ratio of B to A initially is denoted by

$$r = C_B^0/C_A^0,$$

equation (2.3) becomes,

$$X_N = \frac{1 + r}{2C_A/C_A^0 + r - 1}. \tag{2.4}$$

This equation expresses the physically obvious result that if there is an excess of one type of monomer over the other, then the ultimate molecular weight attainable is limited by the blocking of the ends by the monomer in excess. If B—B is in excess, the molecular weight cannot increase beyond the point where all of the A groups have reacted and the ends are all B. The limiting degree of polymerization, when all of the —As have reacted, is given by equation (2.4) with $C_A = 0$ or

$$X_N = \frac{r + 1}{r - 1}. \tag{2.5}$$

For example, if $r = 1.01$ (1% excess of B over A), the limiting degree of polymerization is 201. Thus, if high molecular weight is to be achieved, a rather close balance of the monomers must be attained initially. Among other things, this requires high purity of the starting materials. In the case of polyamides the monomer balance can be achieved by taking advantage of the fact that the amine is a base and forms salts with carboxylic acids, thus the salt below can be made and purified and hence a close stoichiometric balance ensured

$$[-CO_2-(CH_2)_6-CO_2^-][H_3^+N-(CH_2)_6-NH_3^+].$$

The number-average molecular weight is related to the degree of polymerization (equation (2.1) and equation (2.2)) by the conversion from moles of monomer units present to sample weight. For the example of the A—A, B—B step polymer, the total moles of A—A and B—B are converted to weight by means of their respective monomeric unit molecular weights, m_{AA} and m_{BB}. If the molecular weight is high enough that corrections due to the end-groups are negligible then for this example, half of the units are A—A and half B—B and the sample weight is just

$$(m_{AA} + m_{BB})/2$$

times the total moles of monomeric units present, and the approximation

$$M_N = [(m_{AA} + m_{BB})/2]X_N$$

is justified. For example, in nylon 66 the monomeric weights m_{AA}, m_{BB} are calculated as,

$6 \times 12 = 72$	$6 \times 12 = 72$
$2 \times 16 = 32$	$2 \times 14 = 28$
$8 \times 1 = 8$	$14 \times 1 = 14$
$m_{AA} = 112$	$m_{BB} = 114$

Notice that the molecular weights of the monomer units have been calculated as they appear in the chain and not as free monomer, i.e., H_2O is lost in polymerization.

In order to avoid confusion it is important to point out that in step polymers of type A—A, B—B the monomeric units are not identical with the *repeat* unit. The repeat unit is the smallest chemical structure that will represent the structure of the chain. Thus, in this case, the repeat unit is made up of two monomeric units, but, in general, the repeat unit can be larger than, equal to or smaller than the monomeric units (Figure 2.5).

2.2 Molecular weight distribution

The concept of an average molecular weight was introduced in the expectation that real polymerizations may introduce a distribution of chain lengths. In any given polymerization the way in which the lengths are distributed is determined by the chemical and kinetic parameters operative. The nature of the distribution can vary widely from one

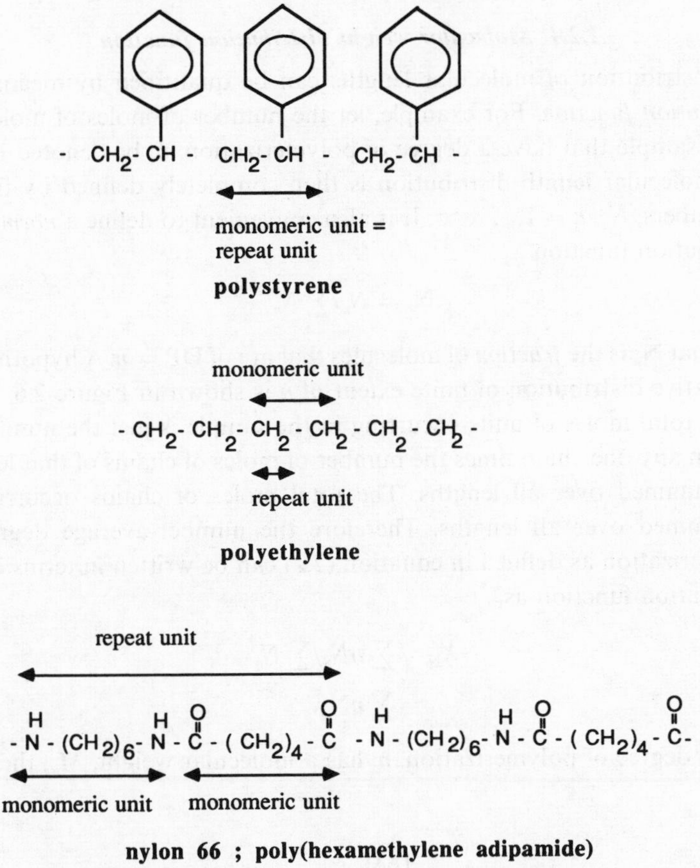

- CH$_2$- CH - CH$_2$- CH - CH$_2$- CH -

monomeric unit =
repeat unit

polystyrene

monomeric unit

- CH$_2$- CH$_2$- CH$_2$ - CH$_2$ - CH$_2$ - CH$_2$ -

repeat unit

polyethylene

repeat unit

$$- N - (CH_2)_6 - N - C - (CH_2)_4 - C - N - (CH_2)_6 - N - C - (CH_2)_4 - C -$$

monomeric unit   monomeric unit

nylon 66 ; poly(hexamethylene adipamide)

Fig. 2.5 Examples of monomeric and repeat units.

polymerization type to another. In some cases it is possible to deduce theoretically from the chemical and kinetic parameters what the distribution should be. In other cases where they are not known, or the situation is too complicated, this is not possible. But very often the distribution can be determined empirically, at least approximately, by means of fractionation experiments. Thus it is appropriate to introduce general concepts and definitions that may be used to describe molecular weight distribution. It will be very instructive to illustrate these and also to go through the process of theoretical deduction of a distribution by means of a simple example, that of step polymerization. A few other distributions will be discussed in later sections and also when polymerization kinetics is taken up (Chapter 4).

2.2.1 Molecular weight distribution function

The distribution of molecular lengths can be quantified by means of a *distribution function*. For example, let the number of moles of molecules in the sample that have a degree of polymerization, n, be denoted by N_n. The molecular length distribution is then completely defined by the set of numbers, N_n, $n = 1, \ldots, \infty$. It is also convenient to define a *normalized* distribution function

$$\bar{N}_n = N_n / \sum N_n \tag{2.6}$$

such that $\bar{N}_n$ is the *fraction* of molecules that are of DP $= n$. A hypothetical illustrative distribution of finite extent of n is shown in Figure 2.6.

The total moles of units occurring in the sample is just the number of units in any one chain times the number of moles of chains of that length, nN_n, summed over all lengths. The total moles of chains occurring is N_n summed over all lengths. Therefore the number-average degree of polymerization as defined in equation (2.2) can be written in terms of the distribution function as,

$$X_N = \sum nN_n / \sum N_n$$
$$= \sum n\bar{N}_n. \tag{2.7}$$

If each degree of polymerization, n, has a molecular weight, M_n, then the

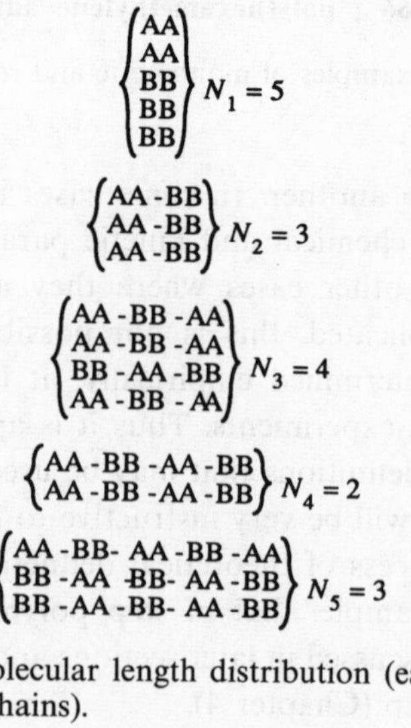

$$\left.\begin{matrix} AA \\ AA \\ BB \\ BB \\ BB \end{matrix}\right\} N_1 = 5$$

$$\left.\begin{matrix} AA\text{-}BB \\ AA\text{-}BB \\ AA\text{-}BB \end{matrix}\right\} N_2 = 3$$

$$\left.\begin{matrix} AA\text{-}BB\text{-}AA \\ AA\text{-}BB\text{-}AA \\ BB\text{-}AA\text{-}BB \\ AA\text{-}BB\text{-}AA \end{matrix}\right\} N_3 = 4$$

$$\left.\begin{matrix} AA\text{-}BB\text{-}AA\text{-}BB \\ AA\text{-}BB\text{-}AA\text{-}BB \end{matrix}\right\} N_4 = 2$$

$$\left.\begin{matrix} AA\text{-}BB\text{-}AA\text{-}BB\text{-}AA \\ BB\text{-}AA\text{-}BB\text{-}AA\text{-}BB \\ BB\text{-}AA\text{-}BB\text{-}AA\text{-}BB \end{matrix}\right\} N_5 = 3$$

Fig. 2.6 An illustrative molecular length distribution (each chain is assumed to represent a mole of such chains).

Table 2.1 *The number-average DP and molecular weight of the distribution of Figure 2.6.*

n	N_n	nN_n
1	5	1×5
2	3	2×3
3	4	3×4
4	2	4×2
5	3	5×3
Number of molecules $= \sum N_n = 17$		$\sum nN_n = 46$
		$=$ number of monomer units

$X_N = 46/17 = 2.71 =$ number of monomer units/number of molecules

$M_N = (23m_{AA} + 23m_{BB})/17$

$\quad = (46/17)(m_{AA} + m_{BB})/2$

$\quad = X_N(m_{AA} + m_{BB})/2.$

number-average molecular weight is

$$M_N = \sum M_n \bar{N}_n \tag{2.8}$$

In Table 2.1 the use of the distribution displayed in Figure 2.6 in computing the number average DP and M_N is shown.

2.2.2 *Molecular weight distribution in step polymerization*

The molecular weight distribution in a step polymerization may be deduced as follows (Flory, 1936). For simplicity it is assumed that the monomeric units are perfectly balanced, $C_A^0 = C_B^0$ (or $r = 1$). Then the distinction between A and B units disappears and the chains may be regarded as all A units (—A—A—, —B—B—, = '—A—'),

$$A—A—A—A—A—A—A—A—A—A.$$

Central to the derivation is the assumption that all unreacted end-groups (**A**), regardless of the length of the chain to which they belong, have the same chemical reactivity and hence may be regarded as occurring randomly through the collection of monomeric units. The probability that any monomeric unit selected at random is a chain end (**A**) is C_A/C_A^0, i.e., the fraction of units that are chain ends. The probability, P, that any monomeric unit selected at random is in the interior of a chain (**A**) is

$$P = 1 - C_A/C_A^0. \tag{2.9}$$

P is also the extent of reaction. The distribution function, N_n, is found by selecting a chain and inquiring as to the chance of its having n monomeric units hooked together. For convenience select the chain by one end, the left end (*). Then this end either has a unit bonded to it on the right, with probability, P, above, or is terminated on the right with probability $1 - P$. In order for the chain to continue it must be the first option. The probability that this end has n bonded units linked together, including the original end, is given by P^{n-1}, since the occurrence of bonded and end groups is assumed to be random and therefore the product of probabilities of independent events. In order to have only n units the chain must be terminated, with probability $1 - P$. Therefore the chance of the selected chain having n units is $P^{n-1}(1 - P)$.

$$A^*—A—A—A—A—A—A—A—A—A$$
$$P \cdot P \cdot P \cdot P \cdot P \cdot P \cdot P \cdot P \cdot P \cdot (1 - P).$$

The normalized number distribution function, $\bar{N}_n = N_n/\sum N_n$, of equation (2.6) is also the probability that a selected chain has n units in it. Therefore,

$$\bar{N}_n = P^{n-1}(1 - P). \tag{2.10}$$

It is to be noticed that this distribution function is, in fact, properly normalized, that is,

$$\sum \bar{N}_n = \sum P^{n-1}(1 - P) = 1.$$

To show this, write

$$\sum P^{n-1} = 1 + P + P^2 + P^3 + \cdots$$

but

$$1/(1 - P) = 1 + P + P^2 + P^3 + \cdots. \tag{2.11}$$

From equation (2.4) with $r = 1$, it is seen, *without recourse to the distribution function*, that

$$X_N = C_A^0/C_A = 1/(1 - P). \tag{2.12}$$

This also follows from the distribution function, as it must, using equation (2.9) and the definition of equation (2.2). From equation (2.7),

$$X_N = \sum nN_n/\sum N_n = \sum n\bar{N}_n$$
$$= \sum nP^{n-1}(1 - P) = (1 - P)(1 + 2P + 3P^2 + 4P^3 + \cdots), \tag{2.13}$$

but when the sum is found by differentiating both sides of equation (2.11)

$$d[1/(1 - P)]/dP = 1/(1 - P)^2 = 1 + 2P + 3P^2 + 4P^3 + \cdots$$

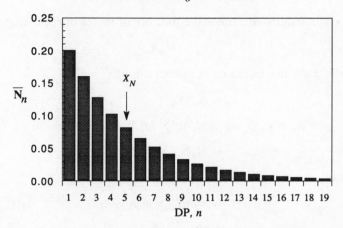

Fig. 2.7 Fractional number of moles of polymer chains of various lengths versus DP for the 'most probable' distribution occurring in step polymerization. Number-average DP is 5.

equation (2.13) reduces to (2.12). The distribution expressed in equation (2.10) has been called by Flory (1953) the 'most probable' distribution. Under certain conditions it can also occur in chain polymerization (Chapter 4).

It is instructive to examine the nature of the particular distribution function expressed in equation (2.10). If P is replaced by $1 - 1/X_N$, see equation (2.12), equation (2.10) becomes

$$\bar{N}_n = \frac{1}{X_N}\left(1 - \frac{1}{X_N}\right)^{n-1}. \tag{2.14}$$

This is a monotonically decreasing function of increasing n. The most likely value of n is $n = 1$, with the number of molecules of DP $= n$ decreasing as n increases. In Figure 2.7 a plot of $\bar{N}_n$ vs n for $X_N = 5$ is displayed.

2.2.3 Weight-average molecular weight

The conclusion that there is a preponderance of smaller molecules in the step polymerization example of the previous section is deceptive in some respects. The *amount* of polymer (as expressed, for example, as its weight) of each chain length, n, is distributed quite differently. If we imagine such a polymer to be fractionated into the various chain lengths, the amount (weight) of polymer in each fraction, w_n, is given by the molecular weight of each chain length, M_n, times the number of moles of each length

$$w_n = M_n N_n,$$

or, if each chain has the same monomeric units,

$$w_n = m_0 n N_0,$$

$$\bar{w}_n = w_n / \sum w_n = M_n N_n / \sum M_n N_n$$

$$= M_n \bar{N}_n / \sum M_n \bar{N}_n \qquad (2.15)$$

and, if each chain has the same monomeric units,

$$= m_0 n N_n / \sum m_0 n N_n$$

$$= n \bar{N}_n / \sum n \bar{N}_n. \qquad (2.16)$$

Just as the number distribution function can be used to define a number-average DP as in equation (2.7), so can the weight distribution function be used to define a *weight-average DP*. That is, in defining the average molecular weight, or DP, each DP, n, is weighted according to the weight of the fraction rather than the number of moles,

$$X_W = \sum n \bar{w}_n \qquad (2.17)$$

and, if each chain has the same monomeric units,

$$X_W = \sum n^2 \bar{N}_n / \sum n \bar{N}_n$$

and a *weight-average molecular weight*

$$M_W = \sum M_n \bar{w}_n$$

$$= \sum M_n^2 \bar{N}_n / \sum M_n \bar{N}_n. \qquad (2.18)$$

In Table 2.2 the weight-average molecular weight of the distribution shown in Figure 2.6 is computed as an example.

2.2.4 Weight-average molecular weight in step polymerization

The appropriate number distribution function, equation (2.10) or (2.14), when substituted into the weight distribution function, $\bar{w}_n$, equation (2.16), gives

$$\bar{w}_n = (1 - P)n P^{n-1} / X_N$$

$$= (1 - P)^2 n P^{n-1}$$

$$= (1/X_N^2) n (1 - 1/X_N)^{n-1}. \qquad (2.19)$$

Table 2.2 *The weight-average DP and molecular weight of the distribution of Figure 2.6.*[a]

n	weight fraction $\bar{w}_n = nN_n / \sum nN_n$	$n\bar{w}_n$
1	5/46	$1 \times 5/46$
2	6/46	$2 \times 6/46$
3	12/46	$3 \times 12/46$
4	8/46	$4 \times 8/46$
5	15/46	$5 \times 15/46$
	$\sum \bar{w}_n = 1$	$X_W = \sum n\bar{w}_n = 160/46 = 3.48$

[a] Assuming $m_{AA} = m_{BB}$.

Equation (2.19) has a maximum at $n = X_N - 1$ and X_N, the fraction with the greatest amount (weight) of polymer, occurs at the number-average DP. See Figure 2.8.

The weight-average DP from equation (2.17) using equation (2.19) is

$$X_W = \sum (1 - P)^2 n^2 P^{n-1}$$
$$= (1 - P)^2 (1 + 4P + 9P^2 + \cdots).$$

The series in parentheses can be summed by noting that it is equal to

$$d(P + 2P^2 + 3P^3 + \cdots)/dP = d[P(1 + 2P + 3P^2 + \cdots)]/dP$$

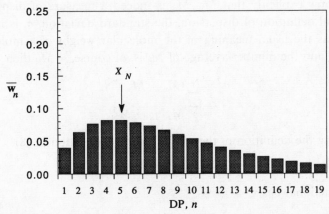

Fig. 2.8 Weight fraction of polymer versus DP for 'most probable' distribution occurring in step polymerization. Number-average DP is 5, as in Figure 2.7.

and therefore (see equation (2.13)) equal to

$$d[P/(1 - P)^2]/dP = (1 + P)/(1 - P)^3.$$

Thus

$$X_W = (1 + P)/(1 - P). \tag{2.20}$$

Since for polymers of any reasonable molecular weight $P \approx 1$, equation (2.20) becomes

$$X_W = 2/(1 - P) = 2X_N. \tag{2.21}$$

Thus the useful rule of thumb is found that the weight-average molecular weight is twice the number-average in a step polymerization.

2.2.5 M_W/M_N as a measure of dispersity

It was deduced above, that for polymers for which the 'most probable' distribution (equation (2.10)) holds, the ratio of weight to number average molecular weight is 2. It is important to realize that this is not generally true and that the ratio M_W/M_N is a convenient index of the breadth or *dispersion* of the molecular weight distribution. If there is but one DP present, say, $n = X$, then $M_W = M_N = m_0 X$ and $M_W/M_N = 1$. This, of course, is the most narrow distribution possible. The value for the 'most probable' distribution $M_W/M_N = 2$ is a convenient base from which to judge dispersity. Values less than 2 indicate a rather 'narrow' distribution. A distribution with $M_W/M_N \gg 2$ could be said to be a 'broad' one.

To show explicitly that M_W/M_N is indeed a measure of dispersity, a statistical definition of dispersion, the standard deviation, σ, is invoked. If M_n has the usual meaning of the molecular weight of a molecule of DP $= n$ and the number-average of M_n is, of course, $= M_N$ then

$$\sigma^2 = \sum (M_n - M_N)^2 \bar{N}_n$$
$$= \sum M_n^2 \bar{N}_n - M_N^2 \tag{2.22}$$

and using the definition of the weight-average, equation (2.18)

$$\sigma^2 = M_W M_N - M_N^2.$$

On a fractional basis,

$$\frac{\sigma}{M_N} = \left(\frac{M_W}{M_N} - 1\right)^{\frac{1}{2}}. \tag{2.23}$$

This relation shows especially clearly that the breadth of the distribution of molecular weight is gauged by the ratio of the weight-average to number-average molecular weights.

2.2.6 *The exponential approximation to the 'most probable' distribution*

The performing of summations to obtain averages using the 'most probable' distribution (equation (2.10) or equation (2.14)) is sometimes simplified by regarding n as a continuous variable and replacing summations by integrations. For the values of X_N typical of technologically useful high molecular weight polymers ($X_N > 100$), $\bar{N}_n$ is a very smooth function of n. That is, it does not change much as n increases from n to $n + 1$. For $X_N = 100$, it changes by just 1% for an increase in n of 1. Therefore it is an excellent approximation to regard n as a continuous variable rather than an integer. Further, it is also an excellent approximation to replace

$$P^{n-1} = \left(1 - \frac{1}{X_N}\right)^{n-1}$$

by

$$P^{n-1} \cong (e^{-1/X_N})^{n-1}$$

since the power series expansion of e^{-1/X_N} is

$$e^{-1/X_N} = 1 - 1/X_N + 1/2X_N^2 - \cdots$$

and the terms beyond the first two are very small for large X_N ($\geq \sim 100$). Thus the 'most probable' distribution can be well approximated as

$$\bar{N}_n = \frac{1}{X_N} e^{-n/X_N}, \tag{2.24}$$

where n is regarded as a continuous variable ranging from *zero* to infinity. Averages such as

$$M_W = \sum M_n^2 \bar{N}_n / \sum M_n \bar{N}_n$$

are performed as

$$M_W = m_0 \int_0^\infty n^2 e^{-n/X_N} \, dn \bigg/ \int_0^\infty n e^{-n/X_N} \, dn. \tag{2.25}$$

Since,

$$\int_0^\infty u^q \, e^{-au} \, du = \frac{\Gamma(q+1)}{a^{q+1}}$$

$$= q!/a^{q+1} \qquad \text{(for integral values of } q\text{)}, \qquad (2.26)$$

equation (2.25) reduces to

$$M_W = m_0 \frac{2!}{(1/X_N)^3} \bigg/ \frac{1}{(1/X_N)^2} = 2m_0 X_N, \qquad (2.27)$$

in agreement with equation (2.21) based on the discrete distribution.

2.2.7 *The 'Schulz–Zimm' distribution*

As is demonstrated in the discussion on the kinetics of chain polymerization (Chapter 4), if termination dominates the chain producing process and is by disproportionation, or if chain transfer dominates chain production, the instantaneous molecular weight distribution of polymer produced at a given conversion is also of the 'most probable' type,

$$\bar{N}_n = (1 - P)P^{n-1} \approx \frac{1}{X_N} e^{-n/X_N}. \qquad (2.28)$$

However, if termination dominates chain production but is by recombination, then the instantaneous distribution is of the form found by Schulz (1939),

$$\bar{N}_n = \frac{n}{[(X_N - 1)/2]^2} \left\{ \frac{1}{1 + 1/[(X_N + 1)/2]} \right\}^n \approx \frac{n}{(X_N/2)^2} e^{-n/(X_N/2)}. \qquad (2.29)$$

When these two distributions are written in terms of the exponential approximation they both become

$$\bar{N}_n = Cn^Z e^{-n/y}, \qquad (2.30)$$

where $y = X_N/(Z + 1)$ and C is a constant independent of n, thus, $Z = 0$ for equation (2.28) and $Z = 1$ for equation (2.29).

It was suggested by Zimm (1948) that equation (2.30) not be restricted to just $Z = 0$ or $Z = 1$ but be used as an *empirical* function for other values of Z. The constant C is determined by normalizing equation (2.30) using equation (2.26),

$$\int_0^\infty Cn^Z e^{-n/y} \, dn = C\Gamma(Z + 1)y^{Z+1} = 1 \qquad (2.31)$$

or

$$C = 1/[\Gamma(Z + 1)y^{Z+1}].\qquad(2.32)$$

Since

$$X_N = \sum n\bar{N}_n$$

$$= \int_0^\infty n^{Z+1}\,e^{-n/y}\,dn/\Gamma(Z + 1)y^{Z+1}$$

$$= \Gamma(Z + 2)y^{Z+2}/\Gamma(Z + 1)y^{Z+1},$$

and using a standard recursion relation for gamma functions,

$$\Gamma(Z + 2) = (Z + 1)\Gamma(Z + 1),$$

then it follows that

$$X_N = (Z + 1)y.\qquad(2.33)$$

The ratio M_W/M_N is found to be

$$M_W/M_N = (Z + 2)/(Z + 1).\qquad(2.34)$$

It is apparent from equation (2.34) that Z is an index of dispersion. For large values of Z, where M_W/M_N approaches 1, the distribution is narrow. Of course for $Z = 0$, the 'most probable' distribution value of $M_W/M_N = 2$ is obtained. Negative values of Z, but larger than -1, correspond to a broad distribution. Thus this empirical distribution function is convenient because it allows the complete range of breadth of distribution from very narrow to very broad to be represented by a function that leads to averages that are simple to perform mathematically. It should be remembered, of course, that it is empirical and that even if Z is chosen to match M_W/M_N of some real polymer, the real distribution and that expressed in equation (2.30) could be different.

2.2.8 The distribution of molecular weight in 'living' polymers

Polymers made with ionic initiators can sometimes be polymerized under conditions where there is no termination or transfer (Section 4.4). The molecular weight then is simply determined by the ratio of the amount of monomer utilized to the number of chains initiated (amount of initiator used). Such polymers are called 'living' because in the polymerization vessel each chain contains a reactive center at the end and can continue to polymerize if more monomer is added. This is in contrast to the usual chain polymerizations (Section 4.2) where termination or transfer of the growing chain containing the reactive center results in a finished unreactive polymer molecule (i.e., by contrast, 'dead'). The molecular weight distribution in such living polymers is quite different from either the 'most probable' distribution or that in combination terminated chain polymerization (Flory, 1940a). The distribution is determined by the random competition of growing ends for monomers. To deduce this distribution, imagine the following situation. In a small time interval Δt a small amount of monomer reacts with the fixed number of growing ends of the chains. Let Δt be small enough that the amount of monomer that reacts is less than the number of ends, so that some ends receive monomer and some do not. At the end of a finite time, t, the resulting situation can be illustrated as in Figure 2.9. The interval Δt is chosen such that three out of four chain ends receive new monomer units in each Δt. A dark circle indicates a monomer that reacted, an open circle indicates one was not received by that chain during Δt. The actual degree of polymerization of one chain is given by the number of dark circles in the sequence of that chain. An open one does not indicate a break in the chain, only that it did not receive a monomer in that particular Δt interval.

The distribution of molecular lengths can be deduced under the assumption of random distribution of the light circles among the dark ones. The question is, what is the probability that a chain has n units (dark circles) in it. If Δt is made small enough that only a small fraction of the chains receives a monomer in Δt, then there will be a vanishing chance of a chain receiving two or more monomers in Δt. Under these conditions, the probability, P, that a given position is occupied by a dark

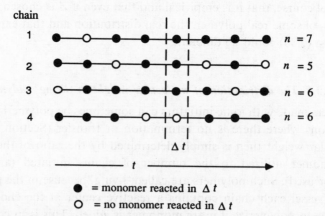

Fig. 2.9 'Living' polymerization. Time history of four growing chains competing for monomer. Each time interval has enough monomer to react with three of the four chains.

circle is given by the ratio of monomers reacting in Δt to the total sizes available ($=$ number of growing chains) or

$$P = r_p \Delta t / \sum N_n, \qquad \text{for } \Delta t \text{ chosen to make } P \ll 1, \qquad (2.35)$$

where r_p is the rate of polymerization (moles monomer/s). The probability of an open circle occurring is $1 - P$. After the polymerization has proceeded for a length of time corresponding to m intervals of Δt, the probability of n occupied circles and $m - n$ unoccupied circles occurring in a specific sequence is $P^n(1 - P)^{m-n}$. However, the particular distribution of open circles among the dark ones is not important, the degree of polymerization of a given chain depends only on the total number of closed circles and not upon their sequence. For example chains 3 and 4 of Figure 2.9 are identical at the end of time t and differ only in the *temporal* history of how they were built. If the above probability is multiplied by the number of ways of mixing n closed circles with $m - n$ open ones, then it would become the probability of n closed circles occurring with $m - n$ open ones regardless of sequence. Or, more simply stated for the problem at hand, it is the probability of n monomers being connected together. Therefore the probability distribution function is,

$$\bar{N}_n = \frac{m!}{n!(m-n)!} P^n(1 - P)^{m-n}, \qquad (2.36)$$

since $m!/n!(m - n)!$ is the number of ways of performing the mixing of n and $m - n$ distinct objects. This function is the classical binomial distribution that results for the probability of n successful events occurring in m tries when the probability of success is constant for each try (Bernoulli trials). As expressed above, the distribution function contains the parameters m and P which involve the physically undefined quantity Δt. However, the quantity mP is defined. From equation (2.35) it is noticed that

$$mP = mr_p \Delta t / \sum N_n. \qquad (2.37)$$

Since

$$mr_p \Delta t = r_p t, \qquad (2.38)$$

and the latter is equal to the total number of monomers incorporated into the polymer, then,

$$mP = \sum n\bar{N}_n = X_N. \qquad (2.39)$$

Since Δt was made small enough that P is small ($\ll 1$), then for all values of n of interest (several multiples of X_N at most) $m \gg n$. Therefore,

$$(1 - P)^{m-n} \approx (1 - P)^m \approx (e^{-P})^m \approx e^{-X_N}$$

and

$$\frac{m!}{(m - n)!} P^n = m(m - 1)(m - 2) \cdots (m - n + 1)P^n \approx (mP)^n = X_N^n.$$

Thus,

$$\bar{N}_n \approx (X_N^n/n!) \, e^{-X_N}. \qquad (2.40)$$

Equation (2.40) is the *Poisson* approximation to the binomial distribution (see Feller (1968)). The Poisson distribution has a maximum in the number distribution at $n = X_N$ and for high molecular weight polymer is very sharply peaked about this value. In fact as shown in Problem 2.8, the standard deviation of the distribution (see equation (2.22)) is,

$$\sigma = X_N^{1/2}$$

or

$$\frac{\sigma}{X_N} = \frac{1}{X_N^{1/2}}. \qquad (2.41)$$

Thus the width of the distribution becomes very narrow at high DPs and is effectively monodisperse.

2.3 Network formation by non-linear polymers: branching and crosslinking; gelation

Up to now, only polymers consisting of collections of linear chains have been considered. Obviously, linear chains arise when the monomer has a functionality of two and participates in two separate linkages in the chain. In the step polymers discussed earlier, each monomer contained two functional groups, two acids or two amines. However, there is no reason why the functionality cannot be greater than two. For example, a polyester could be made using a trifunctional alcohol (e.g., glycerol). This will lead to a polymer containing many *branches* (Figure 2.10). Such branched polymers have some interesting properties from a molecular weight point of view. It turns out that they have a strong tendency to produce some molecules of extremely large, essentially infinite, size. All of the distributions discussed previously for linear polymers tail off smoothly at high molecular weight so that there is a negligible portion of polymer of molecular weight many times the average value. However, the distribution in branched polymers tends to be bimodal, a significant amount of

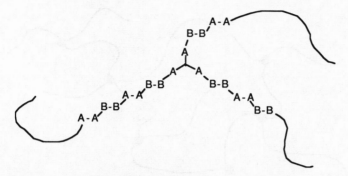

Fig. 2.10 A branch point introduced by a trifunctional monomer.

extremely high molecular weight material being present along with lower molecular weight species. In the initial stages of a step polymerization involving a tri- or higher functionality monomer, the molecular weight builds much in the manner of a linear polymer. However, at a surprisingly low extent of polymerization very large molecules appear. Since these essentially infinitely sized molecules cannot be dispersed molecularly by solvent, their production in the course of a solution polymerization is quite evident physically as the appearance of a solvent-swollen highly viscous *gel*. This gel can be physically separated from the lower molecular weight polymer in solution (the latter is called *sol*) by filtration or centrifugation. Eventually, as the extent of polymerization increases the entire sample becomes gel and there is no longer a sol fraction. The first appearance of gel as the polymerization proceeds is called the *gel point*.

The gelled sample at the completion of polymerization has many random circuitous connections and is an example of a network (Figure 2.11). Networks are usually random but there are examples of regular ones, graphitic planes being illustrative. Phenol–formaldehyde resins (Problem 1.2) are often depicted in structural formulae as regular. Gelation is sometimes an undesirable phenomenon resulting from small amounts of multifunctional impurities or transfer reaction that give rise to heterogeneities from the gel fraction in a product that is desired to be homogeneous or perhaps completely soluble. Other times, the formation of the network is the desired result of the chosen monomers. Extreme resistance to solvents, hardness and lack of melting are some advantages of the dense networks formed by carrying to completion the polymerization of multifunctional monomers. Phenol–formaldehyde and melamine resins are such examples. Historically, elevated temperatures were used

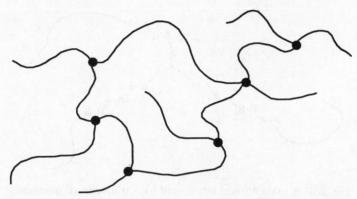

Fig. 2.11 A network formed by a polymer with branches introduced by multi-functional monomers.

to bring about the polymerization and the name *thermosetting* plastics has come into use to describe such insoluble infusible network structures.

Sometimes a polymerization is carried out to low or moderate molecular weight using bifunctional monomers giving linear polymers. The polymerization is then continued using higher functionality monomers or reactions that give rise to branching. The network thus formed is usually described as resulting from *crosslinking* rather than branching. The latter usually implies a result of single-stage polymerization from monomers. Crosslinking is usually a deliberate process carried out to impart insolubility, or more importantly, elastomeric properties. As will be seen (see Chapter 8), rubbery behavior requires the establishment of a network, which will usually be rather sparse compared to the dense networks of thermosetting resins. This sparse network can be conveniently established by the crosslinking of a preexisting linear polymer.

The quantitative theoretical description of molecular weight in branched polymers is naturally more complicated than for linear ones. It is possible to establish in a fairly simple manner the conditions for the onset of gelation as a function of the extent of reaction. This onset has the properties connected with the appearance of a new phase andtherefore of a phase transformation and the associated critical point. If a simplifying assumption is made, detailed results concerning the actual molecular weight distribution and the amount of sol vs gel as a function of reaction extent can be obtained in a straightforward although somewhat tedious manner. This assumption is that no closed loops are formed in the branching process. Thus the molecular architecture is assumed to be that of a *tree* rather than an unrestricted network of branches and closed loops.

A number of results were derived under this assumption by Flory (see Flory (1953) for a discussion and many references) and by Stockmayer (1943, 1944). This approach has sometimes been called the *classical* theory of branching and networks (de Gennes, 1979). More recently interest has centered on approaches where closed loop formation is permitted and the connection with the more general subject of *percolation* phenomena is utilized. The obtainable results have emphasized how certain properties such as molecular weight vary in the vicinity of the critical point and are phrased in terms of scaling laws.

2.3.1 *The gel point in branched polymers from multifunctional monomers*

In order to have a finite probability of molecules with infinite molecular weight, there must be a finite probability of tracing out at least one path through a molecule that continues indefinitely. In Figure 2.12, a portion of such a path along with a number of dead ends and other possible continuing paths is illustrated. Following the treatment of Flory (Flory, 1953, Chap. IX; see also Flory, 1946) it is convenient to define a *branching index*, α, that represents the probability that from a given branch point a selected chain continues on to another branch point rather than terminating in a loose end. In seeking to trace out an indefinitely long path, on arriving at a selected branch point (tri- or higher functional monomer), there are $f - 1$ choices for continuing on the path, where f is the branch point functionality. If α is the probability of successfully proceeding to the next branch point in question, then $(f - 1)\alpha$ is the

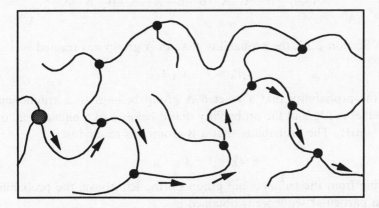

Fig. 2.12 Portion of an indefinitely long path (indicated by the arrows) starting from a selected branch point (the large circle).

probability of successfully continuing to another branch by any of the $f - 1$ paths available for proceeding. For a connected series of i quests for a continuous path, the probability of success is $[(f - 1)\alpha]^i$ and it must remain finite as i increases. Therefore

$$(f - 1)\alpha \geq 1 \tag{2.42}$$

for the formation of very large molecules and the equal sign holds for the *onset* of gelation or the *gel point*.

The criterion of equation (2.42) for the formation of gel is useful because it turns out to be not difficult to express α in terms of composition and extent of reaction. As an illustration of this, the step polymerization of a trifunctional

monomer and a bifunctional A—A monomer with a bifunctional B—B monomer (Figure 2.10) is treated. As in Section 2.2.2, let there be C_A, A groups and C_A^0 of them initially. Also let there be ρC_A^0 of the initial A groups in trifunctional monomers and $(1 - \rho)C_A^0$ of them in bifunctional ones. It is further assumed that all A groups have equal reactivity so that there are ρC_A trifunctional and $(1 - \rho)C_A$ bifunctional groups at any time during the polymerization. There are C_B bifunctional B groups, C_B^0 of them initially. Consider a sequence connecting two branch points,

$$\begin{array}{c} A \\ \diagdown \\ A \diagup \end{array}\!\!-A-B-B-A-A-B-B-A-A-B-B-A-\!\!\begin{array}{c} A \\ \diagup \\ \diagdown A \end{array}.$$

As in Section 2.2.2, the probability that an A group has reacted is

$$P_A = 1 - C_A/C_A^0$$

and the probability that a reacted A group belongs to a trifunctional monomer is ρP_A and the probability that it belongs to a bifunctional one is $(1 - \rho)P_A$. The probability that a B group has reacted is

$$P_B = 1 - C_B/C_B^0.$$

Starting from the trifunctional group on the left above, the probability that a particular sequence is obtained is

$$P_B(1 - \rho)P_A P_B(1 - \rho)P_A P_B \rho P_A.$$

In general, the probability that $n - 1$ A—A and n B—B monomers provide a connection between a selected branch point and another branch point is

$$[(1 - \rho)P_A P_B)]^{n-1} \rho P_A P_B.$$

The probability that a sequence connects the selected branch point with another regardless of length is

$$\alpha = \sum_n [(1 - \rho)P_A P_B]^{n-1} \rho P_A P_B \qquad (2.43)$$

and from equation (2.11),

$$\alpha = \frac{\rho P_A P_B}{1 - (1 - \rho)P_A P_B}. \qquad (2.44)$$

If C_B^0/C_A^0 is again denoted by r (see equation (2.4)), then $rP_B = P_A$ and

$$\alpha = \frac{\rho P_A^2}{r - (1 - \rho)P_A^2}. \qquad (2.45)$$

Thus equation (2.45) provides a means of calculating the branching index in terms of the initial composition (r and ρ) and the extent of reaction P_A. Although the word 'trifunctional' was used in the derivation, actually it is valid for any functionality of the branch point monomer. The onset of gelation can thus be calculated using equation (2.42). For example, if there were no bifunctional As, $\rho = 1$ and if $r = 1$, $\alpha = P_A^2$. If the A monomer is trifunctional, $f = 3$, then gelation starts to take place at (equation (2.42)), $\alpha(f - 1) = 1$, or,

$$\alpha^* = 1/2$$

and therefore

$$P_A^* = (1/2)^{\frac{1}{2}} = 0.707.$$

The onset of gelation is at $\sim 70\%$ extent of reaction, a rather low value considering the very low molecular weight that would be obtained in a linear polymer at the same extent of polymerization. It is also possible to calculate the number-average DP for the above

$$\begin{matrix} \text{A} \\ | \\ \bigwedge \\ \text{A} \quad \text{A} \end{matrix} \quad , \quad \text{A—A,} \quad \text{B—B}$$

step polymerization. The total number of units, N_0, is

$$N_0 = \rho N_A^0/f + (1 - \rho)N_A^0/2 + N_B^0/2, \qquad (2.46)$$

where N_A^0, N_B^0 are total numbers of A and B groups. The total number of chains is the total number of units minus the number of bonds formed during polymerization, or,

$$N = N_0 - N_A^0 P_A = N_0 - N_B^0 P_B, \tag{2.47}$$

and

$$X_N = N_0/N = 1/\{1 - P_A/[\rho/f + (1 - \rho)/2 + r/2]\}. \tag{2.48}$$

Notice for the case $f = 3$, $\rho = 1$, $r = 1$, that

$$X_N = \frac{1}{1 - 6P_A/5}$$

and at the gel point where $P_A^* = (1/2)^{1/2}$, $X_N = 6.60$. Thus gelation starts at an extremely low average DP.

It almost seems to be a paradox that indefinitely large molecules are being formed while the average DP is still very low. This can be rationalized as follows. Suppose a sample consists of two distinct fractions, one of extremely high molecular weight, M_2, has a weight w_2 and the other has molecular weight M_1 and weight w_1. The number-average molecular weight of the blend is

$$M_N = \frac{w_1 + w_2}{w_1/M_1 + w_2/M_2}. \tag{2.49}$$

Suppose that $M_2 \gg M_1$ and if w_2 is comparable to or smaller than w_1,

$$M_N \cong M_1/(1 + w_2/w_1). \tag{2.50}$$

Thus there can be an appreciable amount (w_2) of very high molecular weight material without changing very much the number-average molecular weight from that of the lower fraction. However, consider the weight-average molecular weight in the same circumstance,

$$M_W = \frac{w_1}{w_1 + w_2} M_1 + \frac{w_2}{w_1 + w_2} M_2. \tag{2.51}$$

For $M_2 \gg M_1$ and w_2 not negligible compared to w_1,

$$M_W \cong M_2 w_2/(w_1 + w_2). \tag{2.52}$$

Therefore, the weight-average molecular weight can diverge as the gel point is approached and indefinitely large molecules result, even if the number-average molecular weight remains low. To show in detail how this happens and to compute the relative amount of the gel and sol phases requires deduction of the molecular weight distribution. This was accomplished by Flory and by Stockmayer for a number of circumstances involving branching polymerizations but is much more complicated than in that case of linear step polymers treated in Section 2.2. A summary of the results available is contained in Table 2.3.

The treatments of Flory and Stockmayer in deriving molecular weight distributions differ somewhat. Both make the assumption that the structures are branched and do not contain closed loops. The Stockmayer treatment is analogous to the Mayer theory of condensation. It is found that the molecular sizes in the sol portion remain fixed beyond the gel point and that the effect of further reaction is to create more gel from the fixed property sol much like the conversion of one phase to another in a first order phase transformation. The Flory treatment has no requirement for the structure of the gel portion but assumes that the distribution function for the sol portion is determined in the same manner as before the onset of gelation.

As an illustration of how the gel fraction builds following the onset of the gel point, the following result is quoted (Flory, 1953, p. 376) for a trifunctional monomer polymerization valid for the case where an A_3 monomer condenses with itself ($\rho = 1$, $r = 1$, $P_B = 1$, $\alpha = P$, equation (2.45), $f = 3$, equation (2.42)) or where an A_3 monomer step polymerizes with a B_2 monomer ($\rho = 1$, $\alpha = P_A P_B$, equation (2.45), $f = 3$, equation (2.42))

$$G = 1 - \left(\frac{1-\alpha}{\alpha}\right)^3, \qquad (2.53)$$

where G is the weight fraction of gel and α is the branching index. The gel point according to equation (2.42) occurs at $\alpha^* = 1/2$. The fraction of gel is shown in Figure 2.13 for both the case of A_3 self condensation and A_3 step polymerizing with B_2.

2.3.2 Crosslinking – random

As discussed in the introduction to branching above (Section 2.3), the term crosslinking implies the introduction of network points to an already formed linear polymer. The networks thus differ from those previously discussed only in that the placement of the branch points is controlled by the chemistry of the crosslinking reaction rather than by the nature of the polymerization. A common type of crosslinking occurs by the random reaction of the crosslinking agent with the units of the initial polymer chains. An example would be the vulcanization of rubber by an

Table 2.3 Summary of molecular weight distributions.

conditions	distribution function	averages
general	N_n = number of molecules with n units $\bar{N}_n = N_n / \sum N_n$ $\bar{w}_n = w_n / \sum w_n$ = weight fraction of molecules with n units $= nN_n / \sum nN_n$ $= n\bar{N}_n / \sum n\bar{N}_n$	X_N = units/molecules $X_N = \sum nN_n / \sum N_n$ $X_w = \sum w_n n / \sum w_n = \sum n^2 N_n / \sum nN_n$
Linear polymers		
(1) random distribution of chain ends:	('most probable' distribution, Section 2.2.2)	
(a) step polymerization ($P = 1 - C_A/C_A^0$)	$\bar{N}_n = P^{n-1}(1-P)$ $\cong (1/X_N)\, e^{-n/X_N}$	$X_N = 1/(1-P)$ $X_w = (1+P)/(1-P)$
(b) chain polymerization with disproportionation termination and constant monomer concentration γ = kinetic chain length $\cong X_N$ $P = 1/(1+1/\gamma)$;		
(c) chain polymerization with chain transfer to non-polymer transfer agent at constant concentration (Section 4.2.4)		
	(Schulz distribution, Section 4.2.4)	
(2) chain polymerization with recombination termination at constant monomer concentration	$\bar{N}_n = n/\gamma^2 [1/(1+1/\gamma)]^n$ $\cong n/(X_N/2)^2\, e^{-n/(X_N/2)}$	$X_N = 2\gamma + 1$ $X_w = 3/2 X_N$
	(Schulz–Zimm distribution, Section 2.2.7)	
(3) flexible empirical function capable of representing varying widths of distribution by varying Z:	$\bar{N}_n = C n^Z e^{-n/y}$ $C = 1/[\Gamma(Z+1)y^{Z+1}]$	$X_N = (Z+1)y$ $X_w/X_N = (Z+2)/(Z+1)$
$Z \rightarrow$ large, narrow; $Z = 0$, 'most probable'; $Z = 1$, Schulz; $Z \rightarrow -1$, very broad		
	(Poisson distribution, Section 2.2.8)	
(4) chain polymerization carried out without	$\bar{N}_n = [(X_N)^n/n!]\, e^{-n/X_N}$	$X_w \cong X_N$

Branched polymers

general — branching index α = probability of continuing from a selected branch point to another branch point, at gel point $\alpha^* = 1/(f-1)$
$\alpha = (\rho P_A P_B)/[1 - (1-\rho)P_A P_B]$
ρ = fraction of A groups belonging to A_f units

(1) A_f, AA, BB $\quad A_f = f$-functional monomer with A. A reacts only with B.	see Stockmayer (1952)[a]	$X_N = \left\{1 - P_A \middle/ \left[\left[\dfrac{\rho}{f} + \dfrac{(1-\rho)}{2} + \dfrac{r}{2}\right]\right]\right\}^{-1}$; equation (2.48)
(2) A_f, BB as above but $\rho = 1$	see Flory (1953), p. 370	$r = P_A/P_B$; equation (2.48) $\qquad \alpha = P_A P_B$
(3) A_f, AA $\quad$ A groups capable of reaction with each other, no B groups, as in (1) above, with $P_B = 1$	see Flory (1953), p. 370; Stockmayer (1943)	X_N, see above $\rho = 1$ $\qquad \alpha = \rho P_A/[1 - (1-\rho)P_A]$
(4) A_f as in (3) above and $\rho = 1$	see Flory (1953), p. 370; Stockmayer (1943) $\quad \alpha = P$	$\alpha = P$

Crosslinked polymers

(1) random crosslinking see also Flory (1953), p. 370.		crosslinking index γ = crosslinked units/initial molecule. At the gel point $\gamma = X_N^0/X_W^0$ (one crosslinked unit per weight average molecule) $X_W = X_W^0(1 + \rho_X)/[1 - \rho_X(X_W^0 - 1)]$
(2) crosslinking at end-groups of linear polymer by multifunctional crosslinking agent.	Stockmayer (1944)	at gel point, $P_E P_A(f - 1) = 1$, P_E, P_A extent of reaction of end-groups E, and crosslinking functional groups A.

[a] Stockmayer (1952) worked out the general case of the step polymerization of an arbitrary mixture of A_2, A_3, ..., A_f monomers with B_2, B_3, ..., B_f monomers, where A must react with B.

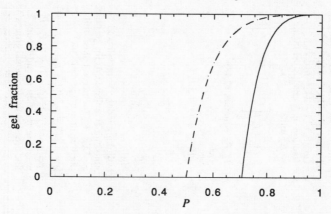

Fig. 2.13 Gel fraction in multifunctional monomer polymerization, plotted against extent of reaction, P. The dashed curve is for a trifunctional monomer, A_3, condensing with itself, $\alpha = P$. The solid curve is for an A_3, B—B step polymerization where the initial monomer groups, A and B, are exactly balanced, $P_A = P_B = P$; $\alpha = P^2$. The curves are based on classical branching theory, equation (2.53). Since $f = 3$, at the gel point $\alpha^* = 1/2$. Thus for the dashed curve $P^* = 1/2$ and for the solid curve, $P^* = \alpha^{*1/2} = 2^{-1/2}$.

introduction of disulfide linkages between the polyisoprene chains (see Chapter 8). It is possible to deduce a criterion for onset of gelation in such systems in a manner very similar to that already introduced in Section 2.3.1 above. Gelation during crosslinking will occur if it is possible to trace out paths of indefinite extent with finite probability. By the same argument as for gelation in branched polymers, that means that the probability of being able to continue on to another chain from an arbitrary one must exceed one. A path could start at the selected crosslink in Figure 2.14. If the fraction of chain units that have undergone crosslinking is ρ_x, then in a chain of n units the probability of further crosslinks, in addition to the one selected, occurring in that chain and therefore the possibility of continuing on with a path is $\rho_x(n - 1)$. Therefore, if the average value of $\rho_x(n - 1)$ exceeds one, the possibility of indefinite networks is present. Therefore, it is desired to compute

$$\varepsilon = \sum_n \rho_x(n - 1)P_n, \tag{2.54}$$

where P_n is the probability that the selected crosslink occurs in a chain containing n units. This probability will be proportional to the number of units n in the chain and, of course, to the number of chains, N_n (before

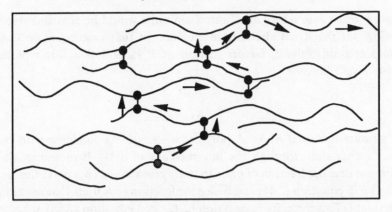

Fig. 2.14 A collection of chains randomly crosslinked. An indefinitely long path (indicated by the arrows) is traced out from a selected crosslink (indicated by the partially shaded circles).

crosslinking), that have n units, and therefore

$$P_n = nN_n / \sum nN_n \qquad (2.55)$$

and

$$\varepsilon = \sum \rho_x n(n-1)N_n / \sum nN_n$$

$$= \rho_x(X_W^0 - 1), \qquad (2.56)$$

where X_W^0 is the weight-average DP before crosslinking. For $\varepsilon = 1$, $\rho_x^* = 1/X_W^0$. Therefore, it is concluded (Stockmayer, 1944) that the critical condition for the onset of gelation is that there is *one crosslinked unit for every initial weight-average chain*. A *crosslinking index*, γ, is defined as the number of crosslinked units per initial chain. Since $\rho_x = v/\sum nN_n$, where v is the number of crosslinked units in the system, then the crosslinking index is

$$\gamma = v / \sum N_n = \rho_x \sum nN_n / \sum N_n$$

$$= \rho_x X_N^0. \qquad (2.57)$$

Since X_N^0 is always less than or equal to X_W^0, the critical value of the crosslinking index required for gelations is always less than or equal to one. For a 'most probable' distribution polymer, it would be one-half.

It is interesting to note that the branching treatment of Section 2.3.1 leads to the same criterion for gelation as that above. The crosslinks can be regarded as tetrafunctional A monomers copolymerized with bifunctional A—A monomer. Equation (2.44) can be modified so that it applies to the case where all of the chain units would be crosslinkable by leaving out the B—B alternate monomers. In that case equation (2.43) would contain only P_A factors instead of $P_A P_B$ and therefore equation (2.44) would become

$$\alpha = \frac{\rho P_A}{1 - (1 - \rho)P_A} \tag{2.58}$$

for a multifunctional A, A—A polymer where the As condense with each other rather than through the intermediary of a B—B monomer. It is apparent that the fraction of units that are crosslinked (ρ_x) plays the same role as ρ in equation (2.44) and hence the fraction of A units that are multifunctional. Therefore, from equation (2.42) and equation (2.58) at the gel point ($f = 4$, $\alpha^* = 1/3$) it follows

$$1/3 = \rho_x P/[1 - (1 - \rho_x)P]$$

or

$$P^* = 1/(1 + 2\rho_x). \tag{2.59}$$

The 'initial' polymer can be regarded as that obtained by severing the tetrafunctional A units into two A—A units. It is apparent from the derivation of equation (2.44) that its molecular weight distribution would be of the 'most probable' type (Section 2.2.2) for which $X_W = (1 + P)/(1 - P)$. Combining this with equation (2.58) it follows that

$$X_W - 1 = 1/\rho_x$$

in agreement with the criterion of equation (2.56). Thus, the equivalence of the crosslinking point of view and the branching polymerization is demonstrated for this special case. Of course, the crosslinking criterion, equation (2.56), is valid for any initial molecular weight distribution, not just the one inferred to be appropriate above for the branching polymerization.

The molecular weight distribution for arbitrary initial distribution resulting from the crosslinking has also been worked out by Stockmayer (1944).

2.3.3 Crosslinking – end-group

There is another crosslinking situation where the chemistry differs from the random reaction at monomeric chain units. Often a linear polymer is crosslinked by reaction at the ends of the chains with a multifunctional crosslinking agent (see Figure 2.15). The criterion for gelation is easily deduced by analogy with the branching case. Let N_E^0 be the number of

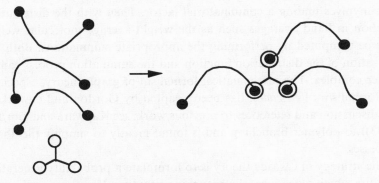

Fig. 2.15 Crosslinking by reacting ends of the initial polymer with a multi-functional crosslinking agent.

reactive end-groups in the initial polymer and let N_E be the number remaining at any point during the crosslinking reaction. If N_A^0 is the initial number of reactive A groups on a multifunctional crosslinking agent and N_A the number remaining during the reaction, then define

$$P_E = 1 - N_E/N_E^0$$

and

$$P_A = 1 - N_A/N_A^0.$$

The probability of two crosslink sites being connected is $P_E P_A$ and from the branching argument of Section 2.3.1, the criterion for indefinitely long paths is

$$P_E P_A (f - 1) \geq 1. \tag{2.60}$$

This equation also follows directly from equations (2.42) and (2.44) for $\rho = 1$. If, for example, the crosslinking agent has been adjusted initially so that $N_A^0 = N_E^0$ and $f = 3$, gelation starts when the extent of reaction of the end-group of $(\frac{1}{2})^{1/2} \approx 70\%$.

2.3.4 Graph theory and branching

To obtain detailed results concerning the various molecular weight averages and how they depend on the degree of reaction, the classical theory of branching proceeds in two steps. First the molecular size distribution function is derived. This involves first setting up a probability model such as was done in Section 2.2.2 for linear ($f = 2$) polymers in equations (2.9) and (2.10). In a branching problem, however, there are always a large number of branching patterns that must be enumerated.

This involves finding a combinatorial factor. Then with the distribution function in hand averages such as the weight-average molecular weight may be computed by performing the appropriate summations. Both the derivation of the distribution function and the summation process can be rather complex. The mathematical formalism of graph theory, especially the formalism of *cascades*, has been adapted by Gordon and coworkers (for discussion and references to previous work, see Kajiwara and Gordon (1973)) to polymer branching and is found greatly to simplify the above processes.

The strategy of cascade theory is to formulate a probability generating function which reproduces the branching situation. If such a function can be formulated, the averages can be directly obtained from it by simple operations, and explicit summations over the complete probability distribution function are not required. It has been found that such generating functions are readily formulated for branched and crosslinked polymers and results for complex situations that would otherwise be quite tedious to obtain can be derived (Kajiwara and Gordon, 1973, Gordon and Ross-Murphy, 1975, 1979, Irvine and Gordon, 1980, Gordon and Torkington, 1981). Figure 2.16 shows the translation of chemical graphs characteristic of a branched structure into the language of cascade theory. The probability generating function is constructed from functions F_0, F_1, F_2, F_3 which express the fertility expectations of each individual (branch points) of a generation (g_0, g_1, g_2, g_3, g_4).

2.3.5 Percolation theory

As indicated above, the classical Flory–Stockmayer approach is based on the structural model of tree-like molecules. Another possible limitation of the model is that the probability of reaction of an unreacted end-group, A, for example, with another group, B, is assumed to be equal to the *overall fraction* of unreacted B groups. Thus this is a *mean-field* assumption, where the overall average extent of reaction is used. There is no accounting for the probability of reaction varying according to the local environment. As far as innate chemical reactivity is concerned this is an accurate assumption. With respect to the effect of the spatial distribution of the reacting groups, it is to be noted that the chemical reactions leading to the branched molecule or network are extremely slow compared to the dynamics of conformational mobility within the molecule. The molecules are extremely dynamic structures over the time scale of the chemistry of reaction. If the molecules were free to pass through each other the

disconnected chemical graph

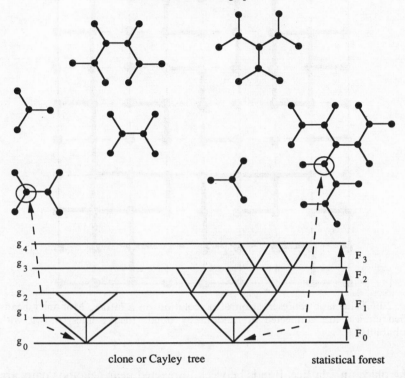

Fig. 2.16 Cascade theory representation of branched chemical structures (after Kajiwara and Gordon, 1973).

mean-field reactivity assumption would therefore be a good one. However, spatial restrictions on molecular beads occupying the same space, i.e., the excluded volume effect (discussed in Sections 3.3.3 and 6.4 and Chapter 9) could to some degree compromise this assumption. Thus there are two features of the classical branching approach, the neglect of loops and the neglect of excluded volume effects, that motivate the investigation of alternative models. Most prominent has been the *percolation* method (see Stauffer, Coniglio and Adam (1982) and Stauffer and Aharony (1991)).

Percolation implies both a *model* for the reacting system *and* a *philosophy* as to what results are to be extracted. The classical branching model emphasizes obtaining a solvable, complete statistical connectivity description, valid within the assumptions of the model. The molecular weight distribution function and the averages of molecular weight are found. The percolation method, as usually applied, allows the reaction to

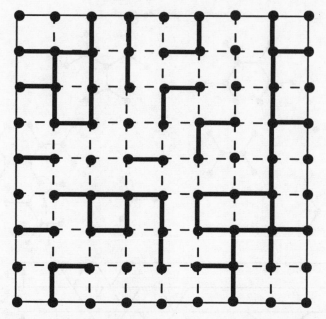

Fig. 2.17 Schematic representation of percolation on a lattice. Monomer beads (filled circles) are connected at random by bonds (heavy lines) drawn with probability, P.

take place on a lattice. Bonds between unreacted near neighbor pairs are drawn at random, Figure 2.17. Such a model is not regarded as solvable in the sense of being able to derive a complete exact theoretical statistical description. However, numerical simulations can be made and some theoretical limiting or approximate results can be obtained. It is found, as in the classical branching model, that at a very definite extent of reaction (fraction of possible bonds drawn) a gelation critical point occurs and indefinitely large molecules appear, i.e., there is a percolation threshold or transition.

The philosophical aspect enters through the kind of results that are to receive attention. The emphasis is on results that are independent of the details of the model and involves properties that change, often dramatically, in the vicinity of the critical point. The results are couched in terms of how the variation takes place. In the immediate vicinity of the critical point, i.e., the gel point, the variation follows a power law expression. For example, the weight-average DP, which diverges at the gel point, is expressed as a function of reaction probability, P, as

$$X_W = c/(P - P^*)^{\gamma_c}; \qquad P \to P^*, \qquad (2.61)$$

where (following the notation of Stauffer *et al.* (1982)) γ_c is a *critical exponent*, $P*$ is the reaction probability at the gel point and c is a proportionality constant or 'amplitude'. Of the three parameters only the critical exponent γ_c is regarded as *universal*. That is, it is independent of the details of the model and represents the variation of X_W in random percolation. The gel point itself, $P*$, was seen earlier, equation (2.42), to depend on the functionality of the groups reacting. It is thus not a universal property of percolation. Similarly the amplitude, c, also depends on the details of the model. Thus attention is centered on the critical exponents. They have been established for several properties. Some examples in addition to X_W follow. The gel fraction is expressed in terms of a critical exponent β_c as:

$$G = c'/(P - P*)^{\beta_c}; \qquad P \to P* \text{ for } P > P*. \qquad (2.62)$$

The above quantities are the result solely of the connectivity of the system. The spatial extent of the molecules is of interest also. The size of one of the molecules containing n mers can be expressed as its radius of gyration (see Section 6.2.2), s. The interest lies in how the size of the molecules depends on the number of mers that have $\text{DP} = n$ in them as the latter increases. The scaling relation is written in terms of the exponent, ρ_c, as

$$\langle s^2 \rangle \to n^{2\rho_c} \text{ as } n \to \infty \text{ at fixed } P. \qquad (2.63)$$

The correlation length, ξ, is the rms value of the distance R between two mers where the average is over the probability that two mers are in the same molecule. Alternatively put, it is a typical 'cluster' size. It is given by the 'z-average' of the molecular weight distribution function, N_n

$$\xi^2 = \langle R^2 \rangle_z = \sum n^2 N_n R_n^2 / \sum n^2 N_n. \qquad (2.64)$$

The scaling relation, with exponent, v_c, is

$$\xi = c''/(P - P*)^{v_c}; \qquad P \to P*. \qquad (2.65)$$

Results found for the above exponents (summarized by Stauffer *et al.* (1982)) are shown in Table 2.4. It is also possible to investigate the behavior of the classical Flory–Stockmayer theory in the vicinity of the gel point. Asymptotic expressions in the form of power laws result. Thus the critical exponents from the classical approach are available as well. These are also listed in the table. It may be seen that, in general, there are major differences in the scaling predictions between percolation and classical theory.

Table 2.4 *Summary of critical exponents from percolation and classical theory.*

property	exponent	percolation	classical
X_w	γ_c, equation (2.61)	1.7	1
gel fraction	β_c, equation (2.62)	0.45	1
radius, $\langle s^2 \rangle^{1/2}$	ρ_c, equation (2.63)		
at $P = P^*$		0.40	1/4
at $P < P^*$		1/2	1/4
at $P > P^*$		1/3	1/4
correlation length	ν_c, equation (2.65)	0.88	1/2

It is worthwhile making some comparisons of the percolation and classical approaches. The percolation results for the critical exponents by virtue of loop inclusion in the model are presumed to be superior to the classical values (Adam, 1991; and references therein). However, these results are confined to the near vicinity of the gel point. The classical method gives detailed results over the entire range of the reaction extent and includes fixing the gel point. The simulations used to establish the percolation critical exponents can, in principle, be used to provide the same information. However, there is no reason to believe that the lattice model *per se* beyond the loop question has advantages over the branching model in representing real materials. The static nature of the lattice in which bonds are drawn among near neighbors, in fact, is not a realistic representation of the chemistry.

2.4 Further reading

The material in Sections 2.1, 2.2 is relatively straightforward. In contrast, the treatment of branching and crosslinking is rather complicated. For the classical theory, in addition to Flory's book (1953) and review article (1946) there are two reviews by Burchard (1981, 1982). The cascade representation by Gordon has references to it in the text. For percolation the review by Stauffer *et al.* (1982) and book by Stauffer and Aharony (1991) are recommended.

Nomenclature

C_A, C_B = concentration of A, B groups (moles/l) (superscript indicates initial values)

DP' = degree of polymerization

G = the weight fraction of gel above the gel point

f = functionality of a monomer unit

m_0, m_{AA}, m_{BB} = molecular weight of a monomeric unit

$M_1, M_2, M_3, \ldots, M_n, \ldots$ = molecular weight of molecules of
  $\mathrm{DP} = 1, 2, 3, \ldots, n, \ldots$

M_N = number-average molecular weight

M_W = weight-average molecular weight

N_A, N_B = number of moles of A and B groups (superscript zero indicates
  initial values)

N_n = number of moles of polymer molecules of degree of polymerization,
  n (number distribution function)

$\bar{N}_n$ = normalized N_n

P = probability of a monomer unit being connected to another, extent
  of reaction of initial monomers = $1 - C/C^0$

P_A, P_B = extent of reaction of A, B groups $= 1 - C_A/C_A^0, = 1 - C_B/C_B^0$

r = ratio of initial concentrations of reactive groups, $= C_B^0/C_A^0$

r_p = rate of polymerization (moles/s)

$\bar{w}_n$ = weight fraction of polymer molecules with degree of polymerization,
  n (weight distribution function)

X_N = number-average degree of polymerization

X_W = weight-average degree of polymerization

Z = parameter specifying the width of the molecular weight distribution
  in Schulz–Zimm distribution

α = branching index, the probability that a chain leaving a branch point
  continues on to another branch point

β_c = critical exponent for gel fraction variation with extent of reaction
  in gelation

γ = crosslinking index, the number of crosslinked units per molecule

γ_c = critical exponent for X_W variation with extent of reaction in
  gelation

ε = gelation criterion in random crosslinking, equation (2.56)

ρ = fraction of reactive groups A that belong to higher functionality
  monomeric units (i.e., fraction of A groups in A_f in an A_f, A—A,
  B—B polymerization)

ρ_c = critical exponent for cluster radius variation with mer content in
  gelation

ρ_x = fraction of units crosslinked in a random crosslinking reaction

σ = standard deviation

ξ_c = critical exponent for correlation length variation with extent of
  reaction in gelation

Problems

2.1 Let ΔH_n = the heat of fusion of an alkane of chain length (number of carbon atoms) = n. Let ΔS_n = the entropy of fusion of an alkane of chain length (number of carbon atoms) = n. The melting point is that temperature where the Gibbs free energy, $\Delta G_n = 0$. Assume that both ΔH_n and ΔS_n are linear functions of the chain length, i.e., they each increase linearly with 'n'. Also assume that ΔH_n and ΔS_n are each independent of temperature (not really highly accurate but good enough for this problem). Show mathematically that the melting point will change with chain length, n, but that it approaches an asymptotic limit as the chain length becomes very long.

2.2 A sample of polyformaldehyde (polyoxymethylene), density = 1.4 g/cm^3, has hydroxyl end-groups on both ends. The hydroxyl group has a characteristic infrared absorption band. The absorbance is measured to be 0.3. A sample, of the same thickness, of 1,10-decanediol, HO—$(CH_2)_{10}$—OH, mixed 10% by weight in a mineral oil (no OH groups) and having a mixture density of 0.90 g/cm^3 has absorbance of 0.4. (Concentrations are proportional to absorbance and sample thickness.)

(a) Calculate the number-average molecular weight and DP.

(b) Would this polymer sample have useful mechanical properties?

2.3 A sample of commercial nylon 66 weighing 1000 g

$$[-(C=O)-(CH_2)_4-(C=O)-NH-(CH_2)_6-NH-]$$

is placed in a tight-fitting metal cylinder and water amounting to 10 g is used to fill the remaining space in the cylinder and the lid is screwed on. It is heated to 250 °C for 24 hours and then rapidly cooled. Calculate approximately the DP of the resulting polymer.

2.4 Derive equations for M_N and M_W of a blend of two fractions 1 and 2 each having $(M_N)_1$, $(M_W)_1$ and $(M_N)_2$, $(M_W)_2$ separately and each weighing w_1 and w_2.

2.5 Find the weight-average molecular weight of Figure 2.6, if $m_{AA} = 2m_{BB}$ (i.e., recompute Table 2.2).

2.6 Suppose a polymer has two kinds of monomers A and B in it that can, in principle, be present in any ratio (copolymer). The distribution function for molecular weight is $N_{n,l}$. That is, $N_{n,l}$ is the number of polymer molecules that have n A groups and l B groups. If m_A and m_B are the molecular weights of the monomeric units, derive equations in terms of N_{nl} for the number-average and weight-average molecular weights.

2.7 Use the exponential approximation to the 'most probable' distribution to calculate M_W, M_W/M_N, and the weight fraction of polymer with DP less than 50, all in a nylon 66 sample with $X_N = 100$.

2.8 Let

$$F_n = X_N^n/n! \, e^{-X_N} \qquad \text{(the Poisson distribution)}$$

(a) Show that since

$$\sum_{n=0}^{\infty} F_n = 1,$$

then

$$\sum_{n=0}^{\infty} n F_n = X_N \sum_{n=0}^{\infty} F_n = X_N.$$

(b) Show that

$$\sum_{n=0}^{\infty} n^2 F_n = X_N^2 + X_N.$$

(c) From equation (2.22) show that

$$\sigma = X_N^{1/2}.$$

2.9 Show that equation (2.48) reduces to equation (2.5) for $f = 2$.

2.10 Derive an equation for the branching index α for the case of both trifunctional monomers A_3 and B_3 present as well as A—A and B—B in a step polymerization.

2.11 Find the scaling exponent, β_c, associated with the gel fraction just above the gel point, equation (2.62), as represented by the classical theory in equation (2.53) for the two cases illustrated in Figure 2.13.

3

Molecular weight determination

There are a number of methods for determining experimentally the molecular weights of polymers. These include both the measurement of number-average and weight-average molecular weights. It will be seen that solution viscosity offers a very convenient method but it is not an absolute method and does not give one of the simple averages. The resolution of molecular lengths into fractions and therefore measurement of molecular weight distribution is also possible experimentally.

3.1 End-group analysis

In linear polymers each molecule has two ends so it is clear that a measurement of total numbers of end-groups in a sample of known weight can result in a determination of number-average molecular weight (=sample weight/moles of chains). There is no general method for accomplishing this and essentially the task embraces organic functional group identification in analytical chemistry. An obvious complication is that the method must be very sensitive since the end-groups are present at very low concentrations in high molecular weight polymers. The available methods can perhaps be classified as chemical or physical. Chemical methods would include acid–base titration of acidic or basic end-groups ($—CO_2H$, for example), reaction of end-groups with determinable amounts of specific reagents, and chemical degradation to identifiable products from end-groups. The most prominent physical method and the most useful method in general is probably infrared vibrational spectroscopy. Diagnostic functional group frequencies are often observed and usually can be measured with great sensitivity. The method does suffer from not being absolute and must be calibrated by measuring band intensities in model compounds to establish absorption coefficients. NMR spectroscopy would also be a highly diagnostic technique but it often lacks the necessary sensitivity.

3.2 Colligative methods for M_N

These are the classic methods that are absolute but depend on extrapolation to an infinitely dilute ideal solution. As will be seen, of the three methods: boiling point elevation, freezing point depression and osmotic pressure, only the last is sensitive enough to be applicable to high molecular weight polymers.

3.2.1 *Boiling point elevation* (*ebulliometry*)

This method is based on the vapor pressure lowering and attendant boiling point elevation of a solvent by a polymer solute. The boiling point of the pure solvent is compared with that of the polymer solution. The temperature difference is related to molecular weight in a known absolute way provided the measurements are carried out in dilute solution and extrapolated to infinite dilution. Assuming that the vapor is ideal, the vapor pressure P_1, over a solution of volatile solvent (1) and non-volatile polymer (2) is related to solvent activity, a_1,

$$P_1/P_1^* = a_1, \tag{3.1}$$

where P_1^* is the vapor pressure of pure solvent at the same temperature. At the specific temperature, T, the solution vapor pressure may be written as,

$$P_1(T) = a_1 P_1^*(T).$$

The condition that this *solution vapor pressure be the same as the pure solvent* at its laboratory boiling point T_b may be written as

$$P_1(T) = a_1 P_1^*(T) = P_1^*(T_b)$$

or

$$a_1 = P_1^*(T_b)/P_1^*(T). \tag{3.2}$$

The effect of temperature on pure solvent vapor pressure may be found from the Clausius–Clapeyron equation (with several attendant approximations) as

$$\ln P_1^*(T_b)/P_1^*(T) = \frac{\Delta H_{vap}^0}{R}\left(\frac{1}{T} - \frac{1}{T_b}\right).$$

Therefore,

$$-\ln a_1 = \left(\frac{\Delta H^0_{\mathrm{vap}}}{RT_b^2}\right)\Delta T, \qquad (3.3)$$

where $\Delta T = T - T_b$ is the boiling point elevation and $T_b T$ is approximated by T_b^2.

The solvent activity may be related to the polymer concentration by means of a power series expansion of $\ln a_1$ in the polymer concentration, C, as

$$-\ln a_1 = A_1' C + A_2' C^2 + A_3' C^3 + \cdots. \qquad (3.4)$$

The first coefficient, A_1', may be determined by making a connection with Raoult's Law. That is, $a_1 \to X_1$ as $C \to 0$ (where X_1 is the solvent mole fraction $= N_1/(N_1 + N_2)$, and N_1, N_2 are the moles of 1 and 2). Therefore,

$$\ln a_1 \to \ln X_1 = \ln(1 - X_2) \text{ as } C \to 0$$

and

$$\ln(1 - X_2) \to -X_2 \to -N_2/N_1 \text{ as } C \to 0.$$

But,

$$N_2 = W_2/M, \qquad (3.5)$$

where W_2 is the polymer weight and M its molecular weight. The polymer concentration is usually defined as weight of polymer/solution volume, $C = W_2/V$. Therefore,

$$-\ln a_1 \to W_2/MN_1 \to CV_1^0/M, \text{ as } C \to 0.$$

Since from equation (3.4)

$$-\ln a_1 \to A_1' C, \text{ as } C \to 0,$$

it is seen now that

$$A_1' = V_1^0/M,$$

where V_1^0 is the molar volume of the pure solvent. Equation (3.4) may be rewritten (with $A_2 = A_2'/V_1^0$, $A_3 = A_3'/V_1^0 \cdots$) as

$$-\ln a_1 = V_1^0\left(\frac{1}{M}C + A_2 C^2 + A_3 C^3 + \cdots\right). \qquad (3.6)$$

Equation (3.6) is a *virial expansion* of $\ln a_1$ in the polymer concentration and $A_2, A_3, \ldots$ are second, third, ... *osmotic virial coefficients*. From equation (3.3) and equation (3.6) it follows that

$$\frac{\Delta T}{C} = K_B\left(\frac{1}{M} + A_2 C + A_3 C^2 \cdots\right), \qquad (3.7)$$

where

$$K_B = \frac{RT_b^2}{\Delta H_{vap}^0} V_1^0.$$

The molecular weight, M, of the polymer is determined from a series of measurements of ΔT at various concentrations by plotting $\Delta T/C$ vs C. If the solutions are sufficiently dilute, a straight line may be drawn through the data where the slope is $K_B A_2$. The intercept at $C = 0$ determines K_B/M and K_B may be determined from pure solvent properties. It is found, in practice, that measurements on solutions in the vicinity of 1% polymer are appropriate for the extrapolation procedure.

The relationship used in equation (3.5) to connect molecular weight with moles of polymer and weight concentration is seen to be the definition of *number-average* molecular weight (weight of polymer/moles of polymer) and thus the method determines M_N. A sample calculation (Table 3.1) indicates that very small temperature differences must be measured (requiring extremely sensitive temperature resolution for moderate accuracy) for even relatively low molecular weights.

A variation of the method called 'vapor pressure osmometry' is based on measuring the temperature difference between pure solvent and solution in an equilibrating chamber. No attempt is made actually to boil the liquids and the method is not regarded as absolute, K_B being determined by calibration. A practical limit of $M_N < 10\,000$ obtains in this and other boiling point methods.

3.2.2 *Freezing point depression* (*cryoscopy*)

This method depends on the relation, under dilute solution conditions, between molecular weight and the difference in the freezing point of a pure solvent and that of a solution. The chemical potential of the solvent (1) at the temperature T in terms of its activity is

$$\mu_1(T) = \mu_1^0(T) + RT \ln a_1. \tag{3.8}$$

At the (initial) freezing point $(=T)$ the solution at this activity is in equilibrium with (assumed to be) pure solvent crystals, of chemical potential $\mu_{1c}^0(T)$ and, therefore,

$$\mu_1(T) = \mu_{1c}^0(T)$$

or

$$\mu_1^0(T) - \mu_{1c}^0(T) = -RT \ln a_1. \tag{3.9}$$

Table 3.1 *Comparison of colligative methods for M_N determination (benzene as solvent)*[a].

quantity measured	boiling point elevation[b] $\Delta T = T - T_b$	freezing point depression[c] $\Delta T = T_f - T$	osmotic pressure[d,e] $\Pi = P'' - P'$	
Magnitude ($C = 1$ g/dl)				
10 000 g/mole	3×10^{-3} K	5×10^{-3} K	2.5×10^{-2} atm = 300 mm	
100 000 g/mole	3×10^{-4} K	5×10^{-4} K	2.5×10^{-3} atm = 30 mm	
Resolution required for $\pm 1\%$ precision ($C = 1$ g/dl)				
10 000 g/mole	30×10^{-6} K	50×10^{-6} K	3 mm	
100 000 g/mole	3×10^{-6} K	5×10^{-6} K	0.3 mm	

[a] $\Delta H_{vap}^0 = 30.75$ kJ/mole, $V_1^0 = 95.8$ cm^3/mole, $T_b = 353.25$ K
$\Delta H_f^0 = 9.94$ kJ/mole, $V_1^0 = 87.3$ cm^3/mole, $T_f = 278.7$ K
[b] $\Delta T = K_b C/M + \cdots$, $K_b = V_1^0 R T_b^2/\Delta H_{vap}^0 = 32.3$ K dl/mole
[c] $\Delta T = K_f C/M + \cdots$, $K_f = V_1^0 R T_f^2/\Delta H_f^0 = 56.7$ K dl/mole
[d] $\Pi = RTC/M + \cdots$, $R = 0.8205$ dl atm
[e] $T = 300$ K

At the freezing temperature T_f of pure solvent, the pure liquid and solid are in equilibrium or

$$\mu_1^0(T_f) = \mu_{1c}^0(T_f).$$

Therefore (from $H = -T^2[(\partial G/T)/\partial T]_0$),

$$\frac{\mu_1^0(T) - \mu_{1c}^0(T)}{T} = -\int_{T_f}^T \Delta H_f^0/T^2 \, dT \qquad (3.10)$$

and, under the assumption (valid for small ΔT) that the molar heat of fusion of pure solvent ΔH_f^0 is temperature-independent, equation (3.10) is integrated and combined with equation (3.9) to give

$$-\ln a_1 = \frac{\Delta H_f^0}{RT_f^2} \Delta T, \qquad (3.11)$$

where $\Delta T = T_f - T$ and $T_f T$ is approximated by T_f^2. Referring again to the virial expansion of $-\ln a_1$ in equation (3.6) it is found that

$$\frac{\Delta T}{C} = K_f\left(\frac{1}{M} + A_2 C + \cdots +\right), \qquad (3.12)$$

where

$$K_f = \frac{RT_f^2}{\Delta H_f^0} V_1^0.$$

As in the boiling point method, measurements of ΔT at concentrations near 1% are used to make a plot of $\Delta T/C$ vs C. The intercept at $C = 0$ gives K_f/M and K_f is determinable from properties of the pure solvent. Sample calculations (Table 3.1) show that the method is somewhat more sensitive than boiling point elevation but it is still restricted to relatively low molecular weights.

3.2.3 Osmotic pressure

The osmotic pressure method is based on the concept of a *semi-permeable* membrane. For such a membrane, solvent is able to transport freely through it, but polymer molecules are completely blocked. A cell can be set up with the membrane separating a compartment containing pure solvent from one containing the polymer solution. The activity or chemical potential of the solvent will be lower on the solution side, thus there will be a driving force for diffusion of the solvent from the pure solvent side into the solution side. The diffusion would, in principle, continue until all of the solvent has migrated into the solution compartment. However, if an extra external pressure is applied to the solution side (and the membrane is rigid) the solvent activity may be increased. In fact, one could imagine applying just the right pressure so that the driving force for solvent diffusion from concentration difference is exactly balanced by the opposite driving force from increased pressure on the solution side. In that case equilibrium would be established and no net solvent diffusion would take place. Such cells can actually be set up in practice (see Figure 3.1).

The system can be treated thermodynamically as follows. The chemical potential of the solvent on the solution side (at the higher pressure, P'') is

$$\mu_1(T, P'') = \mu_1^0(T, P'') + RT \ln a_1, \tag{3.13}$$

where a_1 is the solvent activity and $\mu_1^0(T, P'')$ is the pure solvent chemical potential at the pressure P'' of the solution side. On the pure solvent side (at pressure P'), however, the chemical potential of the solvent is

$$\mu_1(T, P') = \mu_1^0(T, P')$$

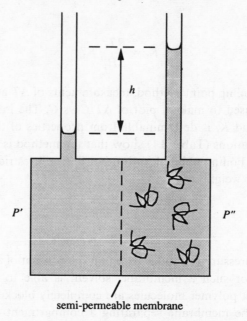

semi-permeable membrane

Fig. 3.1 Self-adjusting membrane osmometer. Diffusion of solvent into solution compartment results in a pressure difference that opposes diffusion. Net diffusion stops and the height difference in the standpipes stabilizes when equilibrium osmotic pressure is reached.

and at equilibrium the chemical potentials of both sides are equal and

$$\mu_1(T, P'') = \mu_1^0(T, P')$$

or

$$\mu_1^0(T, P'') + RT \ln a_1 = \mu_1^0(T, P'). \qquad (3.14)$$

In order to proceed, it is necessary to know the effect of pressure on the pure solvent chemical potential. For the pure solvent, from the general equation for the Gibbs free energy of a single component open system,

$$dG = -S\,dT + V\,dP + \mu_1\,dN_1,$$

a Maxwell-like relation is found,

$$(\partial \mu_1^0/\partial P)_{T,N_1} = \left(\frac{\partial V}{\partial N_1}\right)_{T,P} = V_1^0 \quad \text{(the molar volume of pure solvent)}.$$

Therefore, at constant temperature, the effect of pressure on the pure solvent chemical potential can be found by integrating the above equation

$$\mu_1^0(T, P'') = \mu_1^0(T, P') + \int_{P'}^{P''} V_1^0 \, dP.$$

Using this relation in equation (3.14) results in

$$-RT \ln a_1 = \int_{P'}^{P''} V_1^0 \, dP.$$

Under the excellent approximation that at the small pressure differentials involved V_1^0 will change very little with pressure the above becomes

$$-RT \ln a_1 = V_1^0 \Pi, \tag{3.15}$$

where Π is the *osmotic pressure* $= P'' - P'$. This is an important equation that provides a general method for measurement of solvent activity. The method, of course, depends on the availability of the semi-permeable membrane. However, the large disparity in size between polymer and solvent molecules makes the task of finding suitable membranes a realizable one.

Returning to the question of molecular weight determination, the virial expansion of $\ln a_1$ is again used (equation (3.6)) and when combined with equation (3.15) gives

$$\frac{\Pi}{CRT} = \frac{1}{M} + A_2 C + A_3 C^2 + \cdots. \tag{3.16}$$

A plot of Π/CRT vs C gives, in the concentration range near 1%, a linear plot of slope A_2 and intercept $1/M$. The nature of the extrapolation and the variation of intercept for samples of varying molecular weight are shown in Figure 3.2.

In Figure 3.3 examples are shown of osmotic pressure data for the same polymer in different solvents. Note that the value of the slope, and even its sign, are dependent on the nature of the interactions between the polymer and solvent. As expounded in Chapter 9, a good solvent gives a positive slope and a poor one a negative slope. A zero slope indicates that the intermediate condition of a 'theta' solvent (introduced in the next section) that is neither a good nor a poor solvent. A really poor solvent would show no appreciable solubility so that the range of slopes for poor solvents is restricted to small negative ones. An important practical consequence of this effect is that a poor solvent, with its small negative slope, makes extrapolations easier and requires fewer data points than a good solvent with a high positive slope.

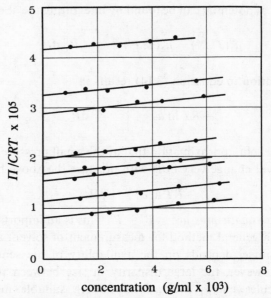

Fig. 3.2 Plots of Π/CRT vs concentration for solutions of cellulose acetate fractions in acetone (after Badgley and Mark, 1949).

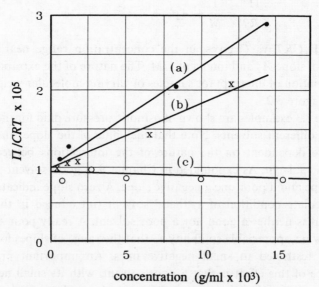

Fig. 3.3 Plots of Π/CRT vs concentration for nitrocellulose in (*a*) acetone, (*b*) methanol, (*c*) nitrobenzene (Gee, 1944).

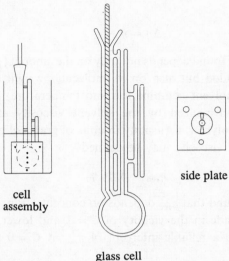

cell
assembly

side plate

glass cell

Fig. 3.4 A schematic drawing of the assembly of a membrane osmometer based on the design of Zimm and Meyerson (1946, with permission from *J. Amer. Chem. Soc.*, © 1946, American Chemical Society). Two semi-permeable membranes are held against the sample cell by channelled metal plates through which the solvent can pass. The osmotic pressure is measured from the height of the liquid in the capillary (hatched area). Automated devices can sense the meniscus electronically or measure the pressure via transducers.

Sample calculations (Table 3.1) show that the method is much more sensitive than boiling point elevation and freezing point depression. That is, liquid heights in the standpipes are readily measured to a fraction of a millimeter and this corresponds roughly to 1% precision at 100 000 g/mol molecular weight. Thus it is quite useful for number-average molecular weight determination of high molecular weight polymers. In practice, in addition to the requirement that the membrane be semi-permeable since solvent diffusion is involved, particular attention must be paid to seeing that equilibrium is attained. This means that the osmotic pressure at each concentration should be followed as a function of time. Considerable effort has gone into instrument and cell design to minimize the time for equilibration (Figure 3.4).

3.3 Solution viscosity

3.3.1 Method

If an increment of polymer is added to a solvent, the viscosity, η, of the solution is increased over that of the pure solvent, η_0, by an

amount,

$$\Delta\eta = \eta - \eta_0. \tag{3.17}$$

The actual increase found depends not only on the amount (weight/volume) of the polymer added but also on its molecular weight (as well as the specific polymer–solvent combination and temperature) (Figure 3.5). If this increase is normalized by the solvent viscosity and the weight concentration of polymer, C (usually in units of grams of polymer/100 ml solvent) a reduced viscosity may be defined

$$\eta_{\text{red}} = \Delta\eta/C\eta_0. \tag{3.18}$$

In practice, it is found that η_{red} depends on concentration, but if measurements of η are made in the vicinity of $C = 1$ and lower, a plot of η_{red} vs C is linear and a reliable intercept of η_{red} at $C = 0$ may be found.

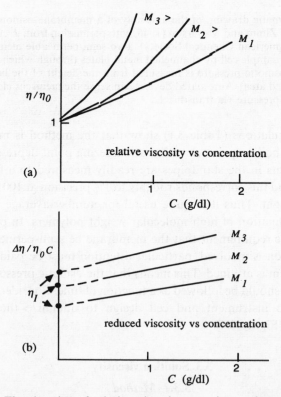

Fig. 3.5 (a) The elevation of solution viscosity at a given polymer concentration depends on molecular weight. (b) The reduced viscosity (secant slope of the relative viscosity in (a)) extrapolated to infinite dilution defines the intrinsic viscosity.

This intercept,

$$\eta_I \equiv \lim_{c \to 0} \frac{\Delta\eta}{C\eta_0}, \qquad (3.19)$$

is called the *intrinsic* viscosity. Being an infinite dilution quantity it is a property of isolated polymer molecules and the solvent. It is not influenced by interactions between polymer molecules. It is found from experiment (and also from theory) that the intrinsic viscosity of a *monodisperse* solution of coiling polymer molecules of molecular weight, M, follows (Figure 3.6) the Mark–Houwink equation,

$$\eta_I = KM^a. \qquad (3.20)$$

The constants 'K' and 'a' are empirically determined and, in general, depend on the solvent–polymer combination and the temperature. Thus the solution viscosity method is not an absolute one and its use depends on its being calibrated by means of a series of polymer fractions of known molecular weight (perhaps determined by osmometry or light scattering). Its popularity results from its relative experimental ease and simplicity (Figure 3.7). The viscosity in a capillary viscometer under conditions of a long tube and relatively slow flow is proportional to the flow time, t, for a given quantity of fluid. Therefore the relative viscosity, η/η_0, determination merely requires measurement of the flow time ratio t/t_0 for solution and pure solvent. There are extensive tabulations of the Mark–Houwink constants (K, a) for a variety of polymer–solvent systems.

3.3.2 Polydispersity

It can be imagined that each molecular weight species, M, is added to the solution as a separate increment. At infinite dilution the increases in

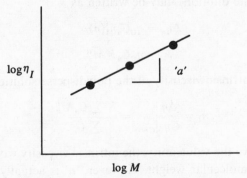

Fig. 3.6 Dependence of intrinsic viscosity on molecular weight.

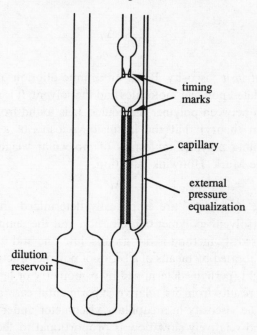

Fig. 3.7 Ubbelohde dilution viscometer. A fixed volume of solution established by the timing marks is allowed to flow through the capillary. The reservoir is of sufficient volume to permit dilution of the solutions in the viscometer. The pressure equalization pipe keeps the head independent of the volume of the solution in the reservoir.

viscosity with each increment will be additive,

$$\Delta\eta = \sum \Delta\eta_n.$$

But from equation (3.19) and equation (3.20) the increment of viscosity change (at infinite dilution) may be written as

$$\Delta\eta_n = \eta_0 C_n \eta_1(n)$$

$$= \eta_0 C_n K M_n^a.$$

Therefore, the intrinsic viscosity of the polydisperse solution is

$$\eta_1 = \left.\frac{\Delta\eta}{C\eta_0}\right|_{c\to 0} = \frac{K \sum C_n M_n^a}{\sum C_n}. \tag{3.21}$$

If 'a' were to have the value unity, the intrinsic viscosity would follow the weight-average molecular weight. However, 'a' is actually found to fall between the values 0.5 and ~ 0.8 for various polymer–solvent combinations.

Therefore, the solution viscosity method does not give one of the simple average (number-, weight-, etc.) molecular weights in polydisperse systems. It is useful to speak of and define a 'viscosity-average' molecular weight by stating that the Mark–Houwink equation applies to polydisperse systems

$$\eta_1 = KM_v^a.$$

The viscosity-average molecular weight is then *defined* as

$$M_v \equiv \left(\sum C_n M_n^a / \sum C_n \right)^{1/a}. \tag{3.22}$$

In practice, it is found and can be demonstrated by illustrative calculations (see Problem 3.1) that M_v differs ordinarily only slightly from the weight-average molecular weight, M_W.

3.3.3 Solution viscosity and polymer structure

Although the solution viscosity method must be regarded in practice as empirical, there is a great deal of theoretical insight and foundation underlying it. In fact, under the proper experimental conditions it can become an absolute method.

The viscosity behavior of an isolated coiling macromolecule in solution can be rationalized on the basis of its behaving like a suspended sphere in a continuous fluid (solvent). The shear field in the solvent flow in the viscosity measurements causes rotation of the effective spherical polymer coil (Figure 3.8). This leads to an additional frictional loss and increases the viscosity of the solvent–polymer suspension. There are two extreme

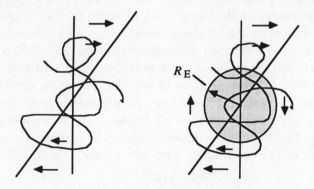

polymer coil in shear field   rotating equivalent sphere of radius R_E

Fig. 3.8 A coiling polymer molecule in shear flow considered as a rotating sphere.

cases one could imagine for the response of the solvent that penetrates the coil. At one extreme, the solvent in the coil can be thought of as experiencing the same shear field as the bulk solvent. This is known as the free draining case. At the other extreme, the solvent in the coil could be considered as completely trapped and, therefore, as riding along with the rotating coil. Theoretical calculations show that normally the latter non-draining or impermeable coil condition is the one that actually obtains. The viscosity can be conceptualized as that due to a suspension of impenetrable rotating spheres. Einstein studied this case and found the suspension viscosity to be

$$\eta = \eta_0(1 + \tfrac{5}{2}\phi_2), \tag{3.23}$$

where ϕ_2 is the volume fraction of suspended spheres (polymer). Thus, ϕ_2 is related to the effective radius of the coil R_E as

$$\phi_2 = \frac{N}{V}\frac{4\pi}{3}R_E^3, \tag{3.24}$$

where N is the number of spheres (molecules) and V the solution volume. The intrinsic viscosity may be written from equations (3.23) and (3.24) as

$$\eta_I = \frac{5}{2}\frac{4\pi}{3}\frac{N_A}{100}R_E^3/M \tag{3.25}$$

where the relation $C = 100NM/N_A V$, N_A = Avogadro number and C in g/dl has been used.

The effective radius is obviously a somewhat nebulous quantity. The average density of polymer segments varies from a maximum at the center of the coil and drops off to zero at large distance. The effective radius of the average sphere can only be determined by detailed calculation of the frictional properties of the distribution of segments. For a spherically symmetric rigid body the rotational inertia that determines the angular velocity behaves as if the mass is concentrated at the radius of gyration, s. This suggests that the radius of gyration of the coil be used as a measure of effective size. Another measure of coil size, related to the radius of gyration, that is somewhat more conveniently amenable to theoretical calculations is the average end-to-end distance, $\langle R^2 \rangle$ (Section 6.1). Thus it is *assumed* that the effective frictional radius R_E is proportional to the mean-square end-to-end distance with proportionality constant, γ,

$$R_E = \gamma\langle R^2 \rangle^{1/2} \tag{3.26}$$

and then from equation (3.25)

$$\eta_I = \Phi\langle R^2 \rangle^{3/2}/M, \tag{3.27}$$

where

$$\Phi = \frac{5}{2} \frac{4\pi}{3} \frac{N_A}{100} \gamma^3.$$

It has been argued (Flory, 1953) that the proportionality constant in equation (3.26) should be independent of detailed structure for coiling polymer molecules and that, therefore, the constant Φ should be a 'universal' one for such molecules.

The theory of the dependence of the end-to-end distance $\langle R^2 \rangle^{1/2}$ of a coiling polymer molecule on molecular weight has been much studied. A special circumstance would obtain if distant parts of the coil did not interfere with the spatial preferences of a given local segment. Under this circumstance, following the chain from one end to the other would be a random walk not perturbed or interfered with by the presence of coil segments from other parts of the chain. The chain would behave as a 'ghost' or 'phantom' chain that can intersect or cross itself freely. An actual chain in a 'poor' solvent would deviate from phantom behavior in the following way. In a poor solvent, polymer– polymer self-contacts are enhanced relative to polymer–solvent contacts. There is a net tendency for segment self-association and the coil is smaller than it would be in the phantom chain case. In a 'good' solvent polymer– polymer segment contacts would be suppressed and polymer–solvent contacts enhanced. The coil is expanded relative to a phantom chain. By the proper choice of solvent, one that is neither a 'good' nor a 'poor' solvent, the tendency for polymer–polymer self-contacts can be exactly balanced by the tendency for polymer–solvent contacts. No preference of a polymer segment for another segment or solvent is exhibited. In this condition, phantom chain behavior can result. The condition is specific to certain solvent–polymer pairs and occurs only at one given temperature. The solvent may become a better solvent as temperature increases and poorer as it decreases. The solvent of such a pair in this condition is often referred to as a θ solvent and the temperature as the θ temperature.

The average-square end-to-end distance of a long coiling polymer molecule under phantom or θ conditions (indicated by subscript zero on $\langle \ \rangle_0$) is given by

$$\langle R^2 \rangle_0 = C_\infty N l^2 \tag{3.28}$$

where N is the number of bonds, l the bond length and C_∞ is a constant (called the *characteristic ratio*, see Section 6.1) that depends on the chain valence angles and the statistics of preference of various bond rotational states of the chain (see Chapter 6). The characteristic ratio may be calculated from theory if the energetics of the bond rotational states are known. Therefore, the average-square end-to-end distance $\langle R^2 \rangle_0$ depends on the molecular weight to the first power under θ conditions,

$$\langle R^2 \rangle_0 = (C_\infty l^2 / m_0) M, \tag{3.29}$$

where m_0 is the monomeric unit molecular weight.

The result in equation (3.29) can be combined with equation (3.27) to give

$$\eta_1 = K M^{1/2} \tag{3.30}$$

with

$$K = \left(\gamma^2 l^2 \frac{C_\infty}{m_0} \right)^{3/2} \frac{5}{2} \frac{4\pi}{3} \frac{N_A}{100} \tag{3.31}$$

or

$$K = (l^2 C_\infty / m_0)^{3/2} \Phi.$$

Detailed theoretical calculation of the frictional properties of coiling chains (dePyun and Fixman, 1966) indicates that under θ conditions

$$R_E \cong 0.87 \langle s^2 \rangle_0^{1/2},$$

where $\langle s^2 \rangle_0$ is the average-square radius of gyration of the coil. In a high molecular weight chain under θ conditions, the segment distribution approaches a gaussian one. For this distribution (see Section 6.2.2),

$$\langle R^2 \rangle_0 = 6 \langle s^2 \rangle_0.$$

Therefore,

$$R_E \cong (0.87 / \sqrt{6}) \langle R^2 \rangle_0^{1/2} \text{ or } \gamma \cong 0.35.$$

In a good solvent, theory indicates that $\langle R^2 \rangle$ rises more rapidly than the first power of M (equation (3.29)), see Section 6.4. The best results indicate that at very high molecular weight

$$\langle R^2 \rangle \rightarrow M^{6/5}$$

is approached. Therefore, if the universality of Φ is accepted equation (3.27) indicates

$$\eta_1 \rightarrow M^{4/5}$$

is approached at very high molecular weight in good solvents. In poor solvents the exponent could fall below 1/2 but this is of little practical importance since precipitation usually occurs at temperatures slightly below the θ temperature.

From the above, two important conclusions can be drawn. First, under θ conditions, K and a in the Mark–Houwink equation are completely determined, in principle, by theory. The constant K depends on the characteristic ratio which, in turn, can be calculated from the bond rotation energetics. In practice, experimental K values are commonly used as an important check on bond rotation energetic models rather than the bond energetics models being used to predict K values for use in an absolute viscometric molecular weight determination. Secondly, it may be concluded that the values of $a > 0.5$ commonly observed (in good solvents) are due to the expansion of the coil resulting from enhanced polymer–solvent interaction and suppressed polymer–polymer interaction.

3.4 Light scattering for M_w

The use of light scattering to study polymer solutions yields the weight-average molecular weight and much additional interesting information on the molecular size and shape. The phenomenon is fairly complicated and places on the experimenter rigorous requirements regarding cleanliness and data processing. The basic phenomenon is familiar to anyone who has entered a dark dusty room in which a beam of light enters from a fairly small window. Since light travels in straight lines, a person standing perpendicular to the beam should not be able to see it. The reason it can be seen is that some of the beam intensity has been deviated into the direction of the viewer by the dust particles. A similar event occurs when a beam of light is passed through a solution of a high polymer. In this case the polymer molecules are responsible for the scattering, even though they are in solution. The phenomenon occurs simply because the molecules have a dimension comparable to or smaller than that of the wavelength of the light and are polarized by the electric field of the passing light ray. In fact, when a light beam is passed through any dust-free gas or a pure liquid, scattering occurs. For small molecules the effect is extremely weak. However, the strength of the scattering does not depend simply on the amount of solute (weight) present but rather is stronger for the same weight of solute partitioned into large molecules than it is partitioned into small ones. Thus scattering can be used to determine molecular weight and is particularly suited to application to large molecules. In order to use the light scattering technique, as well as to understand the phenomenon itself, it is necessary to look at the phenomenon from a basic theoretical viewpoint. The formal detailed treatment of scattering is carried out in Appendix 3.1 but in the next section the basic cause of the phenomenon

is reviewed. Then its application to molecular weight determination starts
with results of that formal treatment in Appendix A3.1.

3.4.1 *The origin of scattering*

The theory of scattering from gas molecules was first derived by Lord
Rayleigh (1871, 1899) who applied classical electromagnetic theory to the
problem and was able to show that the amount of light scattered is
proportional to the fourth power of the wavelength. He also showed that
it was inversely proportional to the number of molecules per unit volume,
provided the molecules were small compared to the wavelength. The most
basic assumption used was that each molecule behaved as a point source
and was not influenced by its surroundings.

To appreciate the basic cause of the scattering, consider a perfectly
spherical molecule having a diameter small compared to the wavelength
of light. Recall that light can be regarded as the product of a pair of
oscillating electric and magnetic fields at right angles to one another.
These fields move through space with a velocity v in such a way that the
field fluctuations are at right angles to the direction of propagation of the
light beam. When the electric field oscillates in the vicinity of a molecule
the motion of the electrons is perturbed generating an induced dipole
moment which also oscillates in magnitude. The magnitude of the induced
dipole moment can be written as

$$p = \alpha E, \tag{3.32}$$

where E is the electric field strength and α is termed the polarizability.
The electric field strength is

$$E(t) = E_0 \cos(2\pi ct/\lambda), \tag{3.33}$$

where t is the time and λ the wavelength. Since the induced dipole moment,
p, varies with time through $E(t)$, it has associated with it a time-dependent
electromagnetic field. This secondary field is propagated through space
about the molecular origin. This is *light scattering*.

The possibility of interference effects arises in light scattering just as in
other scattering experiments such as by x-rays. The size scale which sets
these effects is determined by the *magnitude of the scattering vector, Q*,
which is given by

$$Q = \frac{4\pi n_0}{\lambda_0} \sin \frac{\theta}{2}, \tag{3.34}$$

where θ is the angle between the incident beam and the scattering direction, λ_0 is the wavelength of the light *in vacuo* and n_0 is the refractive index of the medium. At $90°$, $Q^{-1} \approx \lambda_0/20$ for $n_0 = 2$.

3.4.2 Molecular weight and scattering

It is traditional to express the scattered light intensity in terms of the *Rayleigh ratio*, defined as

$$R_\theta \equiv [I(\theta)/I_0](R^2/V_s)F(\theta),$$

where $I(\theta)$ is the scattered intensity at angle θ to the incident beam of intensity I_0 and at distance R and V_s is the scattering volume. Also included here is a geometric factor, $F(\theta)$, that depends on the type of scattering experiment. From Appendix A3.1, equation (A3.27), it is seen that the scattering intensity from a solute in solution, in excess of that from the solvent, can be written in terms of the scattering from within individual molecules and a term representing interaction between the molecules and this leads to a Rayleigh ratio given by,

$$R_\theta = K_s C[MP(Q) + V_s N_A CS(Q)], \tag{3.35}$$

where C is the weight concentration of polymer, M is the polymer molecular weight, where K_s is a factor carrying experimentally known constants and is given by

$$K_s = \frac{1}{N_A \lambda_0^4}\left(2\pi n_0 \frac{\partial n_s}{\partial C}\right)^2 \tag{3.36}$$

and where $\partial n_s/\partial C$ is the experimentally determined change of refractive index of the solution with added solute (polymer). $P(Q)$ is a *scattering function* that arises from intramolecular interference effects. For small Q, equation (3.34), it approaches unity, $P(Q) \rightarrow 1$, $Q \rightarrow 0$. The function $S(Q)$ is a scattering function arising from intermolecular interactions. As may be seen, the term containing it becomes negligible compared to intermolecular scattering as $C \rightarrow 0$.

Clearly the simplest condition encountered is where the molecules are small ($P = 1$) and there is little interaction (S term $= 0$). Under these conditions equation (3.35) reduces on rearranging to

$$R_\theta/K_s = CM. \tag{3.37}$$

From this relation the effect of polydispersity may be assessed. Under dilute conditions the scattering from molecules of differing sizes, M_n, present at weight concentration C_n, will be additive and the total scattering will be given by

$$R_\theta/K_s = \sum C_n M_n = C(\sum C_n M_n/\sum C_n)$$

$$= CM_w, \tag{3.38}$$

where now C is understood to be the total concentration, $C = \sum C_n$. The important conclusion is reached that the light scattering method determines the *weight-average molecular weight*, M_w.

Although equation (3.38) forms the basis for molecular weight determination, in practice neither the effect of intramolecular scattering interference, through $P(Q)$ becoming <1, nor concentration effects, the $CS(Q)$ term, may be ignored. They are eliminated via extrapolation in the case of $P(Q)$ to Q, i.e. θ (equation (3.34)) $= 0$, and in the case of the intermolecular effects to $C = 0$.

The extrapolation to zero scattering angle, or $Q = 0$ can be carried out much more effectively if an explicit model exists for how $P(Q)$ varies with Q. There are many scattering functions that have been developed (for a review, see Burchard (1982)). As a bonus, it will be found that additional information concerning molecular size (spatial dimensions) and shape may be attained. Three commonly cited cases are listed below. They are all expressed in terms of a dimensionless variable, v. The latter is a product of the magnitude of the scattering vector, Q, times a length, x, or, $v = Qx$. The physical interpretation of the length, x, differs according to the case. The three example cases are:

(*a*) random coils (Debye, 1947)

$$\left.\begin{array}{l} P(v) = (2/v^4)(e^{-v^2} + v^2 - 1) \\ v = Q\langle s^2\rangle^{1/2}, \qquad s = \text{radius of gyration;} \end{array}\right\} \tag{3.39}$$

(*b*) rigid rods

$$P(v) = I(v)/v - (\sin v/v)^2 \tag{3.40}$$

where

$$I(v) = \int_0^{2v} \frac{\sin v'}{v'} \, \mathrm{d}v'$$

$$v = QL/2, \qquad L = \text{length of rod;}$$

(*c*) spheres

$$\left.\begin{array}{l} P(v) = [3(\sin v - v \cos v)/v^3]^2 \\ v = QR, \qquad R = \text{sphere radius.} \end{array}\right\} \tag{3.41}$$

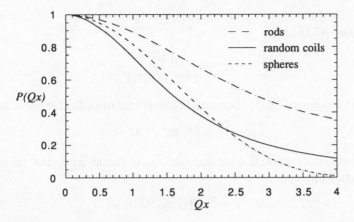

Fig. 3.9 Scattering functions for rods ($x = L/2$, L = length of rod), random coils ($x = \langle s^2 \rangle^{1/2}$ = radius of gyration), and spheres ($x = R$, radius). Q is the scattering vector magnitude, equation (3.34).

The three functions are displayed in Figure 3.9. For use in an extrapolation procedure the scattering function can be expanded in a series. For the important case of random coils, equation (3.39) gives

$$P(v) = 1 - v^2/3 + \cdots = 1 - Q^2 \langle s^2 \rangle /3 + \cdots.$$

The other functions also contain Q^2 as the leading term in the expansion. Equation (3.35) can be rearranged and expressed at low concentration ($C \rightarrow 0$) as

$$K_s C/R_\theta = 1/[M(1 - \langle s^2 \rangle Q^2/3 + \cdots)]$$

$$= M^{-1}(1 + \langle s^2 \rangle Q^2/3 + \cdots). \tag{3.42}$$

Thus this suggests that $K_s C/R_\theta$ should be plotted at low concentration against

$$Q^2 = (4\pi n_0 \lambda^{-1})^2 \sin^2(\theta/2)$$

in order to determine $1/M$ from the intercept and $\langle s^2 \rangle$ from the slope $= M^{-1} \langle s^2 \rangle /3$.

Turning to the concentration dependence, it is necessary to rephrase the development of the dependence of α_s equation (A3.16) that leads to equation (A3.19). It is assumed that at low Q, the concentration dependence of the scattering is entirely a thermodynamic effect that shows up in the ability of the solution to scatter through the effect of concentration fluctuations on the local polarizability. The quantity α_s^2 in equation

(A3.20) is found as a statistical mechanical average, $\langle \alpha_s^2 \rangle$, through equation (A3.16) as,

$$\langle \alpha_s^2 \rangle = \frac{\langle (\Delta\varepsilon)^2 \rangle}{[4\pi(N_p X_s V_s]^2}. \tag{3.43}$$

Then the average, $\langle \Delta\varepsilon^2 \rangle$, is related to concentration fluctuations, ΔC, as

$$\langle \Delta\varepsilon^2 \rangle = (\partial\varepsilon/\partial C)^2 \langle \Delta C^2 \rangle. \tag{3.44}$$

The latter fluctuational average, $\langle \Delta C^2 \rangle$, is found from the theory of fluctuations as

$$\langle \Delta C^2 \rangle = k_B T [(\partial^2 G/\partial C^2)_{T,P}]^{-1}, \tag{3.45}$$

where G is the Gibbs free energy. Conceptually the scattering volume V_s is subdivided into parcels δV about each of the scattering entities, i.e., the N_p polymer molecules, and thus $\delta V = V_s/N_p$. The free energy second derivative is found in terms of solvent chemical potential, μ_1, from the relation

$$(\partial^2 G/\partial C^2)_{T,P} = -\delta V (\partial\mu_1/\partial C)_{T,P}/(CV_1^0), \tag{3.46}$$

where V_1^0 is the solvent molar volume. Use of the virial expansion in equation (3.6) with $\mu_1 = \mu_1^0 + \ln a_1$ gives

$$\langle \Delta C^2 \rangle = \frac{C}{(V_s/N_p)N_A(1/M + A_2 C + \cdots)}. \tag{3.47}$$

Substitution of equation (3.47) into (3.44) gives, when $\partial\varepsilon/\partial C = 2n_0(\partial n/\partial C)$ is invoked, for $\langle \alpha_s^2 \rangle$ in equation (3.43),

$$\langle \alpha_s^2 \rangle = \frac{[2n_0(\partial n/\partial C)]^2 C}{(4\pi N_p X_s/V_s)^2 (V_s/N_p) N_A (1/M + A_2 C + \cdots)}. \tag{3.48}$$

Substitution of equation (3.48) into equation (A.320) and rearranging gives

$$K_s C/R_\theta = M^{-1} + A_2 C + \cdots \tag{3.49}$$

for the extrapolation equation against concentration at low Q. It is seen that the reciprocal of the intercept again gives the molecular weight, cf. equation (3.42), and the slope then gives the second virial coefficient, A_2.

In practice the dual extrapolations against scattering angle, equation (3.42), and concentration, equation (3.49), are usually attempted on the same plot, a Zimm (1948) plot, see Figure 3.10.

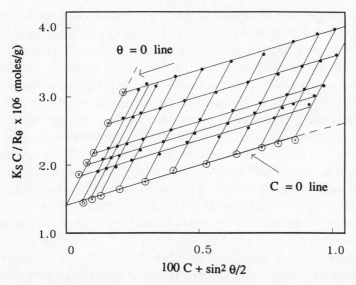

Fig. 3.10 Zimm plot for solutions of polystyrene in benzene at 25 °C (concentration in units of g/cm³). The lower slope set of lines represents constant concentration data and the steep lines represent constant scattering angle data. Note the convergence of the two extrapolated lines to the same point on the vertical axis. A lack of convergence would have indicated some form of specific association between polymer molecules (after Marserison and East, 1967).

To summarize, the following parameters are found and information obtained:

(a) Weight-average molecular weight as the reciprocal of the intercept on the $K_s C/R_\theta$ axis.

(b) The second virial coefficient from the limiting slope of the $\theta = 0$ line.

(c) The mean-square radius of gyration of the molecule from the limiting slope of the $C = 0$ line.

(d) More detailed information on shapes and interactions using curve fitting of plots of P and/or S versus Q (a subject not treated here).

3.5 Molecular weight distribution from chromatography

All of the aforementioned molecular weight determination methods generate an average molecular weight. Estimation of the actual distribution of molecular weights is a more involved matter necessitating a fractionation into the different components.

A method used in the past, fortunately no longer necessary, required that any sample first be separated into the components using techniques derived from solution theory (Chapter 9) involving one of several solvent/ non-solvent precipitation/dissolution methods. Once fractionated the average molecular weights would be determined using the techniques described earlier in this chapter. In practice, such separation techniques are still used when more complicated analyses are required. For instance, the more modern types of branched polyethylene are the 'linear low density polyethylenes' and 'ultra-low density polyethylenes' which are copolymers of ethylene with one or more α-olefins. The properties of these semi-crystalline polymers are determined in a complicated way by the relation between the branching (i.e., comonomer) distribution and the molecular weight distribution. It is therefore very important to have a three dimensional relationship involving molecular weight distribution and branching distribution. These copolymers are first fractionated according to level of branching and then each fraction is characterized for its molecular weight distribution. The most common modern method for such fractionations is *temperature rising elution fractionation* (TREF). This method of separation utilizes the influence of molecular weight and chain microstructure on crystallization and dissolution processes. In this process a polymer is first deposited on beads by cooling a hot solution at a constant cooling rate, which causes different fractions to crystallize on the beads at different temperatures. The beads are collected and placed into long glass columns. Solvent is passed through the columns, which are now being heated at a constant rate, and the solvent progressively dissolves the crystalline material. The solvent fractions are collected and characterized for molecular structure (e.g., by infrared or nuclear magnetic resonance spectroscopy) before being characterized for molecular weight distribution using the method that will now be described.

The most convenient technique for study of molecular weight distribution is known as gel permeation chromatography (GPC) or size exclusion chromatography. The principle behind GPC is fairly simple, but difficult to develop into an exact theory. Gel particles contain pores of various sizes and tortuosities which reach inward from the surface. If a polymer solution is placed in contact with gel particles the solution penetrates the gel, but the penetration of the gel pores by any polymer molecule depends on the hydrodynamic volume of the molecule, see Figure 3.11. This, in turn, depends on the polymer's molecular weight, its flexibility and the nature of its interactions with the solvent. Generally, it can be expected that the larger the hydrodynamic volume the lesser the degree of passage into the gel particle's interior, if it can be assumed

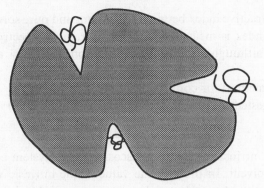

Fig. 3.11 Conceptual illustration of gel permeation or size-exclusion chromatography. Porous particles absorb polymer molecules. The smaller ones penetrate the pores more deeply and thus are eluted by flowing solvent more slowly than the larger ones.

that the pores' diameters decrease with depth below the surface. Theories of the phenomenon have assumed that the pore can be approximated by a cone with its apex pointing to the center of the gel particle.

Essentially the polymer solution is injected into a continuous stream of solvent which is passing through gel particles into a column. The polymer molecules diffuse into the pores as described above. As more of the solvent stream passes through the system polymer molecules are eluted and pass out of the column. Since the penetration of the pores by any polymer molecule is inversely proportional to its hydrodynamic volume, the larger molecules are eluted first and the smallest ones last. Molecules having a hydrodynamic radius larger than that of the gel pores cannot be separated by this technique.

The technique is, in fact, a form of liquid chromatography in which strong interactions between the polymer and the gel substrate must not occur. Should specific interaction occur then the separation would not be determined by hydrodynamic volume alone. The method is not absolute, but depends on a calibration curve to convert experimental data into molecular weights. In fact each column has its own characteristics and must be calibrated individually. The commonly used standard is polystyrene for which virtually monomolecular weight fractions are available. Continuous experimental curves are generated of some parameter which measures concentration as a function of elution volume. The conversion of such an experimental curve into a curve of mass fraction vs molecular weight is necessary. In practice, the measured quantity is usually a difference in refractive index between the elutant and pure solvent. Usually the refractive index is measured using an ultraviolet spectrophotometer (200–600 nm) although infrared spectrophotometers may also be used occasionally.

A common method for conversion of the data is to apply the relation, originally suggested by Benoit, Rempp and Grubisc (1967)

$$\log \eta_1 M_1 = \log \eta_2 M_2, \tag{3.50}$$

where η is the intrinsic or limiting viscosity of the system considered in the standard solvent. In practice the value of the intrinsic viscosity can be replaced by the Mark–Houwink equation, $\eta = KM^a$, leading to

$$\log M_2 = \frac{(1 + a_1) \log M_1 + \log(K_1/K_2)}{1 + a_2}. \tag{3.51}$$

Although automated systems are now available, there are many optimization approaches that need to be applied for any system to be used to its maximum potential. In some advanced systems the molecular weight of each fraction is determined automatically as it elutes using laser light scattering, generating greater versatility. This approach eliminates the need for calibration and the chromatography serves only as a separation method and not as a method for the estimation of molecular weight. Additionally, many rigid or semi-rigid molecules are studied, some are beginning to be of commercial importance. Their viscosity molecular weight behavior cannot be handled by using the above method and a direct molecular weight determination method is thus desirable.

3.6 Further reading

A comprehensive review of experimental methods is contained in a book by Rabek (1980). For a review of the experimental aspects of the GPC technique and its applications, the book by Schroder, Muller and Arndt (1989) is recommended. The best source of experimental Mark–Houwink constants and also solution dimensions is the *Polymer Handbook* (Brandrup and Immergut, 1989). The theory of light scattering is given in the books by Chu (1974), Berne and Pecora (1976) and Schmitz (1990) cited in the text. A review by Burchard (1982) is recommended.

Appendix A3.1 Scattering of light

There have been many recent advances in the light scattering field including the introduction of dynamic light scattering (Chu, 1974; Berne and Pecora, 1976;

Schmitz, 1990) (i.e., time-dependent scattering which can give information on the time scale of molecular motions). The modern development of the field emphasizes the treatment of light scattering in the common language of many other scattering phenomena such as by x-rays, neutrons, electrons, etc. The case taken up here starts from a description of the three dimensional propagation of the radiation to be scattered.

A3.1.1 Light scattering

In three dimensions and in modern notation, the electric field associated with a propagating light wave that is represented in equation (3.33) is written as

$$\mathbf{E}_0(\mathbf{r}, t) = \mathbf{E}_0 \exp(i\mathbf{K}_0^T \cdot \mathbf{r})\, e^{-i\omega_0 t}, \tag{A3.1}$$

where $\mathbf{E}_0(\mathbf{r}, t)$ is a vector quantity denoting the spatial orientation of the oscillation of a field strength having E_0 magnitude. λ_0 is the wavelength of light and $\mathbf{K}_0$ the *wave vector* whose magnitude is given as

$$|\mathbf{K}_0| = 2\pi n_0/\lambda_0, \tag{A3.2}$$

n_0 being the index of refraction of the medium, see Figure 3.12. The polarization of the system is also written in terms of vector quantities as follows,

$$\mathbf{P}(\mathbf{r}, t) = \alpha(\mathbf{r}, t)\mathbf{E}_0(\mathbf{r}, t). \tag{A3.3}$$

In practice, the direction of response of the material to the incident light beam is not restricted to the direction of polarization of the light and so α is a tensor having nine components, that relate the three laboratory directional coordinates of the $\mathbf{P}$ and $\mathbf{E}$ vectors.

It should also be noted that there is the possibility of the photon and the system exchanging energy. Should this occur then the energy of the emitted photon would differ from that of the incident photon and a frequency change would occur. Should there be no exchange of energy then the process is referred

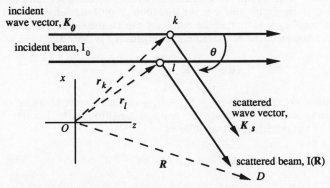

Fig. 3.12 Scattering diagram. An incident wave with intensity I_0 and wave vector $\mathbf{K}_0$ is scattered by centers at k and l. The scattered beam has intensity $I(\mathbf{R})$ and wave vector $\mathbf{K}_0$ at the detector, D. The positions of k and l are defined by vectors, $\mathbf{r}_k$, $\mathbf{r}_l$ from origin at L. The detector is located at $\mathbf{R}$.

to as *elastic light scattering*. If there is an exchange of energy the process is known as *inelastic light scattering* and clearly additional information on the process can be obtained from the change in frequency that occurs.

The polarization at point **r** arises from distortion of the local electric field and can be described as a damped harmonic oscillator with a sinusoidal driving force acting over a volume element $(dr)^3$. The time dependence is often described using $\cos(\omega t)$ and the response has two components, one in-phase and one out-of-phase. The in-phase component is generally the only one of significance unless the applied frequency is close to a natural frequency of the material. It represents absorption.

Assuming that only the in-phase solution is important, the radiated electric field can be written as

$$d\mathbf{E}_s(\mathbf{R}, t) = \frac{n_0^2}{Rc^2} \frac{\partial^2 \mathbf{P}(\mathbf{r}, t')}{\partial t^2} d^3 r, \tag{A3.4}$$

where **R** is the location of the detector and t is the time of detection of light radiated at an earlier time t' given by

$$t' = t - \frac{|\mathbf{R} - \mathbf{r}|n_0}{c}, \tag{A3.5}$$

and c is the velocity of light. The latter part of equation (A3.5) is the time it takes for light to travel from the scattering center at **r** to the detector at **R**. This quantity forms the basis for studies of *dynamic light scattering*.

When equations (A3.1) and (A3.3) are substituted into equation (A3.4) and integrated over the total scattering volume V_s the following is obtained,

$$\mathbf{E}_s(\mathbf{R}, t) = \int_{V_s} d\mathbf{E}_s(\mathbf{R}, t)$$

$$= -\frac{\omega_0^2 n^2}{Rc^2} \int_{V_s} \alpha(\mathbf{r}, t') \cdot \mathbf{E}_0(\mathbf{r}, t') d^3 r. \tag{A3.6}$$

In a laboratory experiment the z-direction is taken as the direction of the propagation of the beam and x is the direction of polarization of a laser beam. If the polarizability is isotropic the equation can be simplified.

The wave vector of the scattered electric field, $\mathbf{K}_s$, is defined by

$$(\mathbf{R} - \mathbf{r})^{\mathrm{T}} \cdot \mathbf{K}_s = \frac{|\mathbf{R} - \mathbf{r}|\omega_0 n_0}{c}. \tag{A3.7}$$

Defining the *scattering vector* $\mathbf{Q} = \mathbf{K}_0 - \mathbf{K}_s$ generates a term **Q** which defines momentum space, which is related to real space **r** through the Fourier transform of the polarizability as follows:

$$\alpha(\mathbf{Q}, t') = \int_{V_s} \alpha(\mathbf{r}, t') e^{i\mathbf{Q}^{\mathrm{T}} \cdot \mathbf{r}} d^3 r. \tag{A3.8}$$

The equation for the scattered field can now be written as

$$E_s(\mathbf{R}, t) = -\frac{\omega_0^2 n_0^2 E_0}{c^2 R} e^{i\mathbf{K}_s^{\mathrm{T}} \cdot \mathbf{R}} e^{-i\omega t} \alpha(\mathbf{Q}, t'). \tag{A3.9}$$

The polarizability can be expressed in terms of an average and the local inhomogeneity, the latter being both space- and time-dependent.

$$\alpha(\mathbf{Q}, t') = \bar{\alpha} \int_{V_s} e^{i\mathbf{Q}^T \cdot \mathbf{r}} \, d^3r = \int_{V_0} \delta\alpha(\mathbf{r}, t') \, e^{i\mathbf{Q}^T \cdot \mathbf{r}} \, d^3r. \tag{A3.10}$$

The first integral, derived from the averaged polarizability, equals zero if $\mathbf{Q}$ is not equal to zero, or equals V_s if $\mathbf{Q} = 0$. The latter case results when $\mathbf{K}_s = \mathbf{K}_0$ and hence only contributes to scattering in the forward direction. Hence scattering in any direction other than the forward direction results from fluctuations in the polarizability of the material. Since $|\mathbf{K}_s| = |\mathbf{K}_0|$ conservation of momentum leads to the expression for the *magnitude of the scattering vector*,

$$|\mathbf{Q}| = 2|\mathbf{K}_0| \sin\frac{\theta}{2} = \frac{4\pi n_0}{\lambda_0} \sin\frac{\theta}{2}, \tag{A3.11}$$

where θ is the scattering angle measured relative to the direction of propagation of the incident light (see Figure 3.12). Scattering is measured as a function of θ thus allowing the variation in $\mathbf{Q}$, the momentum space, to be monitored.

A3.1.2 Light scattering from polymer molecules

In an experiment the quantity measured is the time averaged scattered light intensity, $I(\mathbf{R})$, which is given by

$$I(\mathbf{R}) = f \langle \mathbf{E}_s^*(\mathbf{R}, t')^T \cdot \mathbf{E}_s(\mathbf{R}, t') \rangle, \tag{A3.12}$$

where f is the efficiency of the detector and $\mathbf{E}_s^*$ the complex conjugate of the field scattered at t'.

For a solution both the solvent molecules and the polymer molecules will contribute to the scattering, the total scattering intensity usually being written as the sum of the two contributions

$$I(\mathbf{R}) = I(\mathbf{R})_{\text{solv}} + I(\mathbf{R})_{\text{polymer}}. \tag{A3.13}$$

The problem now comes down to obtaining an expression for the scattering from a polymer molecule in terms of its component parts. If it is assumed that the intensity of the scattered light depends only on the spatial locations of the scattering centers, then the scattering is known as *Rayleigh scattering* (sometimes as Rayleigh–Gans–Debye scattering). If the polarizability is the same for all segments the polarizability can be written as

$$\alpha(\mathbf{r}, t') = \alpha_s \sum_{i=1}^{N_p} \sum_{k=1}^{X_s} \delta(\mathbf{r} - [\mathbf{G}(t')_i + \mathbf{r}(t')_k]), \tag{A3.14}$$

where the center of mass of the ith polymer molecule is $\mathbf{G}(t')_i$ and the relative location of its kth segment is $\mathbf{r}(t')_k$; N_p is the number of polymer molecules in solution, X_s the number of scattering units in each polymer molecule and δ the Kronecker delta which is equal to zero, except for the condition in which the kth segment of the ith polymer is in the scattering volume at time t', when it is equal to one.

The function $I(\mathbf{R})_{\text{polymer}}$ due to N_p macromolecules in the scattering volume

V_s and where the velocity of light is taken as that in the medium, c_m, is given by

$$I(\mathbf{R})_{polymer} = \left(\frac{\omega_0}{c_m}\right)^4 \left(\frac{|E_0|}{R}\right)^2 \alpha_s^2 f$$

$$\times \left(\sum_{i=1}^{N_p} \sum_{j=1}^{N_p} \sum_{k=1}^{X_s} \sum_{l=1}^{X_s} \exp\{-i\mathbf{Q}^T[\mathbf{G}(t')_i - \mathbf{G}(t')_j + \mathbf{r}(t')_k - \mathbf{r}(t')_l]\} \right).$$

(A3.15)

Ideally, this total intensity may be separated into two component parts. The first is obtained by placing $i = j$, and may be regarded as arising from an isolated molecule in which its component parts are interacting only with themselves. The second component of the equation applies when i is not equal to j, and corresponds to interactions between pairs of molecules.

In order to proceed further it is necessary to consider the types of polymer molecules that may be encountered and also how and under what conditions they will interact. Molecules may be rigid, semi-rigid or flexible. They may be small or very large. Additionally they may be in a very dilute solution so that they are separated effectively or they may be in moderately dilute solutions and may be interpenetrated. Other complications arise from their polarity, since this will lead to strong interactions, especially if the molecules are rigid and have a large dipole moment parallel to the molecular direction. Polymer molecules generally behave as if they are large and interactive, even when relatively dilute solutions are used. It becomes essential therefore in molecular weight determinations to account for these two influences. These two complicating factors provide valuable information on the molecule under investigation. If the molecules are interactive then information can be obtained on the nature of the interaction. When the molecule is large, interference effects within the molecule are important and provide information on the spatial dimensions and shape of the molecule.

The polarizability term in equation (A3.15), α_s^2, arises from the difference in refractive indices of the solution and the solvent. It is defined by the equation

$$\alpha_s = \frac{\Delta\varepsilon}{4\pi(N_p X_s/V_s)},$$

(A3.16)

where $\Delta\varepsilon$ is the difference in high frequency dielectric constant between the solution and the solvent. $\Delta\varepsilon$ is equal to $n_s^2 - n_0^2$, which may be regarded as equal to $2n_0(n_s - n_0)$ if the difference is small. Substitution in equation. (A3.16) leads to

$$\alpha_s = \frac{n_0(n_s - n_0)}{2\pi(N_p X_s/V_s)}.$$

(A3.17)

Since the refractive index difference is dependent on (weight) concentration, C, it can be assumed that $(n_s - n_0)/C \cong \partial n_s/\partial C$. The concentration, C, can be written as

$$C = M N_p/(N_A V_s),$$

(A3.18)

where M is the molecular weight of the polymer molecule and N_A is Avogadro's number, leading to

$$\alpha_s = \frac{\partial n_s}{\partial C} \frac{n_0}{2\pi X_s} \frac{\mathbf{M}}{N_0}.$$

(A3.19)

Returning now to the separation of $I(\mathbf{R})_{\text{polymer}}$ into its component parts, the equation for the internal structure effects can be written as

$$I(\mathbf{R})_{\text{int}} = N_{\text{p}}\left(\frac{\omega_0}{c_{\text{m}}}\right)^4 \frac{I_0}{R^2} \alpha_{\text{s}}^2 X_{\text{s}}^2 P, \qquad (A3.20)$$

where

$$P = \left(\frac{1}{X_{\text{s}}}\right)^2 \left(\sum_{k=1}^{X_{\text{s}}} \sum_{l=1}^{X_{\text{s}}} \exp\{i\mathbf{Q}^{\text{T}} \cdot [\mathbf{r}(t')_k - \mathbf{r}(t')_l]\}\right). \qquad (A3.21)$$

Similarly, the equation for the intermolecular effects can be written as

$$I(\mathbf{R})_{\text{ext}} = N_{\text{p}}^2\left(\frac{\omega_0}{c_{\text{m}}}\right)^4 \frac{I_0}{R^2} \alpha_{\text{s}}^2 X_{\text{s}}^2 S, \qquad (A3.22)$$

where

$$S = \frac{1}{N_{\text{p}}^2 X_{\text{s}}^2} \left(\sum_{i=1}^{N_{\text{p}}} \sum_{j=1}^{N_{\text{p}}} \sum_{k=1}^{X_{\text{s}}} \sum_{l=1}^{X_{\text{s}}} \exp\{-i\mathbf{Q}^{\text{T}} \cdot [\mathbf{G}(t')_i - \mathbf{G}(t')_j + \mathbf{r}(t')_k - \mathbf{r}(t')_l]\}\right) \qquad (A3.23)$$

The entire expression given in equation (A3.15) becomes

$$\frac{I(\mathbf{R})_{\text{polymer}}}{I_0} = \frac{I(\mathbf{R})}{I_0} - \frac{I(\mathbf{R})_{\text{solvent}}}{I_0}$$

$$= \left(\frac{\omega_0}{c_{\text{m}}}\right)^2 \left[\frac{n_0(\partial n_{\text{s}}/\partial C)}{2\pi R}\right]^2 \left(\frac{MV_{\text{s}}CP}{N_{\text{A}}} + C^2 V_{\text{s}}^2 S\right). \qquad (A3.24)$$

The *Rayleigh ratio* R_θ is traditionally defined by $(I(\mathbf{R})/I_0)R^2$. It is also convenient to include in it geometrical factors which depend on the type of experiment conducted and are expressed in a function, $F(\theta)$. Thus the Rayleigh ratio is,

$$R_\theta = \frac{I(\mathbf{R})_{\text{polymer}}}{I_0} \frac{R^2}{V_{\text{s}}} F(\theta), \qquad (A3.25)$$

where $F(\theta)$ is the geometrical factor. It is also convenient to lump a number of parameters that occur in equations (A3.20) and (A3.22) into a single parameter K_{s} which is defined as

$$K_{\text{s}} = \frac{1}{N_{\text{A}} \lambda_0^4} \left(2\pi n_0 \frac{\partial n_{\text{s}}}{\partial C}\right)^2 \qquad (A3.26)$$

where the wavelength *in vacuo*, λ_0, has been introduced. This leads to the experimentally useful equation

$$\frac{K_{\text{s}} C}{R_\theta} = \frac{1}{MP + V_{\text{s}} N_{\text{A}} SC}. \qquad (A3.27)$$

The term $F(\theta)$ of equation (A3.25) has a value determined by the experimental conditions, which may be:

(*a*) vertical incident polarization with

  (i) vertically polarized detection (polarized light scattering), $F(\theta) = 1$,
  (ii) horizontally polarized detection (depolarized light scattering), $F(\theta) = 1$;
(b) unpolarized light throughout $F(\theta) = 2/(1 + \cos^2 \theta)$.

Nomenclature

a = exponent in Mark–Houwink equation

a_1 = solvent activity

a_2 = solute (polymer) activity

$A_2, A_3, \ldots$ = osmotic second, third, ... virial coefficients

c = velocity of light (*vacuo*)

c_m = velocity of light in scattering medium

C = solute (polymer) concentration in weight/solution volume, traditionally often g/100 cm^3

C_∞ = characteristic ratio of a long coiling polymer molecule (Chapter 6)

G = Gibb's free energy

$\mathbf{G}$ = center of mass of a scattering polymer molecule

ΔH_{vap}^0 = molar heat vaporization of pure solvent

ΔH_f^0 = molar heat of fusion of pure solvent

K = constant in Mark–Houwink equation

K_b = ebulliometric constant

K_f = cryoscopic constant

K_s = constant in light scattering molecular weight relation

$\mathbf{K}_0, \mathbf{K}_s$ = wave vector of incident, scattered light

l = bond length

M = molecular weight

M_n = molecular weight of a molecule of DP $= n$

M_N, M_W, M_V = number-average, weight-average and viscosity-average molecular weight respectively

n_s, n_0 = refractive index of solution, solvent

N = number of bonds in a polymer chain

N_A = Avogadro's constant

N_p = number of polymer molecules in the scattering volume (light scattering)

N_1 = moles of solvent

N_2 = moles of solute (polymer)

P = pressure

P_1 = solvent vapor pressure

P_1^* = pure solvent vapor pressure

P'', P' = pressure on solution, pure solvent, respectively

$P, P(Q)$ = intramolecular scattering function of a polymer molecule
Q, Q = scattering vector and its magnitude in light scattering
$\langle R^2 \rangle$ = average square end-to-end distance of coiling polymer in solution.
Subscript zero indicates unperturbed or θ conditions
R_E = effective frictional equivalent spherical radius of a coiling polymer
molecule in solution
s = radius of gyration of a coiling polymer molecule
$S, S(Q)$ = intermolecular scattering function of a polymer solution
T_b, T_f = boiling point, freezing point, respectively of pure solvent
V = solution volume
V_1^0 = pure solvent molar volume
X_s = number of scattering centers in a polymer molecule
X_1 = mole fraction component 1 (solvent)
X_2 = mole fraction component 2 (solute, polymer)
η = viscosity
η_0 = viscosity of pure solvent
η_I = intrinsic viscosity
α, α_s = polarizability, excess polarizability of scattering center in solution
μ_1 = solvent chemical potential (superscript zero indicates pure solvent)
μ_2 = solute (polymer chemical potential)
$\Pi = P'' - P'$ = osmotic pressure
ρ = solution density
λ_m = wavelength of light in scattering medium

Problems

3.1 (a) For the data in Figure 3.13 below, determine K and a in the
equation,

$$\eta_I = K M_V^a.$$

(b) Derive an equation for M_V for a polymer with a Schulz–Zimm
distribution in terms of z and M_W.

(c) Using the value of 'a' found in part (a), calculate the ratio
(M_W/M_V) for $z = -0.9$, 0, and 5.

3.2 The fact that the molecular weight from light scattering is the *weight
average* tends to be obscured by the details of scattering theory.
Provide a simplified derivation of this result. Start from the premise
that the intensity of the scattered light is proportional to the
number of polymer molecules, N_p, in the scattering volume of the
solution and that the scattering power of each polymer molecule is

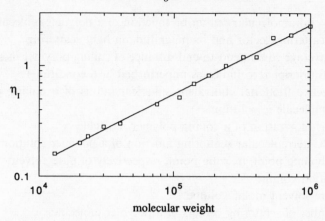

Fig. 3.13 Intrinsic viscosity vs molecular weight for polyisobutylene fractions in diisobutylene, 20 °C. From data of Flory (1943).

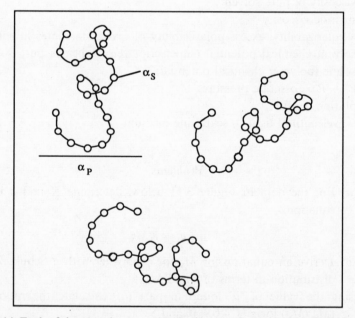

Fig. 3.14 Each of the N_P polymer molecules has an effective polarizability, α_P. The total intensity of the light scattered is proportional to $N_P\alpha_P^2$. Each molecule is made up of X_S scattering segments or monomeric units, S. Each unit has an effective polarizability, α_S. Thus $\alpha_P = X_S\alpha_S$ for long wavelength light. The effective segmental polarizability $\alpha_S = \delta\varepsilon/(X_S N_P)$ where $\delta\varepsilon = [2n_0(\partial n_s/\partial C)]C$. The refractive index of the pure solvent is n_0 and the change of solution refractive index, n_s, with concentration, $(\partial n_s/\partial C)$, is an experimentally measured property of the solution; C is the weight concentration of polymer.

proportional to the *square of its effective polarizability*, α_p. Write the effective polarizability of one polymer molecule, α_P, as the product of the effective polarizability of one segment or monomeric unit, α_S, and the number of scattering segments or monomeric units per molecule, X_S (Figure 3.14). Then write the effective polarizability of a segment as the increase of the solution dielectric constant over that of pure solvent divided by the total number of segments in the scattering volume. Write the increase of dielectric constant of the solution in terms of the increase of the refractive index of the solution, n_s, with concentration times the concentration ($\varepsilon = n_s^2$; $\delta\varepsilon = 2n_0 C \partial n_s / \partial C$). Show that the resulting expression for the intensity of the scattered light is proportional to the molecular weight, M, times the weight concentration, C. Generalize this to a polydisperse system where the scattered light is the sum of that from each fraction of DP $= n$.

4

Polymerization: kinetics and mechanism

Although the overall chemistry of a polymerization might be obvious in terms of monomers used and polymer produced, this knowledge does not ensure that the details of the chemistry are understood. Polymerizations, especially chain polymerizations, can have a number of intermediate steps that lead to the final result. These intermediate steps or mechanism need to be known if the polymerization itself is to be fully understood. The kinetics associated with polymerization are the principal manifestation of the mechanism underlying the overall chemical process. If a specific polymerization mechanism is proposed then the kinetic equations that ensue from the mechanism should lead to relations for the progress through the polymerization of observables such as the rate of monomer consumption and molecular weight build-up that are in agreement with experiment. If it is desired to have control over the polymerization process in the sense of being able to manipulate results such as molecular weight and its distribution, rate of polymerization, or composition in copolymerizations, then it is highly desirable to have a successful mechanistic and kinetic formulation of the process that expresses the effects of the possible variables.

4.1 Step polymerization

The description of the kinetics of step polymerization is greatly simplified if it is realized that the reactivities of the as yet unreacted end groups (carboxyl, amine, etc.) are not sensitive to the length of the chain to which they are attached. The rate constant for reaction **4.1** does not depend much

$$HO\text{-}[BB\text{---}AA]_n\text{---}CO_2H + HO\text{---}[BB\text{---}AA]_m CO_2H \rightarrow HO\text{---}[BB\text{---}AA]_{n+m} CO_2H + H_2O$$

4.1

on the values of n and m. It might be supposed that longer chains would be less reactive since they would be less mobile. However, in any chemical reaction with appreciable activation energy only a tiny fraction of the encounters between reactive species actually leads to reaction. If decreased mobility in long chains does decrease the rate of new encounters between reactive ends it also increases the length of an encounter before the ends diffuse away. Thus, by whatever proportion the number of encounters is reduced, the duration of contact and hence fraction of reactive encounters is increased by the same factor.

The rate constants for reaction of given groups will depend on the attached chain just as such rates depend on substituents in simpler organic reactions. Distant groups will have little effect. Although an appreciable difference may exist between monomer and dimer, higher chain lengths should have virtually identical reactivities. Thus the reactions of the end-groups may be monitored as a class and rate equations written for them as a whole.

4.1.1 Self-catalyzed polymerization

In a reaction such as **4.1**, let C_A = concentration of acid end-groups and C_B = concentration of hydroxyl end-groups. A rate equation of the form below might then be written,

$$dC_A/dt = -kC_A C_B \qquad (4.1)$$

if the reaction is not reversible. The latter condition would prevail if the water is efficiently removed throughout the reaction. However, for the particular example of **4.1** it must be recognized that such reactions are acid catalyzed and the carboxyl groups will also act as a catalyst (see **4.2**).

4.2

In this case it is appropriate to write

$$dC_A/dt = -kC_A^2 C_B, \qquad (4.2)$$

where one of the powers of C_A represents the direct participation of the carboxyl group in the reaction and the other one the effect of carboxyl groups as an acid catalyst as indicated by $[H^+]$ in **4.2**. The concentration of B, C_B, may be eliminated in favor of C_A through $C_B^0 - C_B = C_A^0 - C_A$ or

$$dC_A/dt = -kC_A^2(C_B^0 - C_A^0 + C_A). \qquad (4.2a)$$

If the end-groups are perfectly balanced, then, $C_A^0 = C_B^0$ and

$$\frac{dC_A}{dt} = -kC_A^3, \qquad (4.2b)$$

and the equation integrates to a particularly simple form,

$$\left(\frac{C_A^0}{C_A}\right)^2 - 1 = 2(C_A^0)^2 kt.$$

Since under the assumptions of exactly balanced monomers, $C_A^0/C_A = X_N = 1/(1 - P)$, see equation (2.12), it follows that

$$X_N^2 = 1 + 2(C_A^0)^2 kt. \qquad (4.2c)$$

A plot of DP squared vs time should be a straight line. Figure 4.1 shows an example of such a plot and demonstrates that this is, in fact, the case.

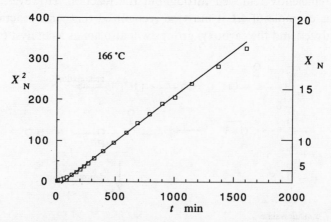

Fig. 4.1 Self-catalyzed polymerization of adipic acid and ethylene glycol. The square of the DP is plotted against time (equation (4.2c)) as left-hand ordinate. A scale for DP itself is shown on right-hand side. Data of Flory (1939).

However systematic deviations at both low and high DP seem to be evident. The rates of acid catalyzed reactions are notoriously sensitive to the medium (solvent) and deviate badly at high acid concentration from simple proportionality to acid concentration. Since polymerizations like reaction (**4.1**) are usually carried out 'neat', i.e., in the melt with no added solvent the polymeric mixture itself is the medium. Therefore, equation (4.2) might be expected to hold well after the reaction has progressed to the point where most of the carboxyls have been used up and the medium is thus not changed much and the acid concentration is low, but it might be expected to be rather poor in the early stages. At the later stages, the simplicity introduced by the assumption that the monomers are exactly balanced has to break down eventually and the molecular weight, rather than increasing indefinitely, must approach a limiting value as suggested by equation (2.5). Thus the observed fall-off of the DP below the straight line at long times is not unexpected.

4.1.2 Added catalyst polymerization

Sometimes an additional strong acid is used as a catalyst and it overrides the catalytic effect of the acid end-groups once the polymerization is underway. If this is the case, the counterpart of equation (4.2) is

$$dC_A/dt = -kC_{HA}C_A C_B, \tag{4.3}$$

where C_{HA} represents the added catalyst and stays constant through the polymerization. If again it is assumed that $C_A^0 = C_B^0$, then,

$$dC_A/dt = -(kC_{HA})C_A^2 \tag{4.3a}$$

and, on integration,

$$C_A^0/C_A - 1 = (C_A^0 C_{HA})kt \tag{4.3b}$$

or,

$$X_N = 1 + (C_A^0 C_{HA})kt. \tag{4.3c}$$

In this case, a plot of DP directly vs time should be a straight line. In Figure 4.2 it may be seen that such a relation is observed and holds over a significantly wide range of DP. It should also be noticed that, in spite of a much lower temperature, the build-up of molecular weight is much faster than in the self-catalyzed case. This is due to the linear nature of equation (4.3) vs time compared to the dependence of DP on the square root of time in equation (4.2c) and also to the effective rate constant, kC_{HA}, in equation (4.3c) being comparable, due to the added catalyst, to the constant k in equation (4.2c), even at the lower temperature.

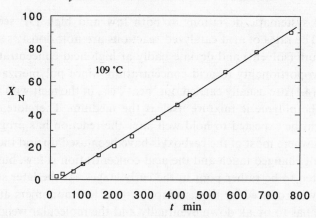

Fig. 4.2 Polymerization of adipic acid and ethylene glycol with added catalyst (0.4 mole% p-toluene sulfonic acid). DP plotted linearly vs time, equation (4.3c). Data from Flory (1940b).

4.2 Free radical chain polymerization

As indicated in Chapter 1 free radical chain polymerizations represent one of the important classes of polymerization methods and a mechanistic description of the processes was formulated there. The consequences of the mechanism are developed here.

4.2.1 Kinetic representation

It is useful to abstract the chemical reactions (1.3–1.6) in Chapter 1 by representing them symbolically and to use them in defining the associated rate constants. *Initiator*, I, *decomposition* into radical, I•, can be written as

$$I \xrightarrow{k_D} 2I\bullet$$

4.3

initiator fragment reaction with monomer, M, in the initiation step to form a radical, $R_1\bullet$, containing one monomeric unit as,

$$I\bullet + M \xrightarrow{k_i} R_1\bullet$$

4.4

propagation, where $R_n\bullet$ is a chain containing n monomeric units and a radical end, as

$$R_1\bullet + M \xrightarrow{k_P} R_2\bullet$$
$$\vdots$$
$$R_n\bullet + M \xrightarrow{k_P} R_{n+1}\bullet$$

4.5

and *termination*, where M_n is a finished chain of $DP = n$ without the radical center as,

(*a*) combination

$$R_n\bullet + R_m\bullet \xrightarrow{k_{T(C)}} M_{n+m}$$

4.6

or,

(*b*) disproportionation

$$R_n\bullet + R_m\bullet \xrightarrow{k_{T(D)}} M_n + M_m$$

4.7

If the symbols above for the various molecular species stand for their concentrations as well, rates for the individual steps can be written as:

initiator decomposition rate $= k_D I$

monomer reaction rate with initiator fragment $= k_I M I\bullet$

polymer radical disappearance(appearance) rate by reaction with monomer $= k_P M R_1\bullet$ (for $R_2\bullet$); $= k_P M R_2\bullet$ (for $R_3\bullet$) etc.

polymer radical disappearance by termination: a radical $R_n\bullet$ can terminate by reaction with any of the other radicals, and the rate of termination is thus

$$k_T(R_n\bullet R_1\bullet + R_n\bullet R_2 + \cdots + R_n\bullet R_{n-1}\bullet + R_n\bullet R_n\bullet + R_n\bullet R_{n+1}\bullet + \cdots)$$

$$= k_T R_n\bullet R\bullet \quad \text{(for } R_n\bullet\text{)},$$

where $R\bullet \equiv \sum R_n\bullet$, the total concentration of polymer radicals and k_T represents either $k_{T(C)}$ or $k_{T(D)}$ or their sum if both types of termination processes were to occur simultaneously.

These steps lead to the following rate equations for the initiator fragment, $I\bullet$, and each radical, $R_1\bullet$, $R_2\bullet$, $R_3\bullet$, ...,

$$dI\bullet/dt = 2k_D I - k_I MI\bullet \tag{4.4}$$

$$dR_1\bullet/dt = k_I MI\bullet - k_P MR_1\bullet - k_T R_1\bullet R\bullet \tag{4.5}$$

$$dR_2\bullet/dt = k_P MR_1\bullet - k_P MR_2\bullet - k_T R_2\bullet R\bullet \tag{4.6}$$

$$dR_3\bullet/dt = k_P MR_2\bullet - k_P MR_3\bullet - k_T R_3\bullet R\bullet$$
$$\vdots$$
$$dR_n\bullet/dt = k_P MR_{n-1}\bullet - k_P MR_n\bullet - k_T R_n\bullet R\bullet \tag{4.7}$$
$$\vdots$$

4.2.2 Rate of polymerization

Since the concentration of radicals and the number of chains that are actually undergoing propagation is small, the rate at which polymer is produced is very close to the rate of consumption of monomer. That is, the rate of monomeric unit incorporation is very close to the rate of disappearance of monomer. The rate of monomer utilization is given, neglecting that used in the initiation step, equation (4.4), by that consumed by the various propagating species, or,

$$dM/dt = -k_P MR\bullet. \tag{4.8}$$

If the rates of change of all the polymer radical species, equations (4.5)–(4.7), are summed, it is found that,

$$\frac{dR\bullet}{dt} = \sum \frac{dR_n\bullet}{dt} = k_I MI\bullet - k_T R\bullet^2, \tag{4.9}$$

i.e., the *total* intermediate concentration can change only through creation from an initiation or destruction by termination.

The *steady-state approximation* is appropriate when a reactive intermediate is present and the rate at which it is formed is closely matched by the rate at which it disappears. Consequently, the *net* rate of accumulation of the intermediate is small compared to the rates of formation and disappearance. In applying the steady-state approximation to equation (4.9), the net rate $dR\bullet/dt$ is neglected compared to either of the separate rates on the right-hand side, and,

$$0 \approx k_I MI\bullet - k_T R\bullet^2$$

or

$$R\bullet = (k_I MI\bullet/k_T)^{\frac{1}{2}}. \tag{4.10}$$

Similarly, applying steady-state to equation (4.4) gives

$$k_I MI\bullet = 2k_D I$$

or

$$R\bullet = (2k_D I/k_T)^{\frac{1}{2}}. \tag{4.11}$$

Thus the rate of monomer utilization and polymer production will be, from equation (4.8),

$$\frac{dM}{dt} = -k_P\left[\left(\frac{2k_D I}{k_T}\right)^{\frac{1}{2}}\right]M. \tag{4.12}$$

It is appropriate to comment on the factors that could be used to control the rate of polymerization. The rate constants for propagation and termination are fixed by the choice of the polymerizing monomer and the temperature (see Section 4.2.5) and perhaps solvent. However, the concentration and the stability of the initiator are subject to choice. A higher initiator concentration, I, would lead to a more rapid rate of polymerization as would a less stable initiator with a higher value of k_D. There are a variety of initiators available that cover a wide range of stabilities. Figure 4.3 is a nomograph displaying the relation between stability expressed as half-life and temperature for a number of peroxides. At a given temperature the half-lives cover a range of at least five orders of magnitude.

4.2.3 *Instantaneous molecular weight*

The number-average DP and molecular weight being formed instantaneously in the time interval, dt, can be found from the definition in equation (2.2), and the rates of polymer unit and chain production. Since the definition below applies only to the polymer being formed through chemical reaction in a time interval, dt, it excludes free monomer from the counting of both units and chains

$$X_N(t) = \frac{\text{rate unit production}}{\text{rate chain production}} + 1.$$

For disproportionation termination there are two chains formed for every

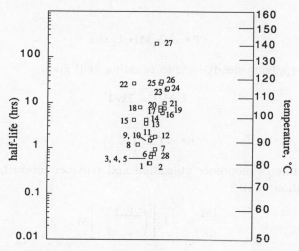

Fig. 4.3 Decomposition rates of initiators. A nomograph relating half-life and temperature. (A line is drawn through the temperature and initiator number and extended to intercept the half-life.) (1) 2,4-dichlorobenzoyl peroxide, (2) t-butyl peroxypivalate, (3) perlargonyl peroxide, (4) decanoyl peroxide, (5) lauroyl peroxide, (6) propionyl peroxide, (7) acetyl peroxide, (8) succinic acid peroxide, (9) t-butyl peroxyoctoate, (10) benzoyl peroxide, (11) p-chlorobenzoyl peroxide, (12) t-butyl peroxyisobutyrate, (13) t-butyl peroxy malic acid, (14) 1-hydroxy-1'-hydroperoxy dicyclohexyl peroxide, (15) bis(1-hydroxycyclohexyl) peroxide, (16) 2,5-dimethylhexane-2,5-diperbenzoate, (17) t-butyl peracetate, (18) methyl ethyl ketone peroxides, (19) di-t-butyl diperphthalate, (20) t-butyl perbenzoate, (21) n-butyl-4,4-bis(t-butyl peroxide) valerate, (22) ketone peroxide, (23) 2,5-dimethyl-2,5-bis(t-butylperoxy)hexane, (24) t-butyl hydroperoxide-70, (25) di-t-butyl peroxide, (26) 2,5-dimethyl-2,5-bis(t-butylperoxy)hexyne-3, (27) t-butyl hydroperoxide-90, (28) azobis(isobutyronitrile) (from trade literature for 'Lucidol'® initiators, Elf Atochem).

two radicals terminated. Therefore the rate of chain production is equal to the rate of termination $= k_T R\bullet^2$. For combination termination, however, two radicals are terminated to form one chain. Therefore the rate of chain production is half the rate of termination $= \frac{1}{2} k_T R\bullet^2$. Thus

$$X_N(t) = \frac{k_P R\bullet M}{\delta k_T R\bullet^2} + 1, \qquad (4.13)$$

where $\delta = \frac{1}{2}$ for recombination and $\delta = 1$ for disproportionation. From equation (4.11) for $R\bullet$,

$$X_N(t) = \frac{k_P M}{\delta (2 k_D I k_T)^{\frac{1}{2}}} + 1. \qquad (4.14)$$

The one is added to equations (4.13) and (4.14) since the rate of polymerization in the numerator of these equations includes formation of dimer, trimer, etc., radicals $R_2\bullet$, $R_3\bullet$, etc., but not $R_1\bullet$ (from $I\bullet + M \rightarrow$). In a polymerization making high molecular weight polymer, the average DP will be one greater than given by the rate of polymerization divided by the rate of termination.

It is to be observed that manipulation of the rate of polymerization (Section 4.2.2) through the initiator concentration and stability leads to opposite effects on the average molecular weight. Increasing I and k_D both lead to decrease of molecular weight. It is also to be noted that the molecular weight decreases versus time in a batch solution polymerization. Since the monomer concentration, M, in equation (4.14) decreases with time so also will the molecular weight.

4.2.4 Molecular weight distribution

Two important parameters of the polymerization, the rate of polymerization and the number-average molecular weight, have been expressed above in terms of the kinetic constants without actually solving the *individual* rate expressions, equations (4.5)–(4.7). It will be necessary to do so in order to determine the distribution of molecular weight. Applying the steady-state approximation to equations (4.4)–(4.7) individually, and solving for the R•s gives

$$R_1\bullet = \frac{2k_D I}{k_P M + k_T R\bullet} \tag{4.15}$$

$$R_2\bullet = \frac{k_P M}{k_P M + k_T R\bullet} R_1\bullet$$

$$\vdots$$

$$R_n\bullet = \frac{k_P M}{k_P M + k_T R\bullet} R_{n-1}\bullet$$

$$\vdots$$

By induction it follows that,

$$R_n\bullet = \left(\frac{1}{1 + 1/\gamma}\right)^n \frac{2k_D I}{k_P M}, \tag{4.16}$$

where

$$\frac{1}{\gamma} = \frac{k_T R\bullet}{k_P M}.$$

The quantity

$$\gamma = \frac{k_P M R \bullet}{k_T R \bullet^2} = \frac{\text{rate of polymerization}}{\text{rate of species termination}}$$

is often referred to as the *kinetic chain length*. According to equations (4.13) and (4.14), the number-average DP, is related to γ as,

$$\left. \begin{array}{l} X_N = \gamma + 1 \quad \text{(disproportionation)}, \\ X_N = 2\gamma + 1 \quad \text{(recombination)}. \end{array} \right\} \tag{4.17}$$

In order to find the instantaneous distribution of molecular lengths being produced, it is necessary to find the rate of production of each length from the rate of termination of each radical species. For the case of *disproportionation termination* the rate of production of finished chains of DP, n, is given by,

$$\frac{d[M_n]}{dt} = k_{T(D)} R_n \bullet R \bullet. \tag{4.18}$$

The square brackets are added to M_n ($[M_n]$) to emphasize the use as the *molar concentration* of molecules of DP $= n$ as opposed to the use of M_n in Chapter 2 as the *molecular weight* of a species of DP $= n$. Except for the volume factor, $[M_n]$ is exactly the same as the distribution function for molecular lengths N_n introduced in Chapter 2. From equation (4.16) it follows that

$$\frac{d[M_n]}{dt} = \left(\frac{2k_D I}{\gamma} \right) \left(\frac{1}{1 + 1/\gamma} \right)^n. \tag{4.19}$$

The distribution defined by equation (4.19) may be normalized by summing over both sides, using equation (2.11), and dividing both sides by the result to obtain for the molecular weight distribution for *disproportionation terminated* free radical polymerization,

$$\bar{N}_n(t) = \frac{d[M_n]}{dt} \bigg/ \frac{d \sum [M_n]}{dt} = \frac{1}{\gamma} \left(\frac{1}{1 + 1/\gamma} \right)^n$$

$$= \frac{1}{1 + \gamma} \left(\frac{\gamma}{1 + \gamma} \right)^{n-1}. \tag{4.20}$$

The time dependence or instantaneous character of the normalized distribution function, through γ, has been recognized by writing $\bar{N}_n$ as a function of t.

Since $\gamma + 1 = X_N$, equation (4.17), it is seen that equation (4.20) is identical with equation (2.14) and therefore the distribution is identical to the 'most probable' one arising in step polymerization. Physically, the basis for the same distribution being produced by seemingly quite different circumstances is the following. This distribution will result whenever the molecular ends are *randomly* distributed among a set of connected monomer units. In the case of linear step polymers this was the case. In the case of the disproportionation terminated chain polymerization this is obscured by the detailed kinetic treatment. Nevertheless, the physical process still is the random distribution of chain ends (by termination) among a set of connected 'mers'. The random distribution of chain ends is a consequence of the fact that the probability of a polymer molecule (chain end) being created (by termination) is independent of the molecular weight of the growing chain.

It is appropriate to note that although a 'most probable' distribution with $M_W/M_N = 2$ is generated instantaneously, in a batch polymerization where monomer concentration decreases continuously, the average molecular weight distribution blended over the batch will be broader than 'most probable'.

If *recombination* is the termination process, then the rate of chain production resulting from termination is given by

$$\frac{d[M_n]}{dt} = k_{T(C)} \sum_{(p+q=n)} R_p{}^\bullet R_q{}^\bullet, \tag{4.21}$$

where the $p + q = n$, notation on the summation indicates that only values of p and q that add to the finished chain length, n, are to be counted. From equation (4.16) it follows that

$$\frac{d[M_n]}{dt} = k_{T(C)} \left(\frac{2k_D I}{k_P M}\right)^2 \sum_{p+q=n} \left(\frac{1}{1 + 1/\gamma}\right)^{p+q}. \tag{4.22}$$

For a given value of n, the sum in equation (4.22) has $(n-1)/2$ equal terms if n is odd, and $n/2$ if n is even. For the large values of n typical of a high molecular weight polymer, it makes negligible difference which is used and therefore,

$$\frac{d[M_n]}{dt} = k_{T(C)} \left(\frac{2k_D I}{k_P M}\right)^2 \frac{n}{2} \left(\frac{1}{1 + 1/\gamma}\right)^n. \tag{4.23}$$

Summation of both sides of equation (4.21) over n gives the total rate of chain production in terms of the total rate of termination. Carrying this out explicitly by summation of both sides of equation (4.23), using the sum evaluation as in equation (2.13), using equation (4.11) and the approximation $1 + 1/\gamma \approx 1$ results in

$$\frac{d \sum [M_n]}{dt} = \frac{k_{T(C)}}{2} R^{\bullet 2} \tag{4.24}$$

in agreement with the denominator of equation (4.19). Dividing equation (4.23) by equation (4.24) gives the normalized distribution function for *recombination terminated* free radical polymerization,

$$\bar{N}_n(t) = \frac{d[M_n]}{dt} \bigg/ \frac{d \sum [M_n]}{dt} = \frac{n}{\gamma^2} \left(\frac{1}{1 + 1/\gamma} \right)^n, \tag{4.25}$$

where according to equation (4.17), $2\gamma + 1 = X_N$. This distribution is known as the 'Schulz' (1939) distribution. It has a maximum when plotted against n which suggests it is sharper than the 'most probable' distribution. Indeed explicit calculation shows $M_W/M_N = \frac{3}{2}$. In principle, recombination termination does not lead to a 'most probable' distribution because the random distribution of end-groups is 'scrambled' by the recombination of chains to form lengths different from the terminating radicals.

4.2.5 Thermochemistry and the effect of temperature on radical polymerization

Two related facets of chain polymerization will be addressed here. The first is the question of why chain polymerization takes place. The second deals with the effect of temperature on the rate of polymerization and the average molecular weight. With respect to the first one, it is pertinent to remark that *step* polymerization can take place, in principle, regardless of the thermochemistry, i.e., the heats of formation of the participants, since mass action can be used to displace the equilibrium to the side of the polymer product by removing efficiently the volatile by-product, e.g., water. However, in the case of chain polymerization this opportunity is not available. If the heat of reaction involved in the polymerization were positive, then very likely the polymerization would not proceed as the entropy change in reducing the monomers to polymer (a negative change, $-T\Delta S$ positive) would be unfavorable to the free energy of reaction being negative. Thus the reaction heat in polymerization should be negative for polymerization to take place. It is easy to demonstrate that this is likely to be so for common polymerizations involving carbon–carbon double bonds, i.e., vinyl and similar polymerizations. From a bond energy point of view, such a polymerization converts a double bond to two single bonds. It is a matter of experience that two such single bonds are stronger than one double bond. The former are of the order of 80 kcal/mol (335 kJ/mol) each whereas a double bond is of the order of 140 kcal/mol (586 kJ/mol), the difference being approximately 20 kcal/mol (84 kJ/mol) on conversion to polymer.

The effect of temperature on the rate of polymerization can be assessed by examining the temperature dependence of the effective rate constant resulting from the combination of rate constants appearing in equation (4.12), $k_{eff} = k_P(k_D/k_T)^{1/2}$. Writing each in Arrhenius form, $A\,e^{-\Delta E^*/RT}$, gives

$$k_{eff} = A_{eff}\,e^{-\Delta E_{eff}^*/RT}, \qquad (4.26)$$

where

$$A_{eff} = A_P(A_D/A_T)^{\frac{1}{2}}$$

and

$$\Delta E_{eff}^* = \Delta E_P^* + \Delta E_D^*/2 - \Delta E_T^*/2. \qquad (4.27)$$

To estimate the effective activation energy it is necessary to speculate on the three separate terms in equation (4.27). In Figure 4.4, an idealized energy diagram for the propagation step is shown. As indicated above, the heat of polymerization is exothermic and thus does not contribute

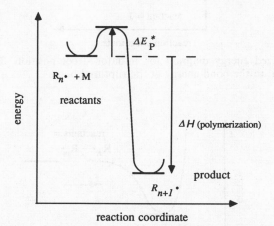

Fig. 4.4 Idealized energy diagram for chain polymerization. Heat of polymerization is the energy (enthalpy) change on adding one more monomer to the chain and is negative. The activation energy for adding the monomer, the propagation step, does not directly involve the heat of polymerization, since the latter is exothermic, but is found to be a small fraction of it.

to the activation energy. It is found that the activation energy for the forward polymerization is a small fraction of the heat of polymerization and thus of the order of 5 kcal/mol (21 kJ/mol). The activation energy for the initiator decomposition is essentially the energy required to break the bond ruptured, in the case of a peroxide, about 40 kcal/mol (167 kJ/mol), Figure 4.5. Termination by combination is an exothermic bond-forming process and thus has a very small or zero activation energy, Figure 4.6. When disproportionation is competitive with combination, the activation energy in that case is presumably also quite small. Therefore the effective activation energy of the rate of polymerization, equation (4.27), is about 25 kcal/mol (105 kJ/mol). This leads to a relatively rapid increase of rate with temperature that is typical of many chemical reactions.

The instantaneous molecular weight is the ratio of two rates and

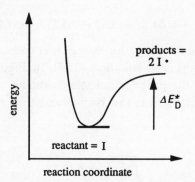

Fig. 4.5 Idealized energy diagram for initiator decomposition. The activation energy is equal to the bond energy of the ruptured bond.

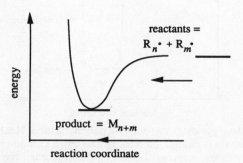

Fig. 4.6 Idealized energy diagram for combination termination. The energetics involve bond formation with zero activation energy.

according to equation (4.14) the ratio has an effective rate constant, $k'_{\text{eff}} = k_P/(k_D k_T)^{1/2}$. By following the procedure above the effective activation energy is

$$\Delta E^{*\prime}_{\text{eff}} = \Delta E^*_P - \Delta E^*_D/2 - \Delta E^*_T/2. \qquad (4.28)$$

With the same estimates of the individual terms as above, the effective activation energy of the temperature dependence of molecular weight produced is -15 kcal/mol (-63 kJ/mol). A *negative* effective activation energy in the Arrhenius expression leads to a *decrease* of molecular weight produced with increasing temperature. Thus the same conclusion is reached as was for using the initiator concentration or its stability to manipulate polymerization rate (Section 4.2.2) and molecular weight (Section 4.2.3). A change, in this case temperature, has opposite effects on the two quantities.

4.2.6 Transfer reactions

In Chapter 1 the concept of chain transfer was introduced in anionic and cationic polymerization as the key step in determining the molecular weight. Transfer reactions often play a role in radical polymerizations as well. There are several possibilities, all of which can serve to broaden the molecular weight distribution. They can also change the detailed chain microstructure and exert a major influence on the crystallinity and the physical properties. All transfer reactions involve hydrogen abstraction from a donor by the growing polymeric free radical. Hence the reaction results in termination of the species and transfer of the active center. These reactions can involve any small molecule such as monomer, initiator or solvent, or may involve a part of the growing polymeric radical itself some distance from the active center.

The first possibility discussed is *transfer to monomer*. Here the first step of the reaction involves abstraction of a hydrogen from the monomer and the reaction competes with the propagation steps. In the second step the monomeric radical initiates a new polymer molecule. Let M_0H denote a monomer, M, to emphasize the capability of donating a hydrogen and $M_0\bullet$ the radical formed on donation. The transfer consists of abstraction (**4.8**) followed by renewed propagation (**4.9**). Thus equations (4.5)–(4.7) for

$$R_n\bullet + M_0H \xrightarrow{k_t} M_nH + M_0\bullet$$

4.8

$$M_0\bullet + M \xrightarrow{k_P} R_2\bullet$$

$$R_2\bullet + M \xrightarrow{k_P} R_3\bullet$$

4.9

$dR_n\bullet/dt$ are modified by adding terms, $-k_f MR_n\bullet$. It is likely that the rates of transfer, $k_f MR_n\bullet$, are small compared to the propagation terms in these equations and the rate of polymerization will be similar. However, finished polymer chains, M_nH, indistinguishable for most purposes from other finished chains, M_n will be formed in **4.9**. Therefore the molecular weight, equation (4.14), is modified due to the rate of chain production from this source, $k_f MR\bullet$, as,

$$X_N(t) = \frac{k_P M}{k_f M + (\frac{1}{2}k_{T(C)} + k_{T(D)})(2k_D I/k_T)^{\frac{1}{2}}} + 1 \qquad (4.29)$$

When a solvent, S, is present transfer to it is also a possibility and can be handled in a manner similar to that for transfer to monomer except that terms will appear in the form, $k_{fS}SR\bullet$ rather than $k_f MR\bullet$. A generalized version of equation (4.13) can be written as

$$X_N = \frac{k_P R\bullet M}{k_f MR\bullet + k_{fS}SR\bullet + k_T R\bullet^2} + 1$$

$$= \frac{k_P M}{k_{fS}S + k_f M + (\frac{1}{2}k_{T(C)} + k_{T(D)})(2k_D I/k_T)^{\frac{1}{2}}} + 1.$$

Since $X_N \gg 1$ the equation can be inverted as,

$$\frac{1}{X_N} = \frac{k_f}{k_P} + \frac{k_{fS}S}{k_P M} + \frac{(\frac{1}{2}k_{T(C)} + k_{T(D)})(2k_D I/k_T)^{1/2}}{k_P M}. \qquad (4.30)$$

In practice, it is found that the rate of transfer to monomer is not very large in most systems and tends to be smaller than the rate of transfer to the initiator (Odian, 1970). Equation (4.30) can be expanded to include a term usually written as $C_I = k_{fI}/k_P$ where k_{fI} represents the analog of k_f for the case of transfer to initiator. If the ratio k_f/k_P is written as C_M, then C_M is usually much smaller than C_I. For instance, in the benzoyl peroxide initiation of styrene $C_M = 0.00006$ and $C_I = 0.055$. The value of C_I obviously depends very much on both the monomer and the initiator. C_M varies from a low value of 0.04×10^{-4} for methyl acrylate to a high value of 6.2×10^{-4} for vinyl chloride. In the polymerization of methyl methacrylate, AZBN has a low C_I value of apparently 0 whereas t-butyl hydroperoxide shows a high value of 1.27 at 60 °C.

It is apparent that S can represent not only solvent but also any organic molecule present that can participate in transfer. If it is present in minor amounts the rate constants in equation (4.30) will be the same as in a bulk polymerization that leads to a DP, X_N^0. Hence the equation can be written as,

$$\frac{1}{X_N} = \frac{1}{X_N^0} + K_S \frac{S}{M}, \tag{4.31}$$

where $K_S = k_{fS}/k_P$. The transfer agent, S, can be intentionally added to obtain close control over the DP. This is of considerable utility where low to medium values of X_N are desired and could be obtained otherwise through the use of high concentration of initiator. When used in this manner, i.e., the deliberate addition of transfer agents, the relation is referred to as the Mayo equation.

There are two types of transfer that involve the polymer itself. One of these results from a radical center on a growing chain attacking a nearby chain and abstracting a hydrogen, Figure 4.7. Then polymerization continues at the newly created center. This mechanism is thus responsible for formation of long branches. At one and the same time it terminates prematurely the initial chain creating a lower-than-average length molecule and creates a molecule having a higher-than-average DP because of the long branch. Such a side reaction will tend to be prevalent when

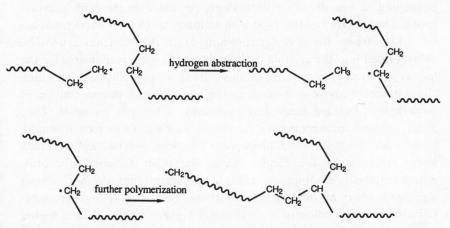

Fig. 4.7 Formation of long branches in polyethylene from *inter*molecular chain transfer.

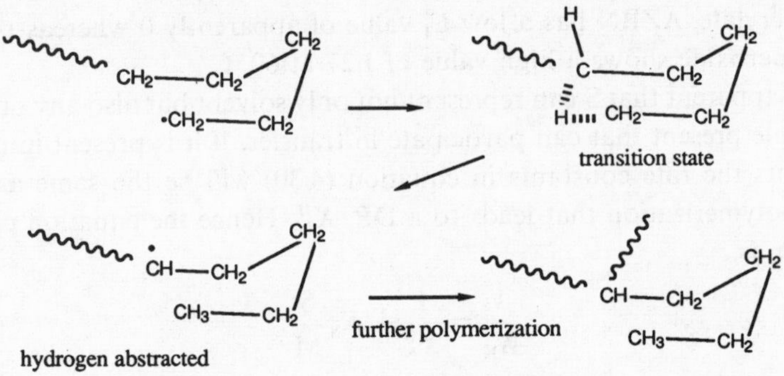

Fig. 4.8 Formation of short branches in polyethylene from *intra*molecular chain transfer.

the concentration of polymer is high and when termination processes are retarded due to diffusional effects. That is, when high conversions are attempted.

Another type of transfer involving the polymer itself results from the active center on a growing chain attacking the *same* chain. The reaction is a special one and occurs only for selected polymers in which a transition state can be achieved at the chain end. This implies ring formation and thus strongly tends to involve a six-membered ring in a flexible chain. It is important in the high pressure gas phase polymerization of ethylene. This polymer is capable of formation of such a transition state, Figure 4.8. This internal transfer reaction creates a short chain branch, C_4 or n-butyl, as pictured as polymerization continues at the new radical site. Branching in low density polyethylene, produced in the high pressure process, is primarily of this type with as many as 15 such butyl branches per 1000 carbon atoms being common. Since the branches cannot be incorporated into the crystals of the polymer they are responsible for the low crystallinity encountered in this variety of polyethylene. The long chain branches above also occur in the high pressure polymerization of polyethylene but are much less numerous, a few per molecule. They affect the rheological properties, the viscosity for a given molecular weight, more than the solid state properties. Ethylene polymerized at much lower pressures by the Ziegler–Natta and other varieties of coordination catalysis, Chapter 1, leads to polymers that do not exhibit significant short or long chain branching. Such *linear*, 'low pressure' polyethylenes crystallize to a significantly higher extent and have higher densities than their high pressure free radical polymerized branched counterparts.

4.2.7 Polymerization to high conversion

The basic scheme for radical chain polymerization above functions well provided that conversions are low. Whenever reactions are taken to high conversion major complications arise from the high viscosities resulting. This problem tends to be severe in the case of commercial operations where high conversions are essential for economic reasons.

The diffusion coefficient of a growing polymer radical is dependent on both its own molecular weight and the resistance to its motion by neighboring, possibly entangled, molecules. As the reaction proceeds to high conversion, especially in bulk polymerizations, the absolute rates of both the propagation and termination reactions decrease due to the decreasing diffusion rates. Especially affected is the termination reaction where two growing polymer radicals must diffuse together. The propagation process is less affected because of the still relatively high diffusion coefficient of the monomer molecules through the viscous medium. Termination processes involving the primary radicals, those produced by initiator decomposition, become important because of the relative ease of motion of these species. However, since both the initiator concentration and the primary radical concentration decrease with increasing conversion, the absolute rates of the propagation and termination reactions decrease. The net result is that the overall rate of reaction, i.e., the rate of consumption of monomer, and the molecular weight of the polymer increase. Transfer reactions to polymer become more important and high degrees of branching can result.

The governing equation of the rate of reaction is, cf. equation (4.12),

$$\text{rate of polymerization} = (k_P/k_T^{1/2})(2k_D I)^{1/2} M$$

and since $k_T^{1/2}$ decreases more rapidly than k_P the ratio $k_P/k_T^{1/2}$ tends to increase. When this ratio and the rate of reaction begin to increase the reaction is described as 'diffusion controlled' (O'Driscoll in Allen and Bevington, 1989). It is sometimes labeled the 'gel effect' and is also often called the 'Trommsdorf effect', although there are several references to the effect previous to the latter oft-cited paper (Trommsdorf *et al.*, 1948; Schulz, 1956). Experimental data for methyl methacrylate solutions of varying concentrations can be seen in Figure 4.9. The effect begins between 10 and 30% MMA but is most evident for concentrations of 60% and above.

The termination reaction can be conceptualized as having three distinct stages (Figure 4.10). They are (*a*) polymeric radicals diffuse together,

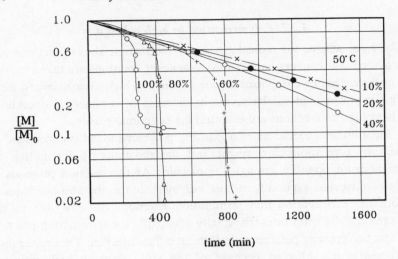

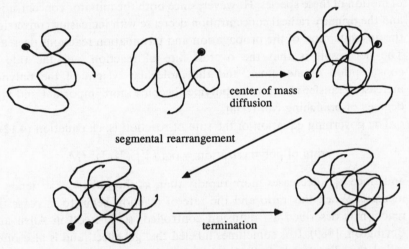

Fig. 4.9 Trommsdorf effect. Monomer consumption is plotted as a function of time for methyl methacrylate in benzene at 50 °C (Schulz, 1956). The percentage figures refer to monomer.

Fig. 4.10 Three stages in diffusion controlled termination.

(b) chain ends diffuse together, (c) chain ends react and termination is completed (Benson and North, 1962). The last process is essentially instantaneous, the first two represent two distinct types of diffusion control. Stage (a) is a macroscopic diffusion process and is governed by Fickian equations. Stage (b) is an internal rearrangement process that is critically dependent on chain flexibility. It has been shown that as temperature is lowered a large decrease in k_T occurs as the molecule becomes more inflexible. The values of k_T for the solution polymerization of several monomers in the diffusion controlled region correlate with the segmental mobility of the final polymers (North and Phillips, 1967).

In industrial syntheses the complications can be severe as there is always the intention, for efficiency, of carrying out the reaction to high conversion. Additional problems can arise since the thermal conductivity of the system also decreases as conversion proceeds due to the reduced diffusion coefficients. Heat of reaction cannot therefore be effectively dispersed and this results in an increase of the system temperature. This, in turn, causes the rate of reaction to increase further, exacerbating the problem. The complications become self-reinforcing, almost autocatalytic in nature, and a situation known as 'popcorn polymerization' can result. It is not unusual for glass reaction vessels to explode under these circumstances. The polymer is a highly crosslinked gel since transfer reactions to the polymer dominate. Extractable polymer tends to have a molecular weight in the million range. Whenever bulk polymerization is retained in commercial synthesis the heat transfer problem is overcome by (*a*) passing the reaction mixture through a cooled tube of small diameter, (*b*) having the medium flow down the walls of a cooled column or (*c*) extruding the material as thin streams which then free fall. Often the polymerization is carried out in two stages, a batch reaction followed by one of the above cooling techniques during the final stages of conversion. Gel effect problems are not as severe in solution polymerization. However, the added expense of solvent recovery often renders such processes uneconomic.

4.2.8 Emulsion and suspension polymerization

Two convenient alternative polymerization methods involve suspending monomer droplets in water. The simplest method, known as *suspension polymerization*, uses monomer droplets 0.1–1 mm in diameter which also contain the free radical initiator. Droplets are usually stabilized by a suspension agent which is often a water soluble polymer. Since the droplets contain both monomer and initiator the kinetic scheme followed is that derived above for bulk polymerization. Heat transfer problems are eliminated but fusion problems still remain.

A popular commercial technique is *emulsion polymerization* (Figure 4.11). The droplets are much smaller than in suspension polymerization

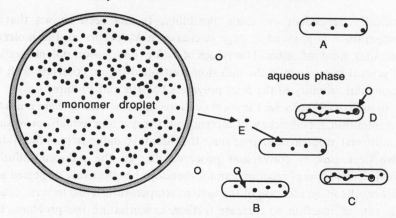

Fig. 4.11 Emulsion polymerization: initial or Phase I stage, stabilized droplets serve as monomer reservoirs. (A) Soap micelles are swollen with monomer also. (B) Initiator radical, O, from aqueous phase enters a micelle and (C) initiates polymerization. (D) The micelles are small enough that when another radical enters the micelle, termination ensues. (E) Monomer in micelles is replenished by diffusion of monomer from droplets.

and are of two sizes. During the initial stages, or phase I, most of the monomer is present in the form of droplets 1–2 μm in diameter stabilized by an emulsifier. The concentration is about 10^{14} droplets per liter. Microdroplets also occur which are 0.1–0.3 μm long and initially about 5 nm thick. They are, in fact, micelles of added soap molecules swollen with monomer. Typically they are found in concentrations of 10^{21} micelles per liter. In contrast to suspension polymerization the initiator is present in the water. The micelles and droplets compete for initiator free radicals and since the micelles are present in concentrations 10^7 times greater than the droplets they tend to dominate. A radical from the initiator system enters a micelle and polymerization ensues. The micelles are so small that whenever two free radicals enter a micelle they terminate rapidly. Hence micelles contain either zero or one free radical at any given time. Under these circumstances classical bulk or solution polymerization kinetics do not apply. The polymerization occurs in the micelles, the monomer supply being replenished by diffusion of the monomer from the droplets to the micelles through the water.

As polymerization proceeds, the micelles are converted into larger latex particles. The diffusion process is governed by the need for equilibrium concentrations of monomer to be maintained in both the micelles and the water. It results in the monomer departing from the droplets which then shrink in size. Phase II occurs when all the micelles have been converted into latex particles and polymerization occurs within them alone. A pseudo-equilibrium is set up in the particles, between monomer and polymer, so monomer continues to diffuse across the water from the emulsion droplets. Phase III begins when the droplets disappear, so as polymerization proceeds within the particles, the monomer is no longer replenished and the process eventually ceases.

Micelles are present only for the first 10% of monomer conversion as they are rapidly converted into latex particles by the polymer generated within them. In practice, most of the polymerization occurs inside the latex particles. Free radicals continue to diffuse into these particles as will monomer from the droplets. The number of radicals need no longer be zero or one because of the larger size of the latex particles.

Clearly a simple model of the process cannot be generated because of the many complications. However, conceptually it is useful to think of an average number of radicals per micelle or latex particle, R_P. The total concentration of radicals is given by

$$R\bullet = N_P R_P / N_A, \tag{4.32}$$

where N_P is the number of particles per unit volume and N_A is the Avogadro number. The rate of polymerization can then be written, cf. equation (4.12), as

$$-dM/dt = k_P M (N_P / N_A) R_P. \tag{4.33}$$

Similarly, the DP can be thought of as

X_N = rate of polymerization/rate of capture of radicals by particles.

$$\tag{4.34}$$

Thus the number of particles times the average number of radicals per particle conceptually plays the role of radical concentration in conventional polymerization and rate of radical capture plays that of termination.

A formal quantitative model has been formulated for the kinetics (Smith and Ewart, 1948). It is necessarily a complicated model but regimes or limiting cases of behavior can be identified and polymerization rates formulated.

Initiation is usually achieved by a redox system. Fenton's reaction is one form,

$$Fe^{2+} + ROOR \rightarrow Fe^{3+} + RO^- + RO\bullet$$

and another common system is persulfate ions with thiosulfate as the reducing agent,

$$S_2O_8^{2-} + S_2O_3^{2-} \rightarrow SO_4^{2-} + SO_4^- \bullet + S_2O_3^- \bullet.$$

There are some complications in practice. These include the following considerations: (*a*) a time-dependent distribution of latex particle sizes is present, (*b*) transfer reactions to soap and other molecules are possible, (*c*) highly water-soluble monomer molecules may react in the aqueous phase, and (*d*) diffusion rates of monomer and the effect of particle size and the DP on them may be important. Despite these complications this synthetic technique is popular because of lack of major control problems and the fact that the product is in the form of easily collected small particles. More details may be found in Napper and Gibert in Vol. 4 of Allen and Bevington (1989).

4.3 Chain copolymerization

Copolymers are of great technological importance because they often have properties significantly different from their respective homopolymers. The most important uses of copolymerization are to lower the melting point and/or degree of crystallinity of crystalline homopolymers, to alter glass transition temperatures, to create phase-separated materials that may or may not be chemically connected and to provide sites for crosslinking reactions. In *step* polymerization (such as polyesterification or polyamidation) the reactivities of the polymerizing species, as has been seen, tend to depend on the functional groups (an aliphatic carboxylic acid and hydroxyl for example) reacting and not so much on which actual monomers. Therefore, the composition and sequencing of monomeric units tend to be random. A mixture of adipic and sebacic acids would tend to be randomly sequenced in a copolyamide. However, in the case of *chain* polymerization, the reactivities of different monomers towards each other can be very different from that of like monomers. This often leads to pronounced fractionation effects where the composition of the polymer being made instantaneously (ratios of the different kinds of monomeric units being incorporated) differs significantly from that of the as yet unreacted monomers in the reactor and the monomer unit sequencing can be far from random.

4.3.1 Composition of copolymers

Consider the monomers A and B, denoted below by subscripts 1 and 2 respectively, that react via a chain mechanism. Let A• denote the total concentration of growing chains that currently have a reactive center on the chain end from the A monomeric unit regardless of the composition of the chain in terms of A and B monomeric units. Similarly, let B• be the chains with reactive centers from B. The rates of A monomer consumption can be written as

$$-dA/dt = k_{11}\text{A•A} + k_{21}\text{B•A}, \tag{4.35}$$

where the first subscript on the propagation constant k refers to the reactive center and the second to the monomer, i.e., k_{11} to

$$\text{—A•} + \text{A} \xrightarrow{k_{11}} \text{—A•}$$

and k_{21} to

$$\text{—B•} + \text{A} \xrightarrow{k_{21}} \text{—A•},$$

and B monomer consumption is

$$-dB/dt = k_{12}\text{A•B} + k_{22}\text{B•B}, \tag{4.36}$$

where k_{12} and k_{22} are defined by

$$\text{—A•} + \text{B} \xrightarrow{k_{12}} \text{—B•}$$

$$\text{—B•} + \text{B} \xrightarrow{k_{22}} \text{—}.$$

As an extension of equation (4.9), recognizing that A• can terminate with either A• or B•, and especially keeping in mind that A•, B• radicals *are created and destroyed by cross-propagation*, it follows that

$$d\text{A•}/dt = k_{\text{I}}^{\text{A}}\text{I•A} - k_{\text{T(AA)}}\text{A•}^2 - k_{\text{T(AB)}}\text{A•B•} + k_{21}\text{B•A} - k_{12}\text{A•B}$$

and similarly,

$$d\text{B•}/dt = k_{\text{I}}^{\text{B}}\text{I•B} - k_{\text{T(BB)}}\text{B•}^2 - k_{\text{T(BA)}}\text{B•A•} + k_{12}\text{A•B} - k_{21}\text{B•A}.$$

The steady-state approximation may be applied to these equations as before, but further simplification is possible. Notice physically that in order for a given chain to contain both A and B monomeric units, the cross-propagation reactions must be present. If a high molecular weight polymer chain is to contain appreciable numbers of both A and B units, then the cross-propagation events must be much more prevalent than initiation and termination. There will be only one initiation and termination event per chain but many cross propagation events if there is incorporation of both types of monomer. Therefore, in a true copolymerization in which the product chains incorporate significant amounts of both monomers, initiation and termination terms may be neglected in the above equations. With this simplification, the steady-state approximation results in (Alfrey and Goldfinger, 1944; Mayo and Lewis, 1944; Simha and Branson, 1944; Wall, 1944)

$$k_{21}B{\bullet}A - k_{12}A{\bullet}B = 0. \tag{4.37}$$

The instantaneous relative rate of monomer incorporation may now be obtained by dividing equation (4.35) by equation (4.36)

$$\frac{dA}{dB} = \frac{A}{B}\left[\frac{k_{11}(A{\bullet}/B{\bullet}) + k_{21}}{k_{12}(A{\bullet}/B{\bullet}) + k_{22}}\right]$$

and using equation (4.37),

$$\frac{dA}{dB} = \frac{A}{B}\left[\frac{r_1(A/B) + 1}{(A/B) + r_2}\right], \tag{4.38}$$

where the *reactivity ratios* r_1 and r_2 have been defined as

$$r_1 \equiv k_{11}/k_{12}, \tag{4.39}$$

$$r_2 \equiv k_{22}/k_{21}. \tag{4.40}$$

The result expressed in equation (4.38) is a very important and useful one (it is known as the *copolymer composition equation*) as it relates the instantaneous incorporation of the monomers to the present monomer concentrations and reactivity ratios. Referring to the definitions of r_1 and r_2 above and back to equations (4.35) and (4.36), it is seen that r_1 expresses the relative preference for A• adding A and compared to B and r_2 for B• adding B relative to A. Table 4.1 lists values of r_1 and r_2 for a number of monomer pairs.

Some interesting observations concerning the dependence of copolymer composition on the reactivity ratios can be made. First of all, it is noticed that, in general, there will be a fractionation effect. The term inside the square brackets in equation (4.38) expresses this fractionation. This term and the ratio of monomers instantaneously incorporated is equal to the ratio of monomers present only in the very special case of $r_1 = r_2 = 1$. Otherwise, for general A/B, the ratio of incorporation is not equal to A/B. Since the ratio of monomers incorporated is not equal to the unreacted monomer ratio, the latter will change through the course of the polymerization and therefore so will the composition of the polymer. The resulting polymer composition over the course of the polymerization must be found by integration of equation (4.38). The composition of the last

Table 4.1 *Monomer reactivity ratios in radical copolymerization.*[a]

M_1	r_1	M_2	r_2	T (°C)
acrylic acid	1.15	acrylonitrile	0.35	50
	0.25 ± 0.02	styrene	0.15 ± 0.01	60
	2 .	vinyl acetate	0.1	70
acrylonitrile	0.35	acrylic acid	1.15	50
	0.02	1,3-butadiene	0.3	40
	0.14 ± 0.04	t-butyl vinyl ether	0.0032 ± 0.0002	60
	0.7 ± 0.2	ethyl vinyl ether	0.03 ± 0.02	80
	0.02 ± 0.02	isobutylene	1.8 ± 0.2	50
	1.5 ± 0.1	methyl acrylate	0.84 ± 0.05	50
	0.150 ± 0.080	methyl methacrylate	1.224 ± 0.100	80
	0.61 ± 0.04	methyl vinyl ketone	1.78 ± 0.22	60
	0.04 ± 0.04	styrene	0.40 ± 0.05	60
	4.2	vinyl acetate	0.05	50
	2.7 ± 0.7	vinyl chloride	0.04 ± 0.03	60
	0.91 ± 0.10	vinylidene chloride	0.37 ± 0.10	60
	0.113 ± 0.002	2-vinyl pyridine	0.47 ± 0.03	60
allyl acetate	0	methyl methacrylate	23	60
	0.00	styrene	90 ± 10	60
	0.7	vinyl acetate	1.0	60
1,3-butadiene	0.3	acrylonitrile	0.2	40
	0.75 ± 0.05	methyl methacrylate	0.25 ± 0.03	90
	1.35 ± 0.12	styrene	0.58 ± 0.15	50
	8.8	vinyl chloride	0.035	50
diethyl fumarate	0	acrylonitrile	8	60
	0.070 ± 0.007	styrene	0.30 ± 0.02	60
	0.444 ± 0.003	vinyl acetate	0.011 ± 0.001	60
	0.12 ± 0.01	vinyl chloride	0.47 ± 0.05	60
diethyl maleate	0	acrylonitrile	12	60
	0	methyl methacrylate	20	60
	0.0 ± 0.1	styrene	5 ± 1.5	70
	0.043 ± 0.005	vinyl acetate	0.17 ± 0.01	60
	0.009 ± 0.003	vinyl chloride	0.77 ± 0.03	60
fumaronitrile	0.01 ± 0.01	methyl methacrylate	3.5 ± 0.5	79
	0.01 ± 0.01	styrene	0.23 ± 0.01	60
	0.000	vinyl acetate	0.14	–
maleic anhydride	0.046 ± 0.052	dodecyl vinyl ether	-0.046 ± 0.054	50
	0.02	methyl acrylate	2.8 ± 0.05	75
	0.02	methyl methacrylate	6.7 ± 0.2	75
	0.015	styrene	0.040	50
	0.003	vinyl acetate	0.055 ± 0.015	75
	0.008	vinyl chloride	0.296 ± 0.07	75

(*continued*)

Table 4.1 (*continued*)

M$_1$	r$_1$	M$_2$	r$_2$	T (°C)
methylacrylic acid	0.526	butadiene	0.201	50
	0.7 ± 0.05	styrene	0.15 ± 0.1	60
	20	vinyl acetate	0.01	70
	0.58 ± 0.05	2-vinyl pyridine	1.55 ± 0.10	70
methyl acrylate	0.05 ± 0.05	acrylamide	1.30 ± 0.05	60
	0.67 ± 0.1	acrylonitrile	1.26 ± 0.1	60
	0.05 ± 0.02	1,3-butadiene[b]	0.76 ± 0.04	5
	3.3	ethyl vinyl ether	0	60
	0.504	methyl methacrylate	1.91	130
	0.15 ± 0.05	styrene	0.7 ± 0.1	60
	9	vinyl acetate	0.1	60
	4	vinyl chloride	0.06	45
	0.20 ± 0.09	2-vinyl pyridine	2.03 ± 0.49	60
methyl methacrylate	0.46 ± 0.026	styrene	0.52 ± 0.026	60
	20 ± 3	vinyl acetate	0.015 ± 0.015	60
	10	vinyl chloride	0.1	68
	2.53 ± 0.01	vinylidene chloride	0.24 ± 0.03	60
α-methylstyrene	0.1 ± 0.02	acrylonitrile	0.06 ± 0.02	75
	0.010 ± 0.01	1,3-butadiene	1.6 ± 0.5	12.8
	0.038 ± 0.003	maleic anhydride	0.08 ± 0.03	60
	0.14 ± 0.01	methyl methacrylate	0.50 ± 0.03	60
	0.38	styrene	2.3	–
methyl vinyl ketone	0.35 ± 0.02	styrene	0.29 ± 0.04	60
	7.00	vinyl acetate	0.05	70
	8.3	vinyl chloride	0.10	70
	1.8	vinylidene chloride	0.55	70
stilbene	0.03 ± 0.03	maleic anhydride	0.03 ± 0.03	60
	0	styrene	11.2 ± 1.2	60
styrene	80 ± 40	ethyl vinyl ether	0	80
	1.38 ± 0.54	isoprene	2.05 ± 0.45	50
	55 ± 10	vinyl acetate	0.01 ± 0.01	60
	17 ± 3	vinyl chloride	0.02	60
	1.85 ± 0.05	vinylidene chloride	0.085 ± 0.010	60
	0.55	2-vinyl pyridine	1.14	60
tetrafluoro-ethylene	1.0	chlorotrifluoroethylene[b]	1.0	60
	0.85	ethylene[b]	0.15	80
	<0.3	isobutylene[b]	0.0	–
vinyl acetate	3.0 ± 0.1	ethyl vinyl ether	0	60
	0.23 ± 0.02	vinyl chloride	1.68 ± 0.08	60
	0.1	vinylidene chloride	6	68

[a] Data are selected values by Odian (1991, with permission, McGraw-Hill, Inc., © 1991) taken from Brandrup and Immergut (1966, with permission, John Wiley & Sons, Inc., © 1966). [b] Emulsion polymerization.

polymer made in a batch polymerization may be quite different from that in the initial stages. In fact, the two extremes could be immiscible and form a phase-separated system.

Some limiting cases of the reactivity ratios are of interest. If r_1 and r_2 both approach zero (corresponding to each radical preferring to react with the unlike monomer) $dA/dB \to 1$, regardless of feed composition. Physically it is seen that an alternating copolymer ABABAB is approached. If one ratio is appreciable (r_1) and one is very small (r_2) then

$$dA/dB \to 1 + r_1 A/B.$$

Thus, if A predominates in the feed and/or r_1 is large, A is incorporated into the polymer greatly in preference to B. Long sequences of A interspersed with B will result.

If both ratios were large, a tendency towards long sequences of each monomeric unit would result. However, this case is not really observed in copolymerization for it would likely result in the homopolymerization of one monomer followed by the homopolymerization of the other. The special case, $r_1 = 1/r_2$ or $r_1 r_2 = 1$ is called *ideal* copolymerization since

$$dA/dB = r_1 A/B$$

and the ratio of the monomeric incorporation is proportional (but not equal) to the feed ratio just as the vapor pressure of a component over an ideal solution is proportional to its concentration in the solution.

It is possible for the parenthetical term in equation (4.38) to be 'accidentally' unity by virtue of the monomer ratio A/B taking a specific value such that the numerator and denominator are equal for the r_1, r_2 values present. In analogy with distillation this is called 'azeotropic' copolymerization. However, there is an important difference. Unlike distillation, it is not possible to approach the azeotropic composition as the result of copolymerization going forward. If the composition differs slightly from the azeotrope it will drift away from it, not towards it.

In illustration of fractionation effects it is instructive to consider an example. Vinylidene chloride (**4.10**) is a highly crystalline polymer that has exceptional properties as a barrier to permeation of many organic

4.10

substances as well as water and thus has formed the basis of commercial films such as Saran® wrap. However, the polymer has a very high melting point, too high for effective melt processing of the film during its manufacture. It has been found that the melting point can be depressed via copolymerization and the barrier properties retained. It is interesting to investigate the composition of such a copolymer in terms of the monomer ratios employed in polymerization. For illustrative purposes only, suppose vinyl acetate (example 9, Appendix A1.1) were chosen as the comonomer. From Table 4.1, letting 1 = vinyl acetate (VAc) and 2 = vinylidine chloride (VDC), $r_1 = 0.1$, $r_2 = 6$. Further, suppose that 20 mol% VAc were desired as the copolymer composition. If the initial feed composition were naively chosen as 20 mol% or A/B = 1/4, equation (4.38) shows that the ratio of monomers initially incorporated into the polymer would be 0.041 or only 3.9 mol%. Solving equation (4.38) for A/B with the ratio dA/dB set to the desired value of 0.25 gives A/B = 1.64. The latter is obviously drastically different from the incorporation ratio, dA/dB. It should be realized that even if the initial monomer ratios were adjusted to the latter value to achieve the desired polymer composition, continued reaction in a *batch* reaction would lead to the monomer ratio changing markedly and thus the polymer composition as well. To maintain the desired composition it would be necessary to be restricted to low monomer conversions or to use a reactor, such as a continuous stirred-tank reactor, where the monomer ratios could be maintained.

The instantaneous incorporation relation, equation (4.38), can be integrated to find the relation between A and B remaining at any point and the initial values of A and B. To accomplish this, define a new variable, $u = A/B$. Then it follows that

$$B\frac{du}{dB} = \frac{dA}{dB} - u$$

and replacing dA/dB in the above with the right-hand side of equation (4.38) leads to separation of variables and,

$$\frac{dB}{B} = \frac{du}{u\left(\dfrac{r_1 u + 1}{u + r_2}\right) - u}.$$

Rearrangement gives

$$\frac{dB}{B} = \frac{(u + r_2)du}{u[u(r_1 - 1) - (r_2 - 1)]}.$$

The right-hand side can be integrated as two standard forms and there results,

$$\ln \frac{B}{B^0} = \frac{r_1 r_2 - 1}{(r_1 - 1)(r_2 - 1)} \ln\left[\frac{u(r_1 - 1) - (r_2 - 1)}{u^0(r_1 - 1) - (r_2 - 1)}\right] - \frac{r_2}{r_2 - 1} \ln\frac{u}{u^0}, \quad (4.41)$$

where B^0 and u^0 are the initial values of B and A/B. Plots of the ratio of A and B ($=u$) remaining unpolymerized as a function of B remaining can be prepared by computing via equation (4.41) values of B/B^0 for various values of u. The overall composition of the polymer at any value of B remaining unpolymerized can be determined from u, u^0 and B/B^0.

$$\frac{A}{B} \text{ (in polymer)} = \frac{A^0 - A}{B^0 - B} = \frac{u^0 - uB/B^0}{1 - B/B^0} \quad (4.42)$$

The use of the integrated equation can be illustrated by means of the VAc/VDC copolymer example above. In a batch reactor, it is obvious that over the course of *complete* reaction the polymer composition would have to match that of the starting monomers. However, the polymer would be a blend of rather different compositions and how different can be determined from the integrated equation. Figure 4.12 shows how the ratio of unreacted monomers, A/B = VAc/VDC, changes with conversion of B ($= 1 - B/B^0$) as calculated from equation (4.41) if the monomer ratio initially is set to that appropriate for the desired 20 mol%, or

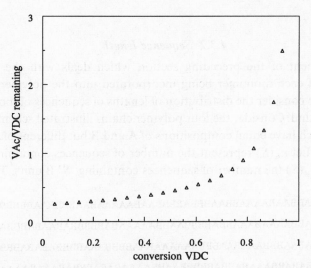

Fig. 4.12 Ratio of unreacted VAc to VDC monomers in a batch reaction versus the conversion of VDC. Initial ratio = 1/4.

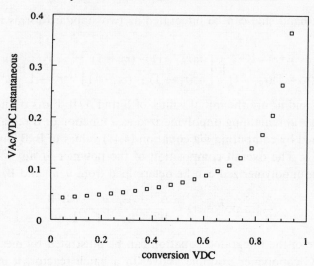

Fig. 4.13 Instantaneous composition of the VAc/VDC polymer being made versus conversion of VDC.

$A^0/B^0 = 1/4$. This 20 mol% will be indeed the final overall composition. However, using the A/B ratio at each point in Figure 4.12 to calculate an instantaneous ratio dA/dB it is seen in Figure 4.13 that the final polymer is a blend of compositions between <4 mol% and over 30 mol%.

4.3.2 Sequence length

The treatment of the preceding section which deals with the relative amounts of each monomer being incorporated into the polymer can be extended to consider the distribution of lengths of sequences of monomers of a given kind. Consider the four polymer chains illustrated schematically in **4.11** which have equal compositions of A and B but different placement sequences. Let $n_A(S)$ represent the number of sequences containing 'S' A units and $n_B(S)$ the number of sequences containing 'S' B units. Then the

AAABBAABAAAABBAABBBABBBBAABAAAAABAABBBBAABBBBB

ABABBAAAABAAABBABBBBBAABBAAAAABABBBBBBABAABABBBA

BAABAABBAAAABAABBBAAAAAABAABBBBBAAABBBBBABABABBB

BBBABABBBAAABABBABBBBBBAABBAAAAABABABBBBABAAABAAAA

4.11

Table 4.2 *Sequence distribution in example* **4.11**

number in sequence S_A or S_B	number of sequences	
	$n_A(S)$	$n_B(S)$
1	18	20
2	12	8
3	6	7
4	4	5
5	2	3
6	1	0
total	43	43

values of $n_A(S)$ and $n_B(S)$ for the above example would be as given in Table 4.2. There are a total of 92 A and B mers each in the example so the average sequence length is 92/43 or 2.14. Within the framework of the assumptions already made concerning the rates of addition of the monomers to the growing chain ends, the distributions $n_A(S)$ and $n_B(S)$ are readily deduced in terms of the rate constant ratios, r_1, r_2 and the monomer concentrations. Consider the probability, P_{AA} that an A unit is followed by another A unit. P_{AA} will be given by the ratio of the rate of production of AA pairs to the total rate of production of both AA and AB pairs. Recall that AA units are produced only by the reaction,

$$\text{---A} \bullet + \text{A} \xrightarrow{k_{11}} \text{---A} \bullet$$

and AB units only by

$$\text{---A} \bullet + \text{B} \xrightarrow{k_{12}} \text{---B} \bullet.$$

Hence,

$$P_{AA} = \frac{k_{11}\text{A} \bullet \text{A}}{k_{11}\text{A} \bullet \text{A} + k_{12}\text{A} \bullet \text{B}}$$

$$= \frac{r_1\text{A}}{r_1\text{A} + \text{B}}. \tag{4.43}$$

Since P_{AB} is given by the ratio of the rate of production of AB pairs to the total rate of production of both AA and AB pairs, then,

$$P_{AB} = \frac{k_{12}\text{A} \bullet \text{B}}{k_{11}\text{A} \bullet \text{A} + k_{12}\text{A} \bullet \text{B}}$$

$$= \frac{\text{B}}{r_1\text{A} + \text{B}}. \tag{4.44}$$

By similar reasoning it follows that,

$$P_{BB} = \frac{r_2 B}{r_2 B + A}, \tag{4.45}$$

$$P_{BA} = \frac{A}{r_2 B + A}. \tag{4.46}$$

Built into the probabilities above through the kinetic representation is the assumption that the probability of addition of a unit of a given kind depends only on that kind of unit and the unit to which it is being added. P_{AB} depends only on B being added to A• and does not depend on the sequencing of As and Bs preceding A•. Such a chain is called mathematically a *Markov chain*. Copolymers obviously are not necessarily required to be Markovian. Previously added nearby units could influence the rate constants so that the simple k_{11}, k_{12}, k_{21}, k_{22} description would be noticeably inadequate. High resolution NMR experiments are in some cases capable of detecting the sequencing of groups of several monomeric units and determining their populations. The Markovian assumption can be checked by comparing the predicted probabilities against the measured ones. It is not uncommon to find significant deviations but the approximation is nevertheless useful and often is rather well fulfilled.

The probabilities are now used to establish the distribution functions $n_A(S)$, $n_B(S)$. In general, the formation of 'S' A units involves, $S - 1$, successive additions of A to a BA unit before a B addition occurs. If an A unit starting an A sequence is selected, then the probability, $\bar{n}_A(S)$, of there being S connected A units is in analogy with equation (2.10),

$$\bar{n}_A(S) = n_A(S)/\sum_S n_A(S) = (P_{AA})^{S-1}P_{AB}. \tag{4.47}$$

The average number of units in an A sequence is

$$\langle S_A \rangle = \sum_S \bar{n}_A(S)S$$

$$= \sum_S (P_{AA})^{S-1}P_{AB} = (1 - P_{AA})(1 + P_{AA} + 2P_{AA}^2 + \cdots). \tag{4.48}$$

From equation (2.13) it follows that,

$$\langle S_A \rangle = 1/(1 - P_{AA}) = 1/P_{AB} = 1 + r_1 A/B. \tag{4.49}$$

By a similar process

$$\langle S_B \rangle = 1/(1 - P_{BB}) = 1/P_{BA} = 1 + r_2 B/A. \tag{4.50}$$

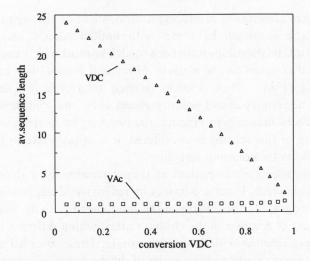

Fig. 4.14 Average sequence lengths for VAc, VDC monomeric units vs conversion of VDC in the VAc/VDC copolymerization of Figures 4.12 and 4.13.

It is apparent that the average sequence length depends on the ratio A/B and that in a batch reactor the lengths will change with conversion. Figure 4.14 shows the variation of $\langle S_A \rangle$ and $\langle S_B \rangle$ with conversion for the VAc/VDC example illustrated in Figures 4.12 and 4.13. It is apparent that the average lengths for VAc and VDC start out very disparate due to the great difference in r_1, r_2 (0.1 vs 6) and reinforced by the difference in concentrations (VAc/VDC = 1/4). As VAc/VDC rises with conversion the average length of VDC sequences drops greatly.

4.3.3 Reactivity ratios and chemical structure

It will be evident from the discussions in the last few sections that reactivity ratios are of great value in the prediction of the copolymer composition and sequence length distributions, which together are responsible for the physical properties of copolymers. From the table of reactivity ratios it is seen that the values are typical only of each pair of monomers. An understanding of the factors responsible for the actual values obtained in each case would lead to the possibility of predicting the behavior of any new copolymer prior to its synthesis and would be of great value to industry. Unfortunately reactivity ratios are controlled by three interactive molecular properties, all dependent on the chemical structure of the monomers. These are resonance, polarity and steric effects.

Theoretical attempts at producing a rationalization through the use of a mechanistic approach have met with limited success and so only semi-empirical relationships have been used. The most widely encountered relation is that based on the work of Alfrey and Price (1947) and Alfrey and Young (1964), which is often referred to as the *Q–e* scheme. It considers the resonance and polarity effects only, and considers them as separate effects although, in practice, the two may be interrelated. Before a discussion of this scheme is considered, it is appropriate to relate the different effects to molecular structure.

Resonance effects are important as they determine the stability of any given radical, which, in turn, controls its reactivity. Comparisons can be made by considering the inverse of the monomer reactivity ratio, which gives the rate of reaction of the chosen radical with a different monomer to the rate of reaction with its own monomer. These so-called monomer reactivities when tabulated for a series of different radicals show broadly the same trends for a given set of comonomers (Odian, 1970). Chemical substituents containing double bonds tend to be most effective at resonance stabilization and give the highest values of $1/r$.

A general order of decreasing effective resonance stabilization is as follows:

phenyl ring, vinyl group;
nitrile group, ketone group;
carboxylic acid, carboxylic ester;
chloride group;
acetate group, alkyl group;
ether group, hydrogen.

The unsaturated units are most effective because of the π electrons which are easily available for resonance stabilization, whereas the remaining groups, although polar, can only supply non-bonded electrons from lone pairs on atoms such as oxygen and chlorine.

Exceptions to the above scheme arise when there is a strong alternating tendency between the two monomers under consideration. In such cases the dominant effect is *polarity*, and alternation generally tends to occur when one of the monomers is an electron donor and the other an electron acceptor. In other words monomers of strongly different polarity are involved. Such effects can usually be predicted from the value of the product of the two reactivity ratios of the monomers, $r_1 r_2$. If the value of this product is very small and approaches zero then an alternating copolymer will result. Deviations from this general principle tend to be caused by *steric effects*. The mechanism responsible for producing alternation has been debated for some time. Although the detailed mechanism is not known, several pertinent facts have been established. For instance, the solvent used appears to exert little influence and so the polar interactions responsible must occur in a *transition state* and not in the ground state of the reactants. One suggestion has been that the mechanism is charge transfer between the species, whereas an alternate mechanism suggests homopolymerization of a 1:1 complex rather than copolymerization of monomer molecules. Much recent evidence supports the latter explanation. Examples of this evidence are that (*a*) the rate of polymerization is at a maximum for a 1:1 mixture, and (*b*) the rate is increased by the addition of Lewis acids (Bamford, 1985). However, it should be recognized that there is not unanimity on the mechanism and the subject is still controversial (Ebdon, Towns and Dodson, 1986).

The effects of *steric hindrance* have generally been evaluated using di-, tri- and tetra-substituted ethylenes in copolymerization experiments. Generally, the inclusion of a beta substituent lowers the reactivity. Although this effect occurs in homopolymerization also (see earlier), the effects are greater in copolymerization. A disubstituted ethylene which is incapable of homopolymerization can enter into a copolymerization reaction if there is a lack of beta substitution on the attacking radical (e.g., a styryl radical). There can be differences in reactivity between *cis* and *trans* isomers of the same disubstituted monomer. For instance, *trans*-dichloroethylene is more reactive than *cis*-dichloroethylene because of the inability of the *cis* isomer to attain the completely planar conformation necessary for resonance stabilization in the transition state. This example is one in which the effects of steric hindrance and resonance cannot be separated from one another. Both the *cis* and *trans* isomers are much less reactive than VDC, which has the same chemical formula, because of the polar nature of its double bond.

The Alfrey–Price–Young *Q–e* scheme is an attempt to include the monomer reactivities and the polar effects in one equation and hence to permit some prediction of copolymer structure and properties. Each propagation rate constant is expressed as

$$k_{ij} = P_i Q_j \, e^{-e_1 e_2} \tag{4.51}$$

where P_i represents the reactivity of radical i and Q_j the reactivity of monomer j. The parameters e_1 and e_2 are measures of the polarity of the radical and monomer respectively. Equations (4.39) and (4.40) therefore become

$$r_1 = (Q_1/Q_2) \, e^{-e_1(e_1 - e_2)}, \tag{4.52}$$

$$r_2 = (Q_2/Q_1) \, e^{-e_2(e_2 - e_1)}. \tag{4.53}$$

Table 4.3 *Average Q and e values for monomers using the Alfrey–Price–Young scheme.*

(a) order of polarity			(b) order of resonance stabilization		
monomer	e	Q	monomer	Q	e
t-butyl vinyl ether	−1.58	0.15	butadiene	2.39	−1.05
ethyl vinyl ether	−1.17	0.032	1-vinyl naphthalene	1.94	−1.12
1-vinyl naphthalene	−1.12	1.94	p-nitrostyrene	1.63	0.39
p-methoxy styrene	−1.11	1.36	p-cyanostyrene	1.61	0.30
butadiene	−1.05	2.39	p-methoxystyrene	1.36	−1.11
styrene	**(−0.80)**	**1.00**	**styrene**	**(1.00)**	**(−0.80)**
vinyl acetate	−0.22	0.026	methyl vinyl ketone	1.00	0.70
vinyl chloride	0.20	0.044	methyl methacrylate	0.74	0.40
p-cyanostyrene	0.30	1.61	diethyl fumarate	0.61	1.25
vinylidene chloride	0.36	0.22	acrylonitrile	0.60	1.20
p-nitrostyrene	0.39	1.63	methyl acrylate	0.42	0.60
methyl methacrylate	0.40	0.74	maleic anhydride	0.23	2.25
methyl acrylate	0.60	0.42	vinylidene chloride	0.22	0.36
methyl vinyl ketone	0.70	1.00	t-butyl vinyl ether	0.15	−1.58
acrylonitrile	1.20	0.60	vinyl chloride	0.044	0.20
diethyl fumarate	1.25	0.61	ethyl vinyl ether	0.032	−1.17
maleic anhydride	2.25	0.23	vinyl acetate	0.026	−0.22

Selected data from Brandrup & Immergut (1989), with permission, John Wiley & Sons, Inc., © 1989.

The product of the reactivity ratios becomes

$$r_1 r_2 = e^{-(e_1 - e_2)^2}. \tag{4.54}$$

Since the individual values of Q and e for any monomer cannot be deduced, because all experimental values are ratios, styrene at 60 °C has been chosen as a reference monomer and its characteristic parameters placed at $Q = 1.00$ and $e = -0.80$. Although this procedure may appear to be arbitrary, it should be remembered that a similar procedure had been used for generations in the electrochemical series. Some characteristic average values for a series of monomers are given in Table 4.3, for the reference temperature of 60 °C. Reactivity ratios tend to be only slightly dependent on temperature (e.g., for the styrene–methyl methacrylate system, $r_1 = 0.42$ at 60 °C and 0.59 at 131 °C, $r_2 = 0.46$ at 60 °C and 0.54 at 131 °C) and so the table can be used in a semi-quantitative manner at other temperatures.

There has not been a great deal of research conducted on this subject recently and any of the more recent references can be used successfully to obtain additional information. Chapters in the series edited by Allen and Bevington (1989), *Comprehensive Polymer Science* are particularly recommended as is the book by Cowie for specific information on alternation tendencies.

4.4 Ionic polymerizations

Classical chain reaction polymerization, as discussed in the last few sections, was concerned with initiation of polymerization using free radicals. During the past 30 years other methods of initiating chain reaction polymerization have been developed, that involve the use of ions. There are essentially three basic classes of ionic initiation to be considered. The two to be discussed in this section are anionic and cationic initiation which involve soluble catalysts. The third class involves initiation at an interface by organometallic catalysts utilizing the ability of certain metals to bond covalently to them. It is usually referred to as coordination catalysis and will be considered in a separate section.

The classical free radical polymerization scheme, with all of its complications, results directly from the high levels of reactivity of free radicals. Unique to free radicals is the ability to react with each other and, in so doing, to self-annihilate. It is this characteristic that is not present when polymerization is initiated by anions or by cations and results in major changes in the overall character of the product. Additionally, free radicals can easily abstract hydrogens from organic molecules and transfer the free radical character to the donor molecule. Such transfer reactions cannot occur as a generality in ionic polymerization and can only occur if certain specific types of group are present. It is, therefore, to be expected that the use of ionic initiators should considerably simplify the reaction scheme and permit greater control to be exercised over the character of the product.

A lack of termination and transfer reactions makes it possible, under ideal circumstances, to allow all of the monomer to be consumed and for the active ends to remain stable. A second monomer can then be added and a block copolymer generated.

Another unusual bonus results from anionic polymerization, in particular, which is the ability to control to a reasonable extent, the tacticity of the polymer. Very difficult to understand from a detailed mechanistic point of view, the use of a charged initiator places considerable restraints on the manner in which the monomer and initiator approach one another. The monomers that are amenable to ionic polymerization tend to be capable of resonance stabilization of the charge and the structure of the transition state has to be very specific. These constraints tend to ensure that the monomer must polymerize in a specific geometry and that this results in a specific configuration of the mer in the polymerized state.

4.4.1 Anionic polymerization

Initiation

The initiators used in anionic polymerization tend to be negatively charged organic molecules which have a positively charged, often metallic, counterion associated with them. The initiators can usually be classed as two types depending on the mechanism of initiation: addition of a negative ion or an electron transfer reaction. Monomers that can be polymerized using these types of catalyst tend to have electron withdrawing groups present (e.g., styrene, methyl methacrylate, acrylonitrile, butadiene; a novel example consists of the cyanoacrylate 'superglue' adhesives).

The most commonly used initiators are organometallic compounds based on low atomic mass alkali metals. These compounds are essentially covalently bonded and are soluble in some organic solvents. Compounds of the higher atomic mass alkali metals tend to be insoluble because of their considerable ionic character. They all function through the addition of a negative ion. Other less frequently encountered initiators of this type are Grignard reagents. Useful charge transfer initiators tend to be complexes of alkali metals with organic compounds. Other more exotic systems have been used, including solutions and suspensions of metals in liquid ammonia or in ethers. The most studied examples of the two systems are n-butyl lithium of the negative ion addition type and sodium naphthalenide of the charge transfer type.

1 *Negative ion initiation*
  (a) Predominantly covalent initiator (e.g. n-BuLi in a nonpolar solvent) (**4.12**).

$$CH_2 = \underset{\underset{H}{|}}{\overset{\overset{R'}{|}}{C}} + R\text{-}Li \quad \xrightarrow{[\pi\text{-complex}]} \quad R\text{-}CH_2\text{-}\underset{\underset{H}{|}}{\overset{\overset{R'}{|}}{C}}\text{-}Li$$

4.12

  (b) Predominantly ionic initiator (e.g. n-BuLi in a polar solvent) (**4.13**).

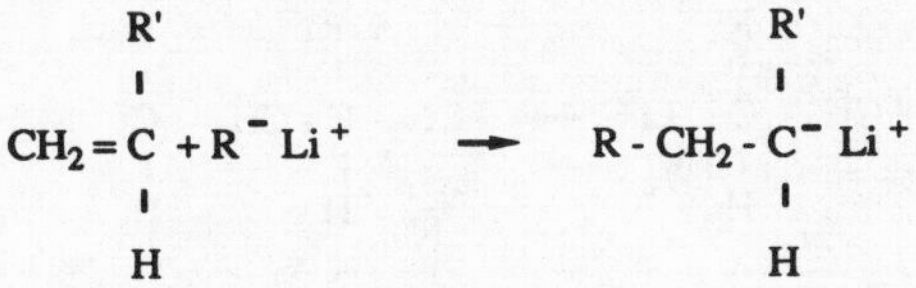

$$\begin{array}{c} R' \\ | \\ CH_2 = C \ + R^- Li^+ \\ | \\ H \end{array} \quad \longrightarrow \quad \begin{array}{c} R' \\ | \\ R - CH_2 - C^- Li^+ \\ | \\ H \end{array}$$

4.13

Note that the carbanion adds to the carbon that is furthest away from the electron withdrawing side group.

(2) *Charge transfer initiation* (e.g., **4.14** and **4.15**)

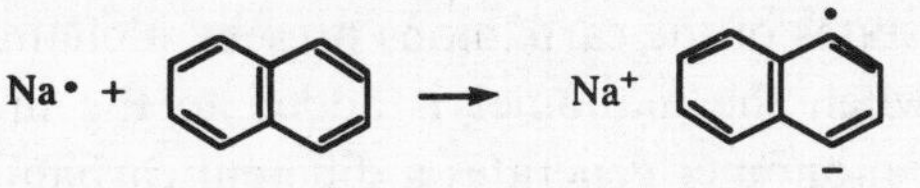

4.14

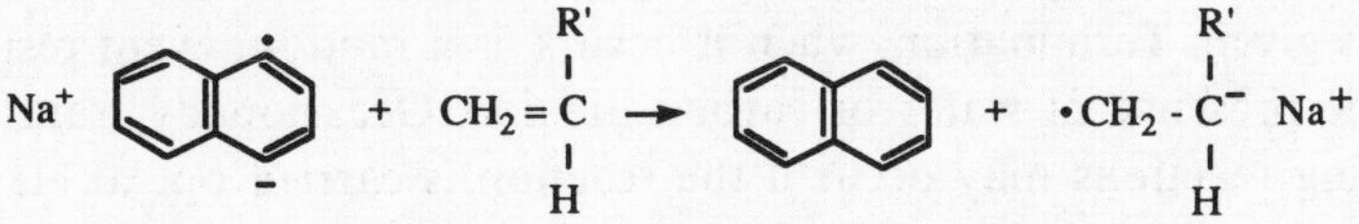

4.15

Note that in this case the initiated monomer is a radical anion and hence has the reactive characteristics of both types of species. Generally, this results in a dimerization, as shown in **4.16**. The species which then proceeds to propagate has two active ends.

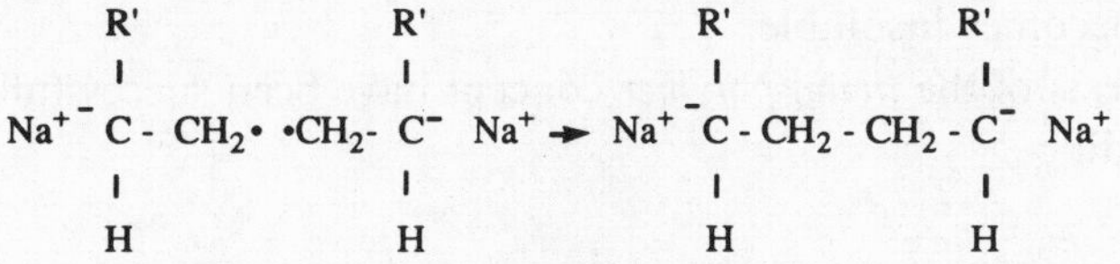

$$\begin{array}{c} R' \\ | \\ Na^+{}^- C - CH_2\bullet \\ | \\ H \end{array} \quad \begin{array}{c} R' \\ | \\ \bullet CH_2 - C^- Na^+ \\ | \\ H \end{array} \quad \longrightarrow \quad \begin{array}{c} R' \\ | \\ Na^+{}^- C - CH_2 - CH_2 - C^- Na^+ \\ | \\ H \end{array} \quad \begin{array}{c} R' \\ | \\ \\ | \\ H \end{array}$$

4.16

Propagation

Propagation consists of the insertion of a monomer molecule at the ionic end of the chain, between the negatively charged carbon and the counterion (**4.17**). In the case of polymers initiated with the charge transfer mechanism, in which dimerization of the first initiated species has taken place, propagation occurs at both ends of the chain.

$$- CH_2 - \underset{\underset{H}{|}}{\overset{\overset{R'}{|}}{C}}{}^- Li^+ \longrightarrow - CH_2 - \underset{\underset{H}{|}}{\overset{\overset{R'}{|}}{C}} - CH_2 - \underset{\underset{H}{|}}{\overset{\overset{R'}{|}}{C}}{}^- Li^+$$

$$CH_2 = \underset{\underset{H}{|}}{\overset{\overset{R'}{|}}{C}}$$

4.17

In most cases of anionic initiation vivid colors are generated because of the electronic states of the carbanions present, a change of color often being observed when the monomer is added to the initiator since the reaction with the monomer generates a different carbanion.

Termination and transfer

There are no intrinsic termination or transfer steps endemic to the reaction schemes given. Termination, when it occurs, is a random event resulting from the presence of water or carbon dioxide. Occasionally transfer or branching reactions may occur if the reaction is carried out at elevated temperatures. In a clean system, from which water has been excluded, the chain ends will remain active, even after all the monomer has been used up. Because of this, the active molecules have often been referred to as *living polymers*. In homopolymerizations small amounts of active ingredients such as alcohols are usually added to generate a termination reaction and so control the molecular weight of the product. If no additives are present, eventually the molecular weight of the polymer becomes so high that the viscosity of the system impedes growth or the polymer becomes insoluble.

Many tests of the *living polymer* concept have been successfully carried out, including:

immediate increase in the viscosity of a stored system when more monomer is added;

formation of block copolymers by sequential addition of two or more monomers; quantitative formation of functional end-groups;

narrow molecular weight distributions (confirming the lack of termination and transfer reactions);

simple relation between degree of polymerization and the monomer/initiator ratio.

Kinetics

Any complexity present in the kinetics of anionic polymerization invariably arises from the rate of dissociation of the initiator into ionic species. In type 1(*a*) initiation shown above the rate depends on the structure of a π-complex formed between the monomer and the covalent alkyllithium.

If the initiator is predominantly ionic, as in the other cases referred to above, then a simple mechanism, based on the rate controlling step being the reaction between carbanion and monomer, can be postulated as follows:

$$R^- + M \rightarrow R\text{—}M^-$$

proceeds at a rate

$$\text{rate of initiation} = k_I R^- M \tag{4.55}$$

propagation occurs as

$$R\text{—}M^- + M \rightarrow R\text{—}M\text{—}M^-$$

$$R\text{—}M\text{—}M^- + M \rightarrow R\text{—}M\text{—}M\text{—}M^-$$

$$R\text{—}M_n^- + M \rightarrow R\text{—}M_{n+1}^-$$

and the rate is given by, cf. equation (4.8),

$$\text{rate of polymerization} = -dM/dt = k_P M M^- \tag{4.56}$$

where M^- represents all polymerizing carbanions in the system.

In an ideal system M^- is a constant since there are no termination or transfer reactions and is equal to the concentration of initiator I. Integration of this equation gives the time dependence of monomer concentration as

$$M = M_0\, e^{-k_P I t} \tag{4.57}$$

For infinite time, the kinetic chain length and the DP will be given by the ratio of the initial monomer concentration to the initial initiator concentration. In the case of initiators such as sodium naphthalenide, where dimerization of the first initiated monomer occurs, the kinetic chain length and DP will be just twice this value. For certain monomers there may be specific chain transfer mechanisms which can operate and more complicated equations result.

It should be recognized that the scheme developed above is highly simplified and assumes that there are no transfer or termination reactions. It also assumes that all of the initiator is ionized in a time shorter than that needed for its first attack on a monomer. As such it is not unique to anionic polymerization, but would be characteristic of any type of polymerization reaction having the aforementioned restrictions placed on it. In practice, there is a narrow distribution of molecular weights present, not a unimolecular species (see Section 2.2.8). Other complicating factors such as the probability of depolymerization being greater than zero, specific transfer reactions to solvent, or rearrangements within the reactive grouping, may cause additional broadening in specific systems.

4.4.2 *Cationic polymerization*

The polymerizing species in this case is a carbocation (the earlier terminology of carbonium ion and carbenium ion are now restricted to CH_5^+ and CH_3^+ species respectively). There are three major classes of initiators used: (*a*) protonic acids, (*b*) a combination of a weak acid with a Friedel–Crafts halide (often referred to as proton–anion complexes) and (*c*) stable carbocations. Examples of type (*a*) include the strong (Brønsted) acids such as sulfuric, phosphoric and hypochloric acids. Examples of type (*b*) are usually weak Brønsted acids (such as H_2O, HCl, HBr, HF, ether etc. used in combination with a Friedel–Crafts halide (often referred to as Lewis acids, but in practice the Friedel–Crafts designation is preferable) such as BF_3, $TiCl_4$, $AlBr_3$, etc. Well-known examples include $BF_3 + H_2O$, BF_3:etherate and $SnCl_4 + H_2O$. In much of the standard seminal literature the halide, e.g., BF_3, has been regarded as the initiator, whereas it is the weak acid (e.g., water) that provides the cation for initiation. Additionally, some strong Lewis acids may function on their own. Type (*c*) includes ionizable organic halides such as tropylium ($C_7H_7^+$) salts and triphenylmethyl halides, which provide stable cations to serve as initiators.

The initiators generate either a proton or carbocation as the initial step;

$$H_2SO_4 \rightarrow H^+ + HSO_4^-,$$

$$BF_3 + H_2O \rightleftarrows F_3BOH^- + H^+,$$

$$SnCl_4 + H_2O \rightleftarrows Cl_4SnOH^- + H^+,$$

$$(C_6H_5)_3CCl \rightleftarrows (C_6H_5)_3C^+ + Cl^-.$$

The second step is usually attack on the monomer, and for this and ensuing steps the initiator will be presented as R^+X^-, where R^+ represents a proton or a carbocation (**4.18**). Note that the proton or carbocation adds to the carbon atom that has attached to it the electron dense groups, ensuring the formation of the most stable carbocation. Propagation takes

$$\underset{\underset{R}{|}}{\overset{\overset{R}{|}}{R^+ X^- + CH_2}} = C \longrightarrow H\text{-}CH_2\text{-}\underset{\underset{R}{|}}{\overset{\overset{R}{|}}{C^+}} X^-$$

4.18

place by a simple insertion of the monomer between the carbocation and its counterion (**4.19**).

$$-CH_2 - \underset{\underset{R}{|}}{\overset{\overset{R}{|}}{C^+}} X^- \longrightarrow -CH_2 - \underset{\underset{R}{|}}{\overset{\overset{R}{|}}{C}} - CH_2 - \underset{\underset{R}{|}}{\overset{\overset{R}{|}}{C^+}} X^-$$

$$CH_2 = \underset{\underset{R}{|}}{\overset{\overset{|}{}}{C}}$$

4.19

Chain transfer can be important if temperatures are near or above room temperature. Termination is important in many of these reaction schemes as it may be easy to transfer a proton to the counterion or to a monomer molecule, thereby ending the kinetic chain. Indeed, in order to generate high molecular weight polymers it is often necessary to carry out the polymerization process at low temperatures. The transfer to monomer step is represented schematically as follows in which R is written as $R' - CH_2$— (**4.20**).

$$-CH_2 - \underset{\underset{R}{|}}{\overset{\overset{CH_2}{|}}{\overset{R'}{|}}{C^+}} + CH_2 = \underset{\underset{R}{|}}{\overset{\overset{R}{|}}{C}} \longrightarrow -CH_2 - \underset{\underset{R}{|}}{\overset{\overset{CH}{||}}{\overset{R'}{|}}{C}} + CH_3 - \underset{\underset{R}{|}}{\overset{\overset{R}{|}}{C^+}}$$

4.20

The monomers that can be polymerized cationically contain electron-supplying substituent groups. A double bond alone is not a sufficient condition and of the common alkenes only isobutylene and butadiene are active. Styrene derivatives are appropriate when p-substituted; but are inactive if o-substituted because of steric problems. However, vinyl ethers and aldehydes are especially reactive towards cationic initiators. The vinyl ethers are sensitive because of the resonance structures that can be generated in the activated state. The mechanism of aldehyde polymerization is not completely established, but is believed to be similar to that represented above for an alkene monomer in that addition of the cation occurs at the carbonyl group.

For many years it was believed that 'living polymers' could not be generated in cationic polymerization because of the ease of transfer to the monomer. However, several systems have now been developed, one of which is based on an initiation scheme using t-butyl acetate–boron trifluoride complexes, which do behave as 'living systems' and can be made to form block copolymers.

The kinetics of cationic polymerization, which might be expected to follow a scheme similar to that of anionic polymerization, are not well understood because of the properties of the initiating species. Unlike anionic polymerization, where it was assumed that all of the initiating species were equally and immediately active, in cationic polymerization the degree of dissociation of the ion pair is of major concern. It should also be recognized that many fewer systems have been investigated in depth for cationic initiation than have been for free radical and anionic initiation. In what follows the initiation problem will be ignored at first and a kinetic analysis carried out simply using an initiator concentration, which represents the concentration of active ion pairs present.

$$\text{rate of initiation} = k_I IM \tag{4.58}$$

$$\text{rate of polymerization} = -dM/dt = k_P MM^+ \tag{4.59}$$

$$\text{rate of termination} = k_T M^+ \tag{4.60}$$

$$\text{rate of transfer to monomer} = k_{tr} MM^+ \tag{4.61}$$

M^+ represents the total molar concentration of cationically charged species. The rate of termination is written as a first-order reaction since the cationic chain end and its counterion remain so close that they can be regarded as a single species. If this were not the case then the reaction would need to contain the product $(M^+)(X^-)$ and the kinetic analysis would be different.

If it is assumed that a steady-state hypothesis, similar to that used earlier for free radical polymerization, equations (4.9) and (4.10), can be applied to cationic polymerization, then

$$\text{rate of initiation} = \text{rate of termination}$$

so

$$M^+ = k_I IM/k_T, \tag{4.62}$$

hence, the rate of polymerization is given as

$$\text{rate of polymerization} = (k_P k_I I / k_T) M^2. \qquad (4.63)$$

In practice, it should be recognized that the rate of initiation will be a complex function of the ion pair concentration, the dissociation constant of the ion pairs, the concentrations of the chemicals that form the initiator, the solvent, the monomer and its concentration. So, a more general expression would be

$$\text{rate of polymerization} = (\text{Rate})_i (k_P M / k_T), \qquad (4.64)$$

where the expression for the rate of initiation, $(\text{Rate})_i$, will vary from system to system and the actual equation that will be applicable to any system cannot be predicted from general principles.

The DP can be predicted, despite the aforementioned mechanistic problems, for situations where transfer is present or absent. In the absence of transfer, the DP equals the kinetic chain length and is given by the ratio of the rate of propagation to the rate of termination

$$X_N = \frac{\text{rate of polymerization}}{\text{rate of termination}}$$

$$= \frac{k_P M M^+}{k_T M^+} = \frac{k_P M}{k_T}. \qquad (4.65)$$

The concentration of cationic ends drops out of the equation, thereby making the DP independent of the initiator. This fact should be contrasted with free radical polymerization where the square root of the initiator concentration appears in the DP equation. This major difference is due to the quite different mechanism of termination.

When chain transfer is present then the equation becomes a little more complex, but still useful.

$$X_N = \frac{\text{rate of polymerization}}{\text{rate of termination} + \text{rate of transfer}}$$

$$= \frac{k_P M}{(k_T + k_{tr} M)}. \qquad (4.66)$$

This equation is seen to be the Mayo equation, first derived for free radical polymerization, equation (4.30), and has been used successfully to study the transfer kinetics in some systems (Kennedy and Chou, 1982). As in free radical polymerization, chain transfer agents can be added and the equation modified accordingly

$$X_N = \frac{\text{rate of polymerization}}{\text{rate of termination} + \text{rate of transfer to monomer} + \text{rate of transfer to X}}$$

$$= \frac{k_P M}{k_T + k_{tr(M)} M + k_{tr(X)} X}, \tag{4.67}$$

where X is the added chain transfer agent.

It has been demonstrated convincingly that the transfer reaction dominates the generation of final polymer, rather than the termination reaction. This fact is in contrast to free radical polymerization, where the termination reaction is dominant. Clearly, an extreme case is possible where it can be assumed that only transfer is of significance to the DP. Under such a circumstance the DP would be given by the ratio of the rate of propagation to the rate of transfer and be independent of monomer concentration.

$$X_N = \frac{k_P M}{k_{tr} M} = \frac{k_P}{k_{tr}}. \tag{4.68}$$

There is often one additional feature that contrasts with free radical polymerization and that is temperature dependence (cf. Section 4.2.5). Since essentially only two rates are significant, the temperature dependence will depend on the relative values of the activation energies of the propagation and transfer reactions. The activation energies for transfer and termination are usually larger than that for propagation, which results in the rate of polymerization increasing as the temperature decreases. An additional consequence of this feature is that DP increases with decreasing reaction temperature.

4.5 Coordination polymerization

There are two major types of heterogeneous anionic catalysis to consider, they being unsupported and supported Ziegler–Natta catalysts and supported activated metal oxide catalysts. Mechanisms of all have been investigated in detail for the past 30 or more years, but little has been reported in the literature for the second class. One of the major reasons for this is that these catalysts are of major importance to the synthesis of modern polyethylenes. The extreme competitiveness of the industry has substantially inhibited dissemination of detailed information on the catalyst schemes developed by different organizations.

4.5.1 *Ziegler–Natta catalysis*

Unsupported systems

Although there are earlier references to the phenomenon, Ziegler and coworkers were the first to recognize the significance of observations that combinations of transition metal halides and organometallic compounds were capable of polymerizing ethylene to produce a polymer which was essentially linear, in contrast to the conventional high pressure free radical product that was highly branched (e.g., Ziegler, Holzkamp, Breil, and Martin, 1955). Shortly thereafter Natta and coworkers reported that this type of catalyst could polymerize α-alkenes to yield highly crystalline polymers (e.g., Natta, 1960; Natta & Danusso, 1967). Such polymers did not result from using any of the other types of initiator system and many now common polymers, such as crystalline polypropylene, exist because of these novel initiation systems. In contrast to other modes of initiation, coordination systems are generally referred to as catalysts, rather than as initiators, since they are not consumed during the process and are more properly regarded as catalysts.

A second important discovery of Natta was that the polymers generated from α-alkenes were stereoregular. The early versions were predominantly 50–70% isotactic, whereas modern polypropylenes are >99% isotactic. The atacticity was distributed between wholly atactic molecules and partially isotactic molecules. The number of Zeigler–Natta catalyst systems that have been developed are too numerous to list in any standard text, because of the variety of metal halides and organometallic compounds that can be used. Generally, a Zeigler–Natta catalyst requires a combination of the following chemicals:

a transition metal compound from groups IV–VIII
an organometallic compound from groups I–III
a dry, oxygen-free, inert hydrocarbon solvent.

The most commonly studied systems (Allen and Bevington, 1989, Vol. 4) have involved aluminum alkyls and titanium halides which are usually added to the solvent at low temperatures. When heated, a brightly colored fine suspension develops which is often deep violet or purple. The polymerization occurs when the monomer is introduced, often in the form of a gas. Polymerization is rapid and heat is evolved, necessitating cooling. The polymer is usually insoluble and precipitates as a gel around the catalyst suspension. Monomer can still penetrate the gel and reaction continues as long as monomer is added to the system. Invariably the reaction is terminated by the addition of a chemical, such as an alcohol, which will react with the catalyst and cause deactivation. Small amounts of water, oxygen, alcohol, etc. may be added to a system to make the reaction controllable. The product tends to have a broad distribution of molecular weights, but has a high degree of stereoregularity.

Despite the volume of studies carried out since the discovery there is still no general kinetic scheme that can be used to predict the rate of reaction or the molecular weight distribution. This is largely because of the complexity of the process. There is no doubt that the polymerization process occurs on the surface of the fine crystalline precipitates of the transition metal halides and that there is some form of complex existing between exposed transition metal atoms at the surface and the organometallic compound. The polymerization process also involves the formation of a complex, this time between the monomer's π-bond and an unfilled d-orbital on the transition metal atom. For additional information on suggested mechanisms, see Keii (1972) or Boor (1979).

Although the molecular weight distribution is broad, the average molecular weight obtained tends to reach a constant value after a pseudo-steady state is achieved. Studies of Natta and Pasquon (1959) showed that the mean lifetime of a polymerizing chain was 16 min for one particular system. Hence, although a reaction system appears to be able to continue polymerizing *ad infinitum*, particular catalyst centers appear to be active for a finite time. Therefore some form of termination or transfer reaction must be occurring.

It was established at an early date that the crystal structure of the transition metal halide was very important in determining the reactivity. Of the four forms of $TiCl_3$, the β-crystal is not very effective as a catalyst since the titanium atoms are effectively encapsulated by the chlorine atoms. Indeed, a series of experiments carried out using electron microscopy demonstrated clearly that the active sites were generally associated with crystal edges where exposed titanium atoms could be found (Arlman and Cossee, 1964). In other experimentation Rodriguez and Gabant (1963, 1966) showed polypropylene to be growing on the spiral growth terrace of $TiCl_3$. Also, Guttman and Guillet (1970) showed polypropylene to leave the surface in the form of fibrils in a vapor-phase polymerization on $AlMe_3$—$TiCl_3$. Any mechanism must therefore of necessity recognize the importance of the detailed chemical structure of the surface. Since there will be a distribution of structures around exposed transition metal atoms, there must be distributions of efficiency and effectiveness amongst catalyst sites. This effect has to be a major contributor to the broad molecular weight distribution encountered.

Additionally, there is another physical complication that is not commonly encountered in other polymerization schemes, that of polymerizations within the precipitated polymer gel. The properties of the swollen gel will be of considerable significance because of the need for the monomer molecule to diffuse to the point of reaction and also because of the need for the growing polymer molecule to diffuse away from the catalyst surface as it is formed. The latter point is of considerable importance since it will be most sensitive to the density of the gel in the vicinity of the surface and to the degree of swelling close to the surface. This introduction of diffusion control into the process must contribute to the breadth of the molecular weight distribution.

Major problems encountered in the modeling of the kinetics of polymerization are therefore dependent on the physics of the precipitate and of the gel which surrounds it. They are very influential in determining the reactivity, but cannot be modeled in any simple fashion. Of course, experimental determinations of the rates of polymerization have been completed and the phenomenology is well established. The rate tends to follow one of three schemes:

rising with time to a constant value;
rising to a peak then decaying to a constant value;
rising to a short plateau then continuously decaying.

Many of the differences between the three schemes arise from the preparation of the catalyst. Generally, though, the first scheme is preferred in production and can be produced with forethought. Schemes for the derivation of kinetic equations have considered adsorption equilibria involving Langmuir–Hinshelwood and other equations, movement of the active center on the surface, and other phenomena, but there does not appear to be any consensus on which of these components are essential to any general scheme (Keii, 1972).

Supported catalysts

In a broad sense, all supported catalysts may be regarded as Ziegler–Natta from a mechanistic point of view. Here, those important catalysts not based on activated metal oxides will be considered, the activated metal oxide type being considered in the next section. All supported ionic catalysts for polymerization purposes can be regarded as descendants of the activated chromic oxides on silica supports of Hogan and Banks

(1954). Although very effective for polymerization of ethylene, supported metal oxides do not function for propylene polymerization. A series of catalysts based on magnesium chloride are the basis of the current polypropylene industry and were first patented by Mitsui and Montecatini-Edison. Later important developments came from Exxon and Shell, as well as the aforementioned companies. All are based on $MgCl_2$ supports activated by a complex of $TiCl_4$ and ethyl benzoate with some type of aluminum alkyl.

The method of preparation of the $MgCl_2$ support is very important, ball milling being used. In addition to reducing the size of the $MgCl_2$ crystallites, the milling generates defect structures at the surfaces involving single and double vacancy magnesium ions. The presence of ethyl benzoate prevents reaggregation of the cleaved $MgCl_2$ crystallites by an adsorption process. Additionally, it has been suggested that the milling alters the crystalline state of the $MgCl_2$ as x-ray diffraction patterns are intermediate between the cubic and hexagonal structures.

The subsequent addition of hot $TiCl_4$ displaces much of the adsorbed ethyl benzoate, possibly at well-bonded sites thus generating the catalytic centers. As in the case of the unsupported Ziegler–Natta catalysts electron microscope studies have provided evidence for the growth being more evident at edges and corners of the crystals.

4.5.2 Supported metal oxides

An alternative route to linear polyethylene and other polyalkenes was discovered about the same time as Ziegler–Natta catalysts, but in the laboratories of Phillips Petroleum (Witt, 1974). The catalyst system employed was chromic oxide on silica or a silica–alumina mixture, and was first activated by treatment at elevated temperature to produce CrO_3. Since that time several inert supports have been used and many new transition metal oxide systems introduced. Indeed much of the research on these systems has concentrated on increasing the efficiency and effectiveness of the catalysts through the substrate. The catalyst is now generally activated through treatment with a reducing agent, such as hydrogen or a metal oxide. Poisons which must be excluded include water and oxygen. The process is now the most important route to linear polyethylene and is the route used for the ethylene copolymers, commonly known as linear low density, very low density and ultra-low density polyethylenes.

Although it has generally been assumed that the mechanism of polymerization is similar to that in Zeigler–Natta catalysis (i.e., through some form of π-complex) a major difference is that the molecules produced through this process have approximately equal numbers of saturated and unsaturated end-groups. It has therefore been assumed that the initial π-complex decays to give two types of bonded adduct, as shown in **4.21**.

$$- Cr -\ \overset{\displaystyle CH_2}{\underset{\displaystyle CH_2}{\|}} \longrightarrow - Cr - \overset{H}{\underset{|}{C}} = \overset{H}{\underset{|}{CH}} \quad OR \quad --Cr\text{-}CH_2CH_3$$

4.21

Clearly this mechanism requires some form of cooperation between centers in order for a hydrogen to be transferred from one attachment site to another. The polymerization then proceeds through insertion of additional ethylene molecules as the Cr—C bond.

When α-alkenes are added as comonomers short branches are added to the chain, but vinyl unsaturation is increased in the final product, suggesting that α-alkenes act as relatively inefficient termination agents.

There has been little progress in recent years in understanding of the mechanisms and kinetics of this type of initiation. Most systems in commercial use at the present time are based on titanium and use silica substrates (polyethylenes).

The production of the silica gel support is of considerable importance. Particle size, pore size and pore size distribution are important variables which influence molecular weight and catalyst efficiency. The particle size effect, of course, functions largely through its effect on the amount of surface area available per unit mass of support. The pore size effect is not adequately explained. However, it is established that a large pore size reduces the molecular weight of the polymer. Chromic oxide is added to the silica support from aqueous solution prior to heating to high temperatures for activation.

Polymerization of the ethylene for many processes is carried out using inert hydrocarbon solvents. However, one of the most significant developments is the gas-based Unipol process by Union Carbide which uses fluidized beds of catalyst systems similar to those described in the preceding paragraph. This system has been licensed to a very great extent throughout the world for the production of polyethylene and its copolymers. It has also been applied to the production of polypropylene and its copolymers.

4.5.3 *Coordination copolymerization*

One of the most important uses of coordination catalysis is the generation of alkene copolymers. Commonly, these copolymers are synthesized from ethylene and one or more α-alkenes. Regardless of the catalyst systems used (Usami, Gotoh and Takayama, 1986), the copolymer molecules generated give the same general relation between comonomer and molecular weight. To a first approximation the product contains two types of molecules:

high molecular weight linear molecules;
low molecular weight branched molecules.

In other words there is a tendency for the catalyst sites to generate either high molecular weight ethylene homopolymer or relatively low molecular weight copolymer. Of course, this generalization is a simplification and the copolymer tends to have a distribution of both molecular weight and comonomer content. As expected, there are differences from one catalyst to another, but the generality is true regardless of whether soluble Ziegler–Natta catalysts are used or the heterogeneous activated metal oxide catalysts. Also it is true regardless of whether the system is liquid or gas based.

In short there appear to be two general classes of catalyst site available and functioning when copolymerization is concerned. One type of site generates the ethylene homopolymer and the second type generates the copolymer. It appears that the lower molecular weight of the copolymer molecules is a direct result of the aforementioned ability of the α-alkenes to act as weak terminating agents. Because of the commercial sensitivity of research in this area, much of the knowledge acquired is in patents or simply not in any literature. However, it is generally known that much of the effort has been expended in recent years into generating catalyst systems that function through only one type of site, in order to generate a homogeneous product with more controllable properties.

Other systems that are in use for copolymerization include the afore-mentioned Unipol process for both ethylene and propylene copolymers as well as certain homogeneous Ziegler–Natta catalysts using vanadium-based systems.

4.5.4 *Homogeneous Ziegler–Natta systems*

Although the vast majority of commercial systems are based on either unsupported Ziegler–Natta catalysts in which a metal halide serves as the support, or on the two types of supported systems already discussed, a small number of soluble Ziegler–Natta catalysts have been generated in the laboratory. They are formed if the transition metal halide and the metal alkyl react to produce a soluble complex.

The most significant groups of these catalyst systems are based on cyclopentadienyl and metalocene complexes and are the subject of intense development efforts by industry. Although known since 1970, their activities were relatively low. High productivity systems were reported by Ewen (1984) and developed in large part by Kaminsky's group (Kaminsky, Kulper, Brintsinger and Wild, 1985; Kaminsky and Sinn, 1988), some being essentially chiral compounds of zirconium with two cyclopentadienes and two additional groups (e.g., Cl or alkyl). Metalocene catalysts, which are being developed currently, are based principally on titanium or zirconium complexed with indene derivatives. Compounds using other transition metals, which do not have a chiral nature, also function, but do not develop isotacticity in polypropylene. They are used in solution with an inert hydrocarbon solvent and need to be activated with methylaluminoxane (generated from the hydrolysis of $AlMe_3$). They exhibit very high activities and generate high molecular weight polymers with a relatively narrow molecular weight distribution. So far ethylene, propylene and their copolymers have been polymerized using these systems. The expected occurrence of only a single site in these systems produces quite different copolymers from those generated using the activated metal oxide systems and linear low density polyethylenes (LLDPEs) produced from them do not exhibit the aforementioned two types of copolymer molecules.

4.6 Further reading

A comprehensive source considering free radical polymerizations of vinyl polymers is a multivolume series edited by Ham (1967). For additional information on anionic polymerization, the reader is referred to the second edition of *Polymer Chemistry* by Stevens (1990) and to the appropriate chapters of Allen and Bevington's fourth volume (Allen and Bevington, 1989). For cationic polymerization the reader should consult Pepper (1954), Plesch (1963), Kennedy (1975), Kennedy and Marechal (1982) and Allen and Bevington (1989), Vol. 3, for more details on systems and mechanisms. In the field of coordination polymerization excellent reviews of the early work are to be found in the books by Boor (1979) and Keii (1972) and in the chapter by Witt (1974). Reviews of several aspects of the subject are to be found in Vols 3 and 4 of Allen and Bevington (1989). The chapter by Tait in Vol. 4 is particularly recommended for its information on the commercial processes and for references on the most significant patent declarations. Chapters by Corradini *et al.* on stereospecificity and by Tait and Berry on copolymerization are very helpful on current research.

Nomenclature

A, B = monomers in a chain copolymerization or their molar concentrations

C_A, C_B = molar concentration of A, B groups in step polymerization

I = initiator or its molar concentration in chain polymerization

I• = radical fragment from decomposition of initiator or its molar concentration

k_I = rate constant for reaction of initiator with monomer (l/(mole s))

k_D = rate constant for initiator decomposition (s^{-1})

k_P = rate constant for propagation (l/(mole s))

k_T = rate constant for termination (l/(mole s))

$k_{T(C)}$ = rate constant for termination by combination (l/(mole s))

$k_{T(D)}$ = rate constant for termination by disproportionation (l/(mole s))

k_{tr} = rate constant for transfer (l/(mole s))

$k_{11}, k_{12}, k_{21}, k_{22}$ = propagation constants in copolymerization (l/(mole s))

M = unreacted monomer or its molar concentration

$M_1, M_2, M_3, \ldots, M_n, \ldots$ = polymer molecules of DP = 1, 2, 3, ... n ...

$[M_1], [M_2], [M_3], \ldots, [M_n], \ldots$ = molar concentration of polymer molecules of DP = 1, 2, 3, ... n .... Unlike other molecular species such as radicals or monomers, brackets are used for concentrations here to avoid confusion with molecular weights as used in Chapters 2 and 3

M_n^-, M_n^+ = a polymer anion or cation of DP, n or its molecular concentration

M^-, M^+ = total concentration of polymerizing anionic or cationic species

M_N = number-average molecular weight

M_W = weight-average molecular weight

N_n = number of polymer molecules (moles) of degree of polymerization, n

N_P = particle number density in emulsion polymerization

$\bar{n}_A(S)$, $\bar{n}_B(S)$ = normalized distribution function for sequences of 'S' A or B units in a copolymer

$P_{AA}, P_{AB}, P_{BA}, P_{BB}$ = probability of addition of an A unit to A•, B to A•, etc. in a copolymerization

r_1, r_2 = reactivity ratios in copolymerization

R• = total polymer radical concentration (moles/liter)

R_1•, R_2•, R_3•, ... = monomeric, dimeric, trimeric radical, ..., or their molar concentrations

R_P = average number of radicals per particle in emulsion polymerization

$\langle S_A \rangle$, $\langle S_B \rangle$ = average number of A or B units in a sequence in a copolymer

X_N = number-average degree of polymerization

X_W = weight-average degree of polymerization

γ = kinetic chain length

Problems

4.1 For the case of a step polymerization in which the initial concentrations of acid and alcohol groups, C_A^0 and C_B^0, *are not equal*:

   (a) integrate the rate equation for self-catalyzed polymerization; and
   (b) find an equation for the number-average DP as a function of time;
   (c) make a plot of X_N vs time for $(C_A^0)^2 k = 0.1 \text{ s}^{-1}$, $C_A^0 = 1.1 C_B^0$.

4.2 Prove mathematically from the integrated rate equations (4.2c), (4.3c) that externally catalyzed step polymerization will always eventually result in higher molecular weight than self-catalyzed when compared at the same time no matter what values the rate constants have.

4.3 Express the rate constant for initiator decomposition, k_D, in reaction **4.3** in terms of half-life, i.e., the time required for half the initiator to decompose unimolecularly.

4.4 The rate of polymerization, equation (4.12), and instantaneous molecular weight, equation (4.14), can be separately experimentally measured. Why cannot the termination constant, k_T, be determined from their ratio? What is determined from their ratio?

4.5 (a) Find the reactivity ratios of your favorite vinyl polymer from the tables. For these values of r_1, r_2 plot the instantaneous polymer composition vs unreacted monomer composition.
   (b) From the integrated copolymer composition equation (4.41), make plots of u versus conversion of B ($= 1 - B/B^0$) for three initial feed ratios, 0.1, 1 and 5.0.

(*c*) from (*b*) and the instantaneous incorporation relation, equation (4.38), make plots of the instantaneous mole fraction of **B** incorporated into the polymer, $f_B = dB/(dB + dA)$, versus conversion of **B** ($= 1 - B/B^0$) for the three ratios in (*b*).

5

Three dimensional architecture: conformation and stereochemical configuration

Up to this point, the long chain nature of polymers has been emphasized but little attention has been paid to the detailed three dimensional character of the molecules. As was apparent in the interpretation of solution viscosity (Section 3.3.3), the shape of molecules in solution characterization methods is of paramount importance. Likewise, the shapes adopted in bulk polymeric materials have a determining role in ensuing properties and these are driven by the preferences of the individual molecules.

5.1 Conformation: the role of bond rotation in determining molecular shape

The shape of a molecule is obviously specified by the connections made by its constituent chemical bonds. Although chemical bonds are fairly rigid with respect to stretching and to bending the valence angles between adjacent bonds, there is usually some degree of freedom of twisting or *internal rotational* movement about at least some of the types of bonds in a typical polymer chain. As a result such a molecule can usually take up a variety of shapes or *conformations*. The classic and simplest example of bond rotational motion is ethane, **5.1**. One methyl group can rotate with respect to the other about the *torsional* or *dihedral angle*, see

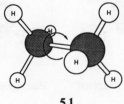

5.1

155

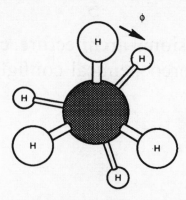

Fig. 5.1 Definition of torsional (dihedral) angle. The angle, ϕ, is measured as a clockwise rotation of the far bond away from the near one looking down the bond being twisted, starting from the eclipsed position. The definition is absolute and independent of the direction of looking down the bond (clockwise does not change to anti- because near and far bonds trade positions).

Figure 5.1 for the quantitative definition of the angle. (It is based on cis planar or eclipsed as the zero position. Some polymer literature in the past has used trans planar for zero. However, there is now general agreement through biochemistry, organic chemistry and polymer chemistry on the cis convention and it has been adopted in an IUPAC convention, conveniently reprinted in Brandrup and Immergut (1989)). The energetics of internal rotation are not directly accessible by spectroscopic methods. However, early studies of the heat capacity and the resulting absolute entropy of ethane showed that it could not be regarded as either (*a*) freely rotating about the C—C bond, or (*b*) completely rigid with respect to such motions. It was found that if it were assumed that some positions with respect to bond rotation were energetically more stable than others, **5.2**, and that a modest energy barrier separated the stable positions

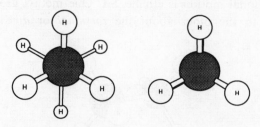

most stable· 'staggered'   least stable 'eclipsed'

5.2

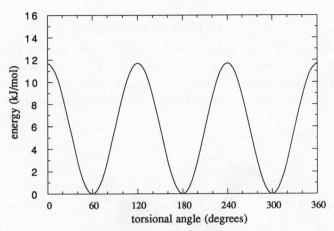

Fig. 5.2 Energy versus rotation about the torsional angle in ethane.

with an angular dependence as shown in Figure 5.2, then such thermo-dynamic data could be fitted well (Kemp and Pitzer, 1936, 1937). The potential goes through three maxima and minima as ϕ goes through 360° since there are three identical eclipsed and staggered positions. It was found that a value of the barrier height equal to 2.8 kcal/mol (11.7 kJ/mol) was required to fit the heat capacity and entropy of ethane. Later work by other methods confirmed this value (Strong and Brugger, 1967; Weiss and Leroi, 1968). The source of this barrier cannot be given a simple explanation in terms of structural chemical concepts. Full explanation requires finding the energy as a function of angle via accurate solution of the Schrödinger equation. However, for present purposes, it may be simply supposed that the H atoms and the C—H bonds at either end repel each other. In the staggered conformation these atoms and bonds are the farthest apart they can be and in the eclipsed position they are the closest. It is to be noted that the possibility of internal rotation with the attendant presence of a rotational barrier arises with the sequential connection of four atoms or, as alternatively described, from connection of three bonds. Near room temperature typical ethane molecules most of the time undergo a torsional oscillation or libration back and forth about the stable values of $\phi = 60, 180, 300°$ corresponding to the stable positions. Occasionally, however, a molecule receives enough thermal energy to traverse the barrier and go from one stable position to another. Nevertheless these occasions are frequent enough that the rate of barrier traversal is very rapid.

5.1.1 Conformational isomerism

Proceeding to a more complicated situation that is appropriate to a polymer chain, consider the four atom C—C—C—C sequence that occurs in polyethylene ($-CH_2-CH_2-)_n$. The smallest alkane to contain this is n-butane, **5.3**. Because of the C atoms at either end there is no

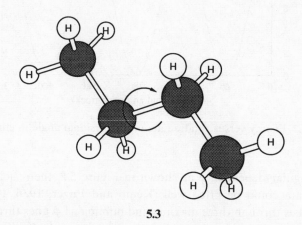

5.3

longer three-fold symmetry about 360° rotation. Of the three eclipsed positions (**5.4**) all are relative energy maxima but the $\phi = 0$ one is the

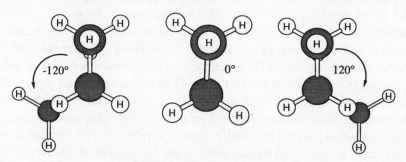

3 eclipsed butane conformations

5.4

least stable. This can be thought of as resulting from the C$\cdots$C repulsion being the largest of the three types C$\cdots$C, C$\cdots$H and H$\cdots$H, and $\phi = 0$ has the shortest C$\cdots$C distance, **5.5**. The other two eclipsed positions ($\phi = \pm 120°$) have the same energy. Similarly, there are three stable positions (energy minima) but the center one (**5.6**), $\phi = 180°$, is the most

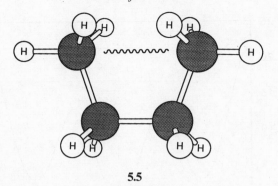

5.5

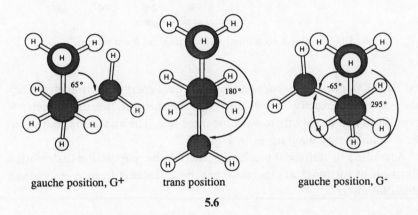

gauche position, G⁺   trans position   gauche position, G⁻

5.6

stable. The 180° position is called *trans* and it is the most stable because it has the greatest C···C distance (**5.7**). The positions at approximately

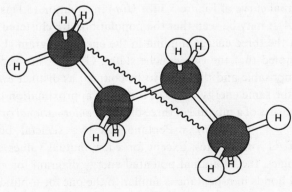

5.7

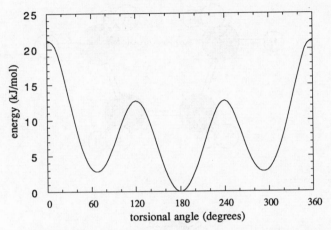

Fig. 5.3 Bond rotational energy diagram for n-butane.

60° and 300° are called *gauche* and are of equal energy by symmetry. They do not occur exactly at 60° and 300° because of the unsymmetrical repulsions and lie slightly toward the trans side. The above considerations lead to an energy diagram as in Figure 5.3.

According to statistical mechanics, the relative population distribution function of torsional angles, $n(\phi)$, can be calculated from a Boltzmann distribution as

$$n(\phi) = \frac{e^{-U(\phi)/k_B T}}{\int_0^{2\pi} e^{-U(\phi)/k_B T} \, d\phi},\tag{5.1}$$

where $U(\phi)$ is the torsional energy as a function of internal rotational angle. The population, $n(\phi)\Delta\phi$, for 20° increments resulting from the use of the potential curve of Figure 5.3 for $U(\phi)$ in equation (5.1) is displayed in Figure 5.4. It may be seen that the populations are clustered about the positions of the three energy minima in the energy diagram (Figure 5.3). It is to be noted that the two gauche states, G^+, G^-, are geometrically non-superimposable and therefore to be counted as distinct even though they are of the same energy. Thus, it is a good approximation to say that n-butane consists of a mixture of three distinct *conformational* or *rotational* isomers, G^+, T, G^- (**5.6**). n-Pentane has two skeletal bonds each exhibiting T, G^+, G^- states. Except for a few mutual values of the two torsional angles, the rotational potential energy diagram for *each* of the two skeletal bonds in n-pentane is similar to the one for n-butane (Figure 5.3). Thus the molecule can exist in any of the states TT, $TG^+ = G^+T$,

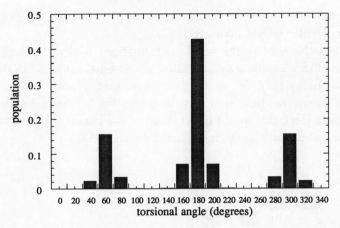

Fig. 5.4 Relative population of torsional angles in butane, 300 K.

$TG^- = G^-T$, G^+G^+, G^-G^- and $G^-G^+ = G^+G^-$. More will be said about the G^-G^+ states later. Although the G^+ and G^- states are distinct, the sequences in the pairs TG^+ vs G^+T, TG^- vs G^-T and G^+G^- vs G^-G^+ are not distinct. They differ only in the direction of writing the structure or, as alternatively stated, TG^+ can be obtained from G^+T simply by a *rigid* rotation of the *entire* molecule. n-Butane, n-pentane or any alkane can be regarded as a mixture of conformational or rotational isomers or, simply, *conformers*. The population of each conformer is determined by the number of conformers and their energy differences in a Boltzmann distribution sense. The concept that the rotational barriers between conformations are sufficiently high and the minima sufficiently sharp to lead to distinct well-characterized species and that the energy differences between species are often sufficiently low to lead to appreciable population of several species is known as the *rotational isomeric state model*.

Values for the differences in energy between trans and gauche states in linear alkanes have been determined from populations from temperature sensitive vibrational Raman bands in n-butane, n-pentane and n-hexane (Szasz, Sheppard and Rank, 1948; Sheppard and Szasz, 1949; Verma, Murphy and Berstein, 1974), from populations from electron diffraction (Bartell and Kohl, 1963), and by fitting heat capacities and entropies of n-alkanes (Pitzer, 1940; Person and Pimentel, 1953). Barriers to internal rotation in simple molecules can be determined from the rotational (microwave) spectra if the molecule is polar. The barriers to rotation in larger molecules are not very precisely known and come from heat capacity and entropy fitting (Pitzer, 1940; Person and Pimentel, 1953) and are inferred from microwave spectroscopy on model compounds of lower molecular weight (Wilson, 1962).

The fact that one of the bond conformations is slightly more stable than the others, i.e. the trans in Figure 5.3 is ~ 600 cal/mol (2500 J/mol) lower in energy than G^+ or G^-, has important implications in that it leads to a *preferred* (lowest energy) *conformation* for the *entire* molecule. For polyethylene this would be when all of the bonds are trans and can be described as the planar zig-zag conformation, **5.8**.

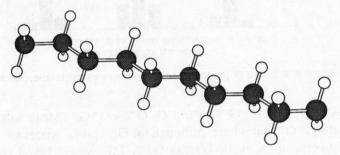

all trans planar zig-zag conformation of polyethylene.

5.8

5.1.2 Helices and crystals

As temperature is lowered, there is an increasing tendency for a molecule to populate its energy preferred conformation. The planar zig-zag conformation of polyethylene has a regularly repeating structure with respect to an axis along the molecular chain. Because of this regular structure, a collection of similar chains with their chain axes parallel can be packed together into a regular array or lattice. Thus the molecules can crystallize. The cross-section of the chain as projected on a plane has a shape that is rather rectangular in appearance (**5.9**). An efficient packing arrangement for rectangles is a 'herringbone' pattern that results in an orthorhombic

5.9

unit cell containing two chains, Figure 5.5. The basic periodicity of the unit cell along the chain axis is related to the repeating structure along the molecular axis and not to the entire polymer molecule. The molecules are so long that the ends are 'lost' (actually incorporated as defects or in residual uncrystallized material). In polyethylene the repeating unit along

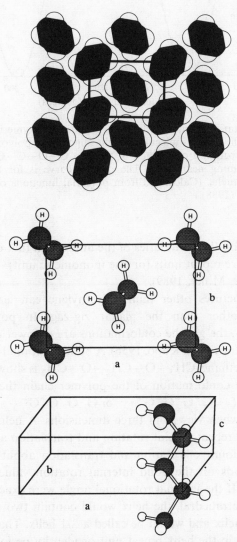

Fig. 5.5 Crystal packing in polyethylene. Upper drawing shows packing pattern in a,b plane using space filling representation of the molecules. Middle drawing is a,b plane and lower is a perspective showing c direction and one chain only.

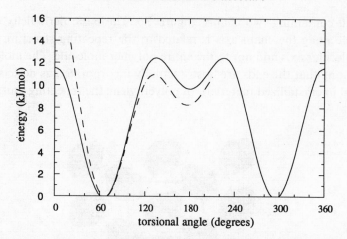

Fig. 5.6 Bond rotational energy diagram for dimethoxymethane, CH_3—O—CH_2—O—CH_3, the lowest molecular weight oligomer of polyoxymethylene. The solid curve is for rotation of the first of the two C—O—C—O bonds with the second one remaining near trans. The dashed curve is for the second bond remaining near gauche. (Calculated from potential functions of Sorensen, Liau, Kesner and Boyd (1988).)

the chain axis, and the boundaries of the unit cell in that direction, is two chemical structure repeat units (or one monomeric unit) —CH_2—CH_2— (Tadokoro, 1979; Miller, 1989).

Chemical structures other than polyethylene can lead to preferred conformations other than the planar zig-zag. In polyoxymethylene [—CH_2—O—]$_n$, the gauche conformations are of lower energy than the trans (Uchida, Kurita and Kubo, 1956). A bond rotational energy diagram for dimethoxymethane, CH_3—O—CH_2—O—CH_3 is shown in Figure 5.6. The most stable conformation of the polymer chain therefore is an all gauche sequence ($\cdots G^+G^+G^+G^+\cdots$ or $G^-G^-G^-G^-\cdots$). This generates a regular helix when viewed in three dimensions. A helix is a structure characterized by repeated joint rotation and translation about and along an axis. The amount of rotation and translation about and along the helix axis depends on the bond internal rotation value (and also the valence angles). If the internal rotational angle were exactly 60° and the valence angles tetrahedral, the helix would contain two monomer units per turn of the helix and would be called a 2/1 helix. That is, it contains two repeat units in the helix repeat unit or identity period and one turn of the helix in the same identity period, Figure 5.7. The reason that G is preferred over T in this instance is not entirely transparent but

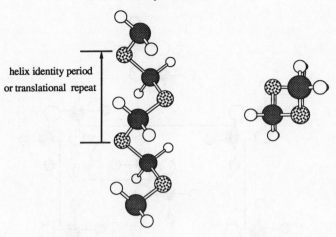

helix identity period
or translational repeat

viewed perpendicular to the helix axis   viewed along the helix axis

Fig. 5.7 Polyoxymethylene. 2/1 Helix resulting when the C—O—C—O torsional angles are 60°.

apparently is related to the fact that the C—O bond is polar. Electrostatistically, the sequence is most stable in the eclipsed conformation and least in the trans (**5.10**). This should contribute to stabilizing relatively

$$\text{trans} \qquad \text{eclipsed}$$

5.10

the gauche over the trans. However, the actual energy difference of ~ 2 kcal/mol (~ 8 kJ/mol), Figure 5.6, between G and T is greater than would result from a purely classical electrostatic energy description of the polarity. An accurate quantum mechanical treatment is required for full explanation. However, it is often found that the repulsions between polar bonds in internal rotation are greatly modified compared to non-polar ones such as the carbon–carbon bond.

Polyoxymethylene (POM) does crystallize in the conformation of the 2/1 helix (Carazzolo and Mammi, 1963). As is the case with polyethylene, the cross-section is rectangular (Figure 5.7) and the resulting crystal packing is orthorhombic as is shown in Figure 5.8.

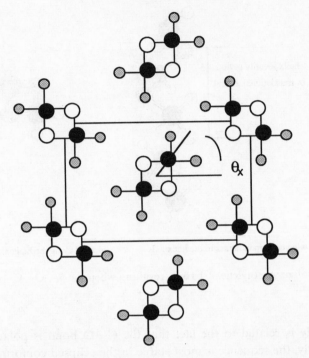

Fig. 5.8 Unit cell of the orthorhombic form of POM containing the 2/1 helix.

It is not uncommon for *polymorphism* to occur in polymers. That is, more than one crystal form or structure can be observed. When encountered, rather than observing a reversible transition with temperature or pressure, often one or more of the structures is metastable. A structure may result from a given method of crystallization but reverts to a more stable form under some condition such as heating. The reason for its occurrence often is that more than one chain conformation exists that has competitive energies and each packs well. Sometimes, however, basically the same conformation finds competitive packing arrangements. An example of the former is poly(vinylidene fluoride), $—(CH_2—CF_2—)_n$. It can exist in a number of crystal forms but two of them are especially interesting and important (Tadokoro, 1979; Lovinger, 1982; Davis, Broadhurst, McKinney and Roth, 1978; Broadhurst, Davis, McKinney and Collins, 1978). In one of them, an all trans ...TT... conformation (**5.11**) is found and the other contains a repeated ...TG^+TG^-... conformation (**5.12**). In both forms the molecule possesses a dipole moment. However, the packing is such that the unit cell has a dipole moment in the all trans crystal form but

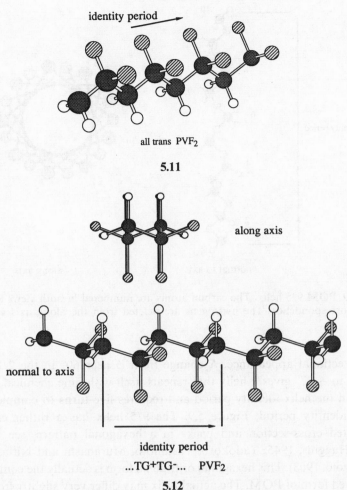

identity period

all trans PVF$_2$

5.11

along axis

normal to axis

identity period

...TG+TG-...   PVF$_2$

5.12

not in the other. Therefore the all trans form crystal is *ferroelectric* and has useful properties as an electromechanical transducer. The other form results from crystallization from the melt and apparently is the more stable form but it can be converted to the all trans form by mechanical deformation.

An important example of basically the same conformation packing in more than one form is represented by POM. As discussed above, the molecule prefers an all gauche helix and was seen to crystallize in an orthorhombic form (Figures 5.7 and 5.8). However a relatively small change in the torsional angles, leaving the chain still basically described as an all gauche helix, results in a rather dramatic change in the

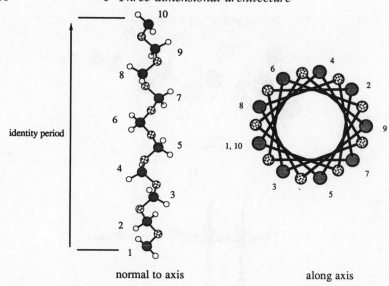

normal to axis   along axis

Fig. 5.9 POM 9/5 helix. The carbon atoms are numbered in both views to show their correspondence. The hydrogens are deleted from the along axis view for clarity.

cross-sectional appearance. A change from ϕ near $60°$ in the 2/1 helix above to $\sim 77°$ gives a helix that repeats itself with nine chemical repeat units in the helix identity period and requires five turns to complete the same identity period, Figure 5.9. The 9/5 helix has a rather circular projected cross-section and packs in a hexagonal pattern, see Figure 5.10 (Huggins, 1945; Tadokoro, Yasumoto, Murahashi and Nitta, 1960; Carazzolo, 1963). This hexagonal or trigonal form is actually the commonly observed form of POM. The actual helix may differ very slightly from the exactly 9/5 one (Takehashi and Tadokoro, 1979; Carazzolo, 1963).

5.1.3 Melts and solutions

The crystalline state of polymers as well as other more simple molecules represents the energetically most favorable state (lowest energy) of the material. However, it is *free energy* that actually determines the stability of forms of matter at constant temperature and pressure. Materials can increase their entropy through disordering processes and may achieve forms that are lower in free energy than the crystalline form. This disordering, of course, is called melting and is a familiar phenomenon. The difference in free energy between the molten or liquid phase and the

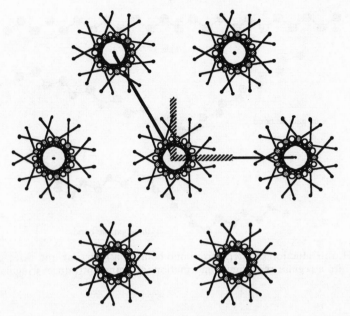

Fig. 5.10 Hexagonal packing of POM in the 9/5 helix form.

crystalline phase can be divided between an enthalpic (energetic) and an entropic contribution,

$$\Delta G_{\text{fusion}} = \Delta H_{\text{fusion}} - T\Delta S_{\text{fusion}}. \tag{5.2}$$

At low temperatures the positive ΔH_{fusion} term dominates, ΔG_{fusion} is positive and the material is crystalline. At high temperatures the positive ΔS_{fusion} term results in $-T\Delta S_{\text{fusion}}$ dominating and ΔG_{fusion} becomes negative and the material is a liquid. The melting point occurs at $\Delta G_{\text{fusion}} = 0$ or $T_{\text{m}} = \Delta H_{\text{fusion}}/\Delta S_{\text{fusion}}$. The disordering upon melting in polymers usually takes place *both* by *disordering the packing of chains* and also by *disordering the shapes of the molecules* from the most preferred conformation to a wide variety of conformations achieved by bond rotations (Starkweather and Boyd, 1960). Dissolution of the individual molecules in a solvent is also a highly disordering process and crystalline polymers can sometimes be dissolved in a suitable solvent (see Chapter 9). For non-specific solvents the process is also usually aided by higher temperatures where the entropy effects are more effective in offsetting the packing energies favorable toward crystallization ($\Delta H_{\text{solution}}$, positive). The best solvents are those that render $\Delta H_{\text{solution}}$ either small or negative by specific interactions with the polymer such as hydrogen bonding.

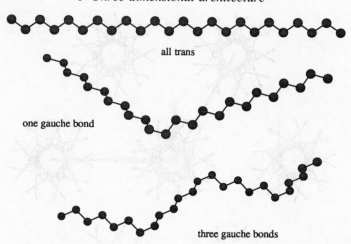

Fig. 5.11 Introduction of alternative bond conformations (one and three gauche bonds) into a regular repeating conformation (all trans) introduces irregularity in shape.

The average shape of a polymer molecule in solution is sensitive to the energetics of the stable rotational states. If some bonds were not in the repeated conformational sequence, for example, a few gauche bonds in a predominantly all trans sequence, the molecules would appear as in Figure 5.11. Since the sequence of local bond conformations is dictated by Boltzmann statistics, if the energy differences between these local conformations are modest, then at higher temperatures there will be appreciable population of the various bond rotational states. The contour of the molecule will become tortuous and coil-like in very long chains. A solution of such disordered coils can be idealized as in Figure 5.12. The common mode of solution of familiar synthetic polymers is in the random coil state. It is also the case that such molecules form random coils in the *melt*. But under this circumstance, the coils must be highly interpenetrating. A single molecule is able to pursue a path dictated by its local bond conformational energetics and yet be intertwined with neighboring molecules pursuing similar courses. In Figure 5.13 a depiction is shown by a section cut out of a melt of polyethylene chains. It was produced by a computer simulation (Boyd, 1989). Just one chain selected from the assembly in the box is shown in Figure 5.14.

If the energy differences between the preferred conformation and competing ones are sufficiently large or if the number of bonds that can show rotational isomerism is small compared to the total number of

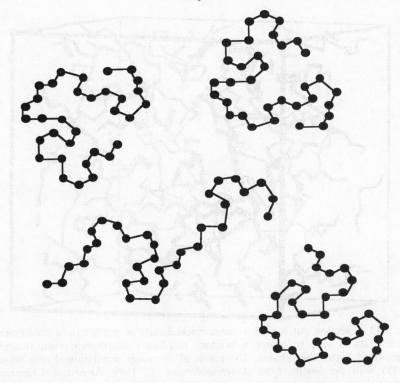

Fig. 5.12 An idealized rendering of a solution of random coils.

skeletal bonds, the entropy gain from forming disordered coiled conformations may not be sufficiently great to offset the energy penalty.
However, dissolution might still take place. If the molecules went into
solution in their most preferred conformation each molecule would be
rod-like (Figure 5.15). Solutions of rod-like molecules can indeed sometimes be formed. Poly(p-phenylene terephthalamide), structure 18 in
Appendix A1.1, is an example. The bonds in the phenyl rings and the
C—N bond in the amide group are not flexible with respect to internal
rotation. This leads to such a reduction in the entropy of conformational
disordering that random coils are not formed. However, the molecules
can be molecularly dispersed as rigid rods in anhydrous sulfuric acid, a
solvent that is highly specific for hydrogen bonding with the amide groups.
The rods also have a strong tendency for spontaneous ordering into
parallel alignment on application or shearing to the solution (Morgan,
1979, 1985).

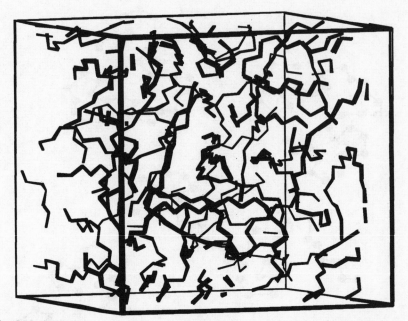

Fig. 5.13 A section cut out of a dense packed melt of polyethylene molecules. Each chain is able to adopt a random coil-like conformation even though intertwined with other chains. The result of computer simulation. From Boyd (1989) with permission from *Macromolecules*, © 1989, American Chemical Society.

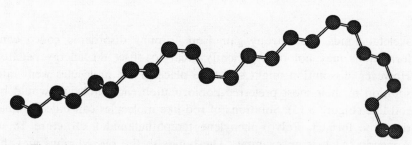

Fig. 5.14 A single chain selected from the assembly in Figure 5.13.

5.2 Stereochemical configuration: the role of stereochemistry in determining molecular shape

In Chapter 1 it was pointed out that in the chain polymerization of vinyl or similar monomers there is a choice of the placing of the substituents with respect to the carbon atom to which they are attached. In free radical polymerization the placement tends to be random, leading to a structure

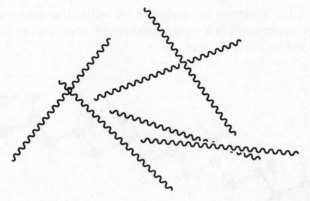

Fig. 5.15 An idealized rendering of a solution of rod-like molecules.

called *atactic*. A section of an atactic chain of polypropylene, depicted in an all trans conformation, is shown in **5.13**. It may be seen that the methyl groups occupy positions on either side of the plane of the zig-zag. In coordination catalysis, however, the placement can be highly regular as in **5.14** and is called *isotactic*. It has also been possible in polypropylene

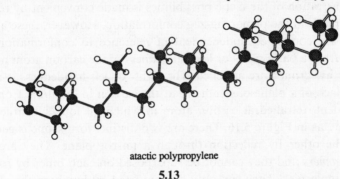

atactic polypropylene

5.13

isotactic polypropylene

5.14

and some other polymers to synthesize an ordered structure where the substituent placement is in a regular alternating manner and this arrangement is called *syndiotactic*, **5.15**.

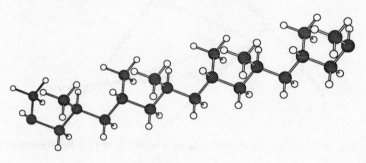

syndiotactic polypropylene

5.15

5.2.1 Stereoisomerism

The illustration of the tactic possibilities is made convenient by reference to the plane in the planar zig-zag conformation. However, these arrangements are completely independent of reference to conformation. They follow from a basic right- or left-handedness of the carbon atom to which the methyl groups are attached. The right- or left-handedness is a result of the lack of a plane of symmetry at the carbon in question. Consider a tetravalent, tetrahedral carbon atom in which all four substituents are different, as in Figure 5.16. There are two distinct forms, one is generated from the other by reflection through a mirror plane. They are called *stereoisomers* and they cannot be superposed on each other by rotations of the molecules. They can only be converted by breaking the chemical bonds and replacing the atoms, for example, by interchanging any pair of substituent atoms (H, Br), (H, Cl), etc. According to a standard-rule-based convention, the isomers are differentiated by calling one of them R (*rectus* or right-handed) and the other S (*sinistra* or left-handed). The stereo sense of an atom or molecule is called its *stereochemical configuration*. The fundamental requirement for existence of stereoisomers is that the object lack a place of symmetry. Applied to a molecule, a substituted atom in the molecule lacking such a plane is said to be *asymmetric* or *chiral*. In the case of tetrahedral carbon atoms, this requires that all four substituents be different. In the molecule in Figure 5.17, which does have a plane of symmetry (Br has been replaced by Cl), rotation

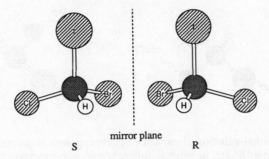

Fig. 5.16 Stereoisomers. When all four substituents about a tetravalent carbon are different, two non-superposable isomers exist. One is generated from the other by reflection through the plane.

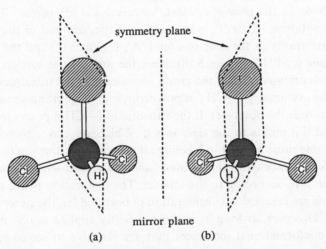

Fig. 5.17 Two substituents (Cl) are the same and a plane of symmetry exists at the central carbon atom. The mirror images are identical and no stereoisomers exist.

about an axis through the C–I bond brings view (*b*) in coincidence with view (*a*) and the central carbon atom is *not* asymmetric. In polypropylene, the two pieces of the chain in either direction from a methyl substituted carbon have to be regarded as different. In general, the lengths of the two pieces will be different and they will not be superposable (Figure 5.18). It follows that all four substituents are different. Each methyl substituted carbon is therefore asymmetric and a chiral center. At this point a notation to be used in this chapter is introduced for labeling of asymmetric carbon atoms in vinyl polymer. There is an absolute 'R, S' system, based on

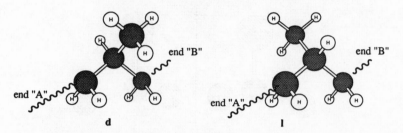

Fig. 5.18 A methyl substituted carbon in polypropylene is asymmetric since the two ends 'A' and 'B' are, in general, different and non-superposable. d, l refer to labeling system described in text.

priorities derived from molecular weights of substituents (Morrison and Boyd, 1966). In the present context, however, it is not useful.[†] Therefore another notation, 'd' or 'l', is resorted to. First, one end of the chain is picked arbitrarily as the near one (end 'A', Figure 5.18) and the other as the far one (end 'B', Figure 5.18). Now the asymmetric carbon atom in question is oriented so that the group considered as the substituent giving rise to the asymmetry (—CH$_3$ in polypropylene) is pointing *upward* (with end A *towards* the observer). If the substituent (—CH$_3$) points to the *left*, it is l and if it points to the *right* it is d. Whichever end is picked as 'A', the near one must be used in labeling all asymmetric atoms in that chain. Figure 5.18 is labeled in this manner and the atactic sequence of **5.13** is labeled in **5.16** according to the scheme. The convention is not absolute. If the ends are reversed in consideration of near and far, the stereo labeling reverses. However, as long as it is consistently applied in one molecule, various conformational sequences that are sensitive to stereo sense can easily be constructed or counted. A further convention is used in which the sequence of bonds is written down on paper as proceeding from

[†] If the asymmetric carbons were labeled in the absolute 'R' or 'S' system, the stereo sense of the carbons would change on proceeding down the chain in an *isotactic* chain. That is, if end 'A' were picked as the starting end it would have a molecular weight less than end 'B'. The R, S system priorities would be 'B' > 'A' > CH$_3$ > H and asymmetric carbons labeled 'd' in Figure 5.18 would be 'R'. However, on proceeding to label the carbons down the chain, eventually the majority weight would switch to 'A' and the labels in an isotactic chain would switch from 'R' to 'S'. This is clearly not a useful system in this instance. It can be circumvented by using a system, to be described below, that labels only the *relative* stereo sense of *adjacent pairs* of asymmetric carbons. However, there is a convenience in placing labels at the carbons individually and the 'd, l' system as defined here serves the purpose. It is to be noted that in current usage, 'd' and 'l' refer to the experimentally determined direction of rotation of polarized light in an optically active compound. However, the notation has had wide historic employment in describing asymmetric centers as done here and this makes it an appropriate choice for the purposes here.

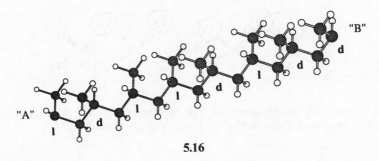

5.16

left to right and the left-hand end is always picked as the near group (end 'A') and the right-hand end as the far one (end 'B').

The placement of the methyl groups in **5.13–5.15** is a manifestation of d or l character of the asymmetric carbon atoms. This persists in *all* conformations (which are achieved by bond rotation). One tactic form cannot be converted into another by bond rotation; the asymmetry of the carbons is not affected by bond rotations. The tactic forms could be converted only by breaking chemical bonds and switching CH_3s and Hs. The various tactic forms have to be assembled chemically (synthesized).

It is appropriate to ask the question whether or not isotactic poly-propylene could be prepared in or separated into pure all-d or pure all-l forms that would be expected to show optical activity. The answer is (almost) no because there is no (appreciable) difference between the two forms. Although a carbon atom lacking a plane of symmetry is asymmetric, the question of whether the entire polymer molecule can be resolved into distinct asymmetric forms depends on whether the entire molecule lacks a plane of symmetry. Referring to **5.14** all the methyl groups are 'l' according to the left to right convention of assigning 'A' and 'B'. A simple rotation of the figure end-to-end out of the plane of the molecule would superpose the all-l figure exactly on an all-d counterpart (except for any end-group effects). This shows that the chain portions are, in fact, identical and indistinguishable. Another way of looking at it is to recognize that the chain possesses a symmetry plane perpendicular to the chain axis at the mid-point of the chain (except for end-group effects). A molecule with such a plane does not have asymmetric forms for the *entire molecule*. Any optical activity based on resolving the two forms through the mismatch of end-groups would be a very weak effect in a high molecular weight polymer due to the low concentration of end-groups. Similar consider-ations hold for any stereoregular form such as the syndiotactic polymer.

In low molecular weight homologs of polymers, however, the resolution

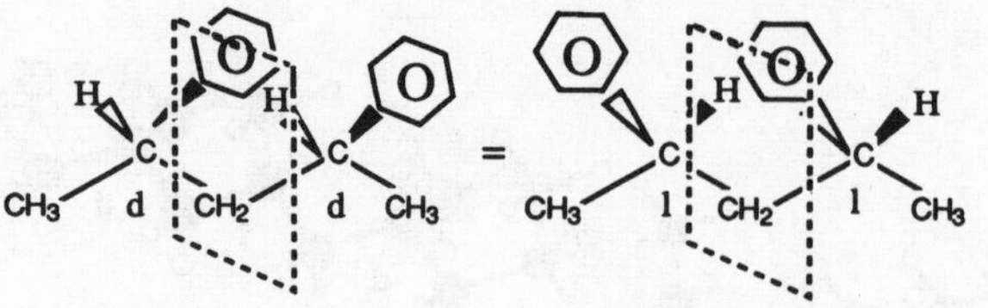

Fig. 5.19 *Meso* 2,4-diphenylpentanes. The two forms (d,d) and (l,l) each possess a plane of symmetry at the methylene group and are thus identical.

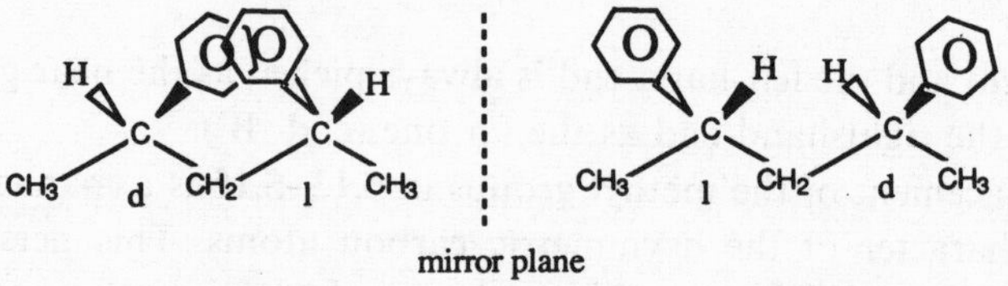

mirror plane

Fig. 5.20 *Racemic* 2,4-diphenylpentanes. The two forms (d,l) and (l,d) do not have an internal plane of symmetry and are thus non-superposable and distinct.

of stereochemical forms can be accomplished. For example, 2,4-diphenyl-pentane (a homolog of polystyrene) has two asymmetric carbon atoms. There are thus four possible combinations of asymmetric carbon atoms (d,d), (l,l), (l,d) and (d,l). The two forms (d,d) and (l,l) possess a plane of symmetry and are identical to each other, Figure 5.19. They are called *meso*. The other two forms (l,d) and (d,l) do not possess a plane of symmetry and are mirror images of each other, Figure 5.20. A mixture of these two forms can, in principle, be resolved by appropriate methods into their separate pure forms that differ in properties only in their direction rotation of polarized light. Such mirror image pairs are called *racemic*. The meso and racemic forms differ, in general, in their physical properties. The meso and racemic forms of 2,4-diphenylpentane have, in fact, been prepared (Overberger and Bonsignore, 1958).

It has become common practice to apply the terms meso and racemic to adjacent pairs or asymmetric carbon atoms in vinyl polymers. Such a pair is called a *dyad*. In this terminology, an isotactic polymer would be sequences of all meso ('m') dyads or '...mmmmm...'. A syndiotactic polymer would be all racemic ('r') dyads or '...rrrrrrrrr...'. An atactic one would be a mixture of them, '...mmrmrrrmmrmr...'. In **5.17** the atactic sequence of **5.13** and **5.16** is labeled according to its dyad ('m' or 'r') content. Notice that the dyad labeling is independent of the absolute configuration (i.e. l,d vs d.l in Figure 5.19) at each asymmetric carbon atom.

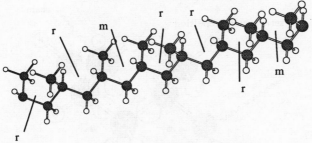

dyad labeled atactic polypropylene

5.17

5.2.2 Tacticity and conformations

The various tactic forms have preferred conformations just as any polymer chain tends to. For the isotactic polypropylene molecule it certainly is *not* the all-trans conformation shown in **5.14**. The adjacent methyl groups are far too close together and a serious steric repulsion results (they 'bump into each other'), Figure 5.21. To explore the alleviation of unfavorable methyl interactions by bond rotation one need only consider the interactions between atoms attached to adjacent asymmetrical carbons and therefore rotations of the pair of bonds between methyls. These are the torsional angles labeled ϕ_1, ϕ_2 in Figure 5.21. Although the energetics are modified by the superimposed effects of the steric interactions of the substituents, there is still a preference for local minima in the internal rotational energy to occur near trans and gauche conformations of each bond. Thus bond conformational pairs near the nine combinations of TT, TG^+, TG^- etc. need to be examined. A number, in addition to TT, have serious repulsions. An example is shown in Figure 5.22 where a

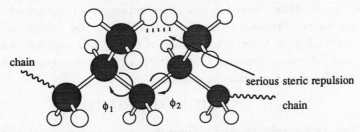

Fig. 5.21 A meso dyad (d,d) in polypropylene. When the conformation in the dyad is TT (ϕ_1, $\phi_2 = 180°$), as shown, the methyl groups are very close together and have a series steric repulsion.

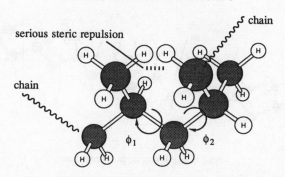

Fig. 5.22 A meso dyad (d,d) in polypropylene. When the conformation in the dyad is TG$^+$ ($\phi_1 = 180°$, $\phi_2 = 60°$), as shown, a methyl group is very close to a chain methylene group and a serious steric repulsion results.

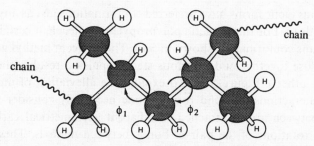

Fig. 5.23 A meso dyad (d,d) in polypropylene. When the conformation in the dyad is TG$^-$ ($\phi_1 = 180°$, $\phi_2 = -60°$), as shown, a strain-free environment results.

methylene CH$_2$ group in the chain is in close juxtaposition to a substituent methyl group. Of the nine pairs, only the combinations TG$^-$ or G$^+$T in a meso (d,d) dyad are strain-free, Figure 5.23.

The preferred conformation of an entire isotactic chain (of d,d meso dyads) then must be built from either TG$^-$ or G$^+$T repeated sequences. The two kinds are not mixed in one chain because a serious across-dyad steric interference results if attempted. Each of these two conformational sequences *leads to a helix, one left-handed* (G$^+$T) *and one right-handed* (TG$^-$) but otherwise identical. Consideration of l,l meso dyads leads to nothing additional. Consider the sequence

$$\cdots /_l TG^+/_l TG^+/_l TG^+/_l TG^+/_l TG^+/_l TG^+/_l,$$

where the slashes refer to partitioning the sequence into pairs of bond conformations internal to the l,l dyad. If it is written in reverse order, the bond formations merely change order but not values but the stereo changes from l to d (see the discussion concerning **5.16**). Therefore

$$\cdots /_l TG^+ /_l TG^+ /_l TG^+ /_l TG^+ /_l TG^+ /_l TG^+ /_l \cdots$$

$$= \cdots /_d G^+ T /_d G^+ T /_d G^+ T /_d G^+ T /_d G^+ T /_d G^+ T /_d \cdots$$

and

$$\cdots /_l G^- T /_l G^- T /_l G^- T /_l G^- T /_l G^- T /_l G^- T /_l \cdots$$

$$= \cdots /_d TG^- /_d TG^- /_d TG^- /_d TG^- /_d TG^- /_d TG^- /_d \cdots .$$

The left-handed helix is illustrated in Figure 5.24. It contains three repeat units and one turn in the helix identity period and is therefore a 3/1 helix. Isotactic polypropylene crystallizes in this conformation (Natta and Corradini, 1960; Natta, Corradini and Ganis, 1960, 1962).

In syndiotactic polypropylene the all-trans conformation **5.15** has no methyl repulsions and should be a low energy conformation. The local conformations $/G^- G^- /$ for an l,d dyad or $/GG/$ for a d,l dyad are also devoid of methyl repulsions. One could therefore consider a repetition of the sequence $/_l G^- G^- /_d G^+ G^+ /_l$ as a low energy chain conformation that would be alternating in stereo sense. This is not possible though because there is a serious across-dyad repulsion involving $G^- /_d G^+$ or $G^+ /_l G^-$. However, it is possible to isolate $G^+ G^+$ or $G^- G^-$ pairs with TT pairs

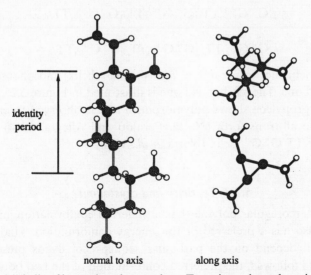

normal to axis   along axis

Fig. 5.24 3/1 helix of isotactic polypropylene. Two views along the axis are shown, one with all hydrogens (upper) and one with main-chain hydrogens blanked out for clarity in seeing how the substituent methyl groups are completely non-interfering sterically.

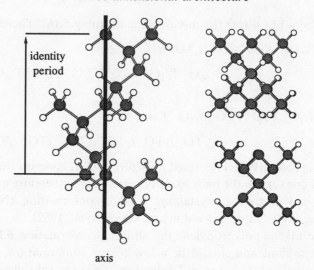

Fig. 5.25 TT/G$^+$G$^+$ helix of syndiotactic polypropylene. Two views along the axis are shown, one with all hydrogens (upper) and one with main-chain hydrogens blanked out for clarity in seeing how the substituent methyl groups are completely non-interfering sterically.

and create an alternative low energy chain conformation as

$$\cdots/_lG^-G^-/_dTT/_lG^-G^-/_dTT/_lG^-G^-/_dTT/_l\cdots$$

or

$$\cdots/_dG^+G^+/_lTT/_dG^+G^+/_lTT/_dG^+G^+/_lTT/_d\cdots.$$

The first sequence generates a right-handed helix and the second a left-handed one. The left-handed one is illustrated in Figure 5.25. Syndiotactic polypropylene shows polymorphism and is observed to crystallize in both the all-trans form (Natta, Peraldo and Allegra, 1964) and the alternating TT G$^+$G$^+$ helix (Natta *et al.*, 1960).

5.2.3 *Tacticity and crystallinity*

Just as a stereoregular polymer has a preferred conformation, an atactic polymer also has a preferred or low energy conformation. That conformation will depend on the particular sequence of dyads present. For example, if l follows l, the preferred conformation of the two bonds being the asymmetric carbons would be that of the all-l chain, either $/_lTG^+/_l$ or $/_lG^-T/_l$. If d follows l, it would be $/_lTT/_d$ or $/_lG^-G^-/_d$ and if l follows d it would be $/_dTT/_l$ or $/_dG^+G^+/_l$ (subject to the restriction of no G$^+$/G$^-$

or G^-/G^+ sequences). The irregular sequencing of preferred local conformations engendered by the irregular sequencing of d and l leads to a non-extended or coiled conformation for the preferred conformation of a particular atactic molecule. Since different individual atactic molecules have different d,l sequences, the preferred coiled conformation will also be different for different molecules. Thus a collection of random d,l sequenced atactic polymers in their preferred conformations will be a collection of randomly coiled molecules. For this reason atactic polymers tend not to crystallize. Further, even if the molecule were constrained to lie in a regularly repeating conformation (cf. all-trans for atactic polymer in **5.13**), it still would tend not to crystallize since the methyl groups would be irregularly oriented down the chain and the molecules could not pack regularly into a lattice. Thus there is a technologically important principle that although stereoregular polymers are fully capable of crystallization, atactic polymers cannot do so and are thus *amorphous*. The latter perhaps should be amended with the proviso that there are all degrees of crystalline order and that packing of some atactic polymers in extended conformations might lead to x-ray diffraction showing *lateral order* but not a fully developed three dimensional pattern.

From the above conclusion that some polymers can crystallize and some must remain amorphous it is obviously important to be able to examine a structural formula and to decide which situation prevails. This sorting out takes place in two steps. First the structure is perused for chiral centers. For the vast majority of situations this merely requires looking for sp^3 tetrahedral carbon atoms that are asymmetric. This translates into searching for those tetravalent carbons that have the two substituents unlike. If no chiral centers are present then the question of stereoregularity (tacticity in polymers from unsaturated monomers) does not arise and the polymer, if of a regularly repeating chemical structure, i.e., not a copolymer, can, in principle, crystallize. If asymmetric carbons are found then the question of stereoregularity does arise. Obviously, it cannot be decided from a chemical formula for the repeat unit what the stereoregularity of a real example of the polymer is. That depends on the method of synthesis. Therefore the further question must be asked. How was the polymer in question polymerized? For most of the common technologically important polymers this is well known. As suggested in Chapter 1, polymers from free radical initiated unsaturated monomers have a strong tendency for atacticity. Polystyrene, for example, is polymerized commercially largely by this method and is atactic and amorphous. However, it can be polymerized by coordination catalysis to the isotactic form. The latter has been much studied scientifically but is not an important commercial material. With polypropylene the situation is somewhat reversed. Commercially, the much more important material is the isotactic form. It is a relatively high melting ($\sim 170\,°C$) crystalline material used as a textile fiber and for shaping into molded objects and extrusion into pipe. The atactic form is an amorphous high viscosity, tacky liquid at room temperature. The latter does find commercial use as an adhesive but in small quantities compared to the isotactic form.

It is appropriate to comment that tacticity is not necessarily an all or nothing proposition, there can be degrees of tacticity. Consider a polymerizing chain incorporating d or l carbons one at a time. Then a probability of self-replication can be visualized. The probability of adding d to d or l to l to form a meso dyad might be only 90% for example. Usually then d would follow d and l follow l but 10% of the additions on the average would result in a reversal to form a racemic dyad

d d d d l l l l l l l l l l d d d d d d d d d···

or in the m, r dyad classification

m m m r m m m m m m m m r m m m m m m m m···

giving a predominantly isotactic chain. 'Perfectly' atactic polymer would come from a self-replication probability of 50%. 'Perfectly' syndiotactic polymer then has a zero self-replication probability. All degrees of tacticity between 100% and 0% self-replication probability would be imagined. The statistics of sequencing developed for copolymers in Section 4.3.2 may be applied. The probability of self-replication is $P_{AA} = P_{BB}$ and $r_1 = r_2$, A = B. Many unsaturated monomers polymerized by chain mechanisms, especially ionic ones, are stereochemically irregular but have a noticeable bias towards one side such as the syndiotactic.

5.2.4 Conformations in melts and solutions

Emphasis has been placed on the preferred lowest energy conformations and the possibility of packing into crystals. Although there is considerable preference for the conformations discussed above, other non-preferred conformations are possible. This would even include local bond conformations rejected in assessing the lowest energy chain conformation as containing serious steric repulsions. In melting or dissolution in the

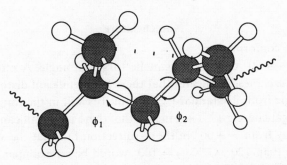

Fig. 5.26 A meso dyad (d,d) in polypropylene, TT(+) conformation. It is identical to the TT conformation of Figure 5.21 except that the second torsional angle ϕ_2 has been increased by 40° to a somewhat distorted value of 220° and therefore labelled as T(+). The methyl groups are now much farther apart and suffer much less steric repulsion. The conformation is a local minimum in energy and balances the methyl repulsion energy against the torsional barrier energy.

normal fashion of random coil formation, it must be that tactically regular polymers populate local bond conformations that involve repulsions. In the case of isotactic polypropylene, the *only* conformation available without interferences judged to be serious was the TG 3/1 helix. However, this polymer melts and dissolves in suitable solvents in the usual random coil manner. It turns out to be the case that although the repulsions in the other conformations are indeed significant they are not as serious as might be supposed (Boyd and Breitling, 1972b). They are greatly relieved by bond rotational adjustments away from exact trans and gauche values. Consider by way of example, the methyl···methyl interference in the TT conformation in an isotactic d,d meso dyad as shown in Figure 5.22. By rotation of one of the skeletal bonds, as demonstrated in Figure 5.26, the distance between the methyl groups is greatly increased. The result is that the energy penalty associated with such conformations is actually rather modest. These effects are taken up in more detail in Chapter 7.

5.3 Further reading

Both the subjects of conformational isomers and stereoisomers are covered in a number of standard texts on organic chemistry. Morrison and Boyd (1969) is a very popular and readable one. For application to polymers there are books by Bovey (1969, 1982), Hopfinger (1973) and Birshstein and Ptitsyn (1966).

Nomenclature

G = gauche conformation

G^+, G^- = $\sim +60°$ or $= -60°$ gauche torsional angle. A plus or minus *in parentheses* indicates that there is a significant distortion of the angle from its standard value but it is still in the range occupied by gauche, e.g. $G^+(+)$, means that there is a significant distortion away from $\sim +60°$ in the $+$ direction but that the value is still less than $120°$. $G^+(+) = 100°$ would be an example

T = trans bond conformation, $\sim 180°$, $T(+)$, $T(-)$ means that the torsional angle is significantly distorted away from the $180°$ value in the direction indicated but that the value is still in the trans range, between $-120°$ and $+120°$

Problems

5.1   For the section of an atactic polypropylene chain shown in **5.16**, write down a sequence of conformations that is 'strain-free' in the sense of the TG^+ 3/1 helix for isotactic polypropylene. Use a notation like that in the text for the 3/1 helix, i.e.,

$$\cdots /_1 TG^+ /_1 TG^+ /_1 TG^+ / TG^+ / TG^+ / TG^+ /_1 \cdots$$

5.2   From all of the polymers in Appendix A1.1, list those that are *inherently* crystallizable in principle; i.e., those that contain no chiral center. Of those that have an asymmetric carbon, find out, for the commercially important form, which are atactic and which are isotactic or syndiotactic. List which are amorphous and which are crystalline.

5.3   Using a set of molecular models of the 'stick' type (as opposed to 'space filling' ones), or better yet, using a computer program such as 'Alchemy', Tripos Associates, or 'Chem-3D', Cambridge Scientific Computing, build the following conformations of polymer chains:

   (a) All T planar zig-zag polyethylene.
   (b) Illustrate the 'pentane interference' effect, Figures 6.4, 6.5, by building one G^+G^- pair in otherwise all T polyethylene. Then partially relieve strain by rotation of one of the gauche bonds to a significantly distorted value, but still identifiable as gauche.
   (c) All G^+ 2/1 helix of polyoxymethylene.
   (d) TG^+TG^- conformation of PVF_2.

(e) Alternating TG$^+$ 3/1 helix of isotactic polypropylene.

(f) TT pair in isotactic polypropylene (partially relieve strain by rotation of one bond to distorted T).

(g) TTG$^+$G$^+$ helix for syndiotactic polypropylene.

6

The statistical behavior of conformationally disordered chains

It is not possible to keep track of the details of the configurations of polymer molecules when they have become disordered or coiled through populating various local bond conformations. An elementary calculation is instructive. Consider a chain with three conformational states for each skeletal bond, a trans and two gauche states for example. Then a chain with N bonds capable of internal rotation will have 3^N total possible conformational states. For $N = 1000$, a modest chain length, there are 10^{477} states possible! Obviously statistical descriptions are called for. This can take the form of directly finding the average value of a desired property or, in more detail, finding a distribution function for the property. For example, in the consideration of the relation between the solution viscosity and molecular weight (Section 3.3.3) it was apparent that a measure of average dimensions or size was needed. Under appropriate conditions, in a 'theta' solvent where phantom chain behavior obtains, the *mean-square* end-to-end distance can be directly calculated. Under these conditions, and where the chain length is long, it is also possible to calculate a distribution function for the probability of a chain having an arbitrary end-to-end extension. In this chapter these particular questions will be taken up, the calculation, under phantom conditions, of mean-square dimensions and the distribution function for end-to-end distance. The effects of non-self-intersection in good solvents will also be considered.

6.1 Characteristic ratio

The mean-square value of the end-to-end distance, $\langle R^2 \rangle$, in long chains and under *unperturbed* or phantom conditions where self-intersection is permitted will be found to be proportional to the number of bonds, N, in the chain. It also, by simple dimensional scaling, must be proportional to the square of the bond length, l. Thus it is appropriate to define a quantity called the *characteristic ratio*, C_N, that expresses the relative or comparative dimensions of a statistically coiling molecule as

$$C_N = \langle R^2 \rangle_0 / N l^2. \qquad (6.1)$$

The subscript zero on the brackets $\langle \ \rangle_0$ is placed to indicate that the average of R^2 is to be taken under unperturbed conditions. The N subscript on C indicates that the ratio may be somewhat dependent on N in relatively short chains, C_∞ will denote the asymptotic long chain limit. As discussed in Section 3.3.3, experimental values of the characteristic ratio can be determined from solution viscosity measurements in theta solutions and also from light scattering measurements under the same condition. Since the characteristic ratio is sensitive to the energetics of local bond conformations, the formulation of methodology for calculating it from the energetics is an important task. Comparison of calculated and experimental values provides an important method for the validation of proposed models for the chain conformational behavior. For purposes of illustrating how such calculations proceed, two simple models for chain conformation are developed below. A more involved treatment of realistic chains is presented separately (Chapter 7).

6.1.1 The persistence length

If each chain bond in a linear polymer with N bonds is represented by a vector, $\mathbf{l}_j$, the end-to-end vector can be expressed as the sum of the bond vectors (Figure 6.1),

$$\mathbf{R} = \sum_{i=1}^{N} \mathbf{l}_i$$

and

$$R^2 = \mathbf{R} \cdot \mathbf{R} = \left(\sum \mathbf{l}_i \right) \cdot \left(\sum \mathbf{l}_j \right) = \sum_j \sum_j \mathbf{l}_i \cdot \mathbf{l}_j$$

$$(6.2)$$

or

$$\langle R^2 \rangle = \sum_i \sum_j \langle \mathbf{l}_i \cdot \mathbf{l}_j \rangle. \qquad (6.3)$$

Consider one typical term in the sum over i in equation (6.3) and divide it by the bond length, l_i, or, $(1/l_i) \sum_j \langle \mathbf{l}_i \cdot \mathbf{l}_j \rangle$. This represents the sum of the average projection of all the bonds in the chain on the selected bond, i (including i itself). In statistically coiling chains this sum will be

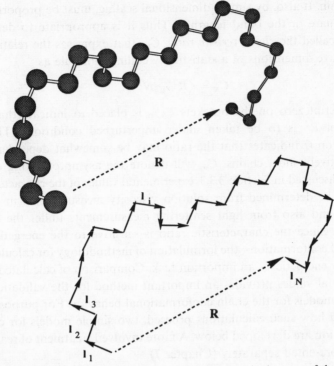

Fig. 6.1 The end-to-end vector, **R**, of a chain (upper) is the sum of the individual bond vectors, $\mathbf{l}_i$ (lower).

convergent, the projections of bonds far away from i will average to zero. The sum of these average projections, in one direction, along the chain from the selected bond, i, is called the *persistence length* (Porod, 1949), L_P, and therefore,

$$L_P = l + (1/l_i) \sum_{j>i} \langle \mathbf{l}_i \cdot \mathbf{l}_j \rangle. \qquad (6.4)$$

Thus in a long chain of uniform bonds of length, l, the persistence lengths at each bond will tend to the same value and comparison of equations (6.1) and (6.3) with equation (6.4) shows the characteristic ratio will be related to the persistence length as $C_\infty = 2(L_P/l) - 1$. The existence of a finite converged persistence length ensures that $\langle R^2 \rangle_0$ will have the linear dependence on N supposed in the definition of the characteristic ratio in equation (6.1).

6.1.2 The characteristic ratio of the freely rotating chain

If the rotational potentials of each bond are independent and do not depend on the conformation of neighboring bonds, then all terms of equation (6.3) for which the indices i and j differ by the same amount are equal, $\langle l_1 \cdot l_3 \rangle = \langle l_2 \cdot l_4 \rangle = \langle l_3 \cdot l_5 \rangle \cdots$. The numbers of each kind of term are easily counted by setting up a square array.

$$
\left.
\begin{aligned}
R^2 = l_1 \cdot l_1 + l_1 \cdot l_2 + l_1 \cdot l_3 + l_1 \cdot l_4 + l_1 \cdot l_5 + \cdots \\
l_2 \cdot l_1 + l_2 \cdot l_2 + l_2 \cdot l_3 + l_2 \cdot l_4 + l_2 \cdot l_5 + \cdots \\
l_3 \cdot l_1 + l_3 \cdot l_2 + l_3 \cdot l_3 + l_3 \cdot l_4 + l_3 \cdot l_5 + \cdots \\
l_4 \cdot l_1 + l_4 \cdot l_2 + l_4 \cdot l_3 + l_4 \cdot l_4 + l_4 \cdot l_5 + \cdots .
\end{aligned}
\right\}
\tag{6.5}
$$

It follows that

$$
\begin{aligned}
\langle R^2 \rangle = Nl^2 &+ 2(N-1)\langle l_1 \cdot l_2 \rangle + 2(N-2)\langle l_1 \cdot l_3 \rangle \\
&+ 2(N-3)\langle l_1 \cdot l_4 \rangle + \cdots .
\end{aligned}
\tag{6.6}
$$

The term $l_1 \cdot l_2$ is calculated from Figure 6.2 as $l^2 \cos \theta'$ where l is the bond length and θ' is the supplement of the bond angle, θ, which is assumed to be rigid. The term $l_1 \cdot l_3$ is the projection of bond 3 onto bond 1. This can be accomplished in two steps. First, in Figure 6.3, notice that bond 3 has a component $l \cos \theta'$ along bond 2. It also has a component $l \sin \theta'$ perpendicular to bond 2 and oriented at a direction determined by the rotational angle, ϕ. For a chain with no hindering rotational

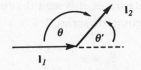

Fig. 6.2 Two adjacent bond vectors define the valence angle θ.

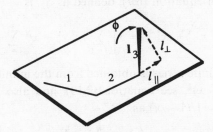

Fig. 6.3 The bond vector l_3 can be resolved into components parallel and perpendicular to the previous bond, l_2.

potential, the perpendicular component will average to zero. Such a chain is called the *freely rotating* chain. Notice that the same would be true for a chain with trans and gauche positions of *equal* energy and symmetrically disposed at 180°, 60° and 300°. Therefore *on the average* for the freely rotating chain bond 3 is merely an extension of bond 2 and adds $l \cos \theta'$ to its length. In turn it may be projected back on to bond 1 and there results,

$$\langle \mathbf{l}_1 \cdot \mathbf{l}_3 \rangle = l \cos \theta'(l \cos \theta')$$

$$= l^2 \cos^2 \theta'.$$

The process may be extended to the other terms

$$\langle \mathbf{l}_1 \cdot \mathbf{l}_4 \rangle = l^2 \cos^3 \theta' \text{ etc.,}$$

and thus

$$\langle R^2 \rangle = Nl^2 + 2(N-1)l^2 \cos \theta' + 2(N-2)l^2 \cos^2 \theta'$$

$$+ 2(N-3)l^3 \cos^3 \theta' + \cdots 2(N-i)l^2 \cos^i \theta' + \cdots + 2l^2 \cos^{N-1} \theta'$$

$$= Nl^2 + 2l^2 \sum_{i=1}^{N-1} (N-i)x^i, \tag{6.7}$$

where $x = \cos \theta'$. This series can be summed by considering the two terms in $N - i$ separately. Defining the first sum as S_1 then

$$S_1 = \sum_{i=1}^{N} x^i = x \sum_{i=0}^{N-1} x^i = x \sum_{i=0}^{\infty} x^i - x \sum_{i=N}^{\infty} x^i = x \left(\sum_{i=0}^{\infty} x^i - x^N \sum_{i=0}^{\infty} x^i \right).$$

Thus although S_1 is a finite series, it is seen that its value can be obtained from the simple infinite geometric series, $1 + x + x^2 + \cdots = 1/(1-x)$, as

$$S_1 = x \frac{1 - x^N}{1 - x}.$$

The second sum in equation (6.7), defined as S_2, is

$$S_2 = \sum_{i=1}^{N} ix^i = x \sum_{i=1}^{N-1} ix^{i-1} = x \left[\sum_{i=1}^{\infty} ix^{i-1} - x^N \sum_{i=1}^{\infty} (N+i)x^{i-1} \right]$$

and therefore the finite series is obtained from the infinite series $1 + 2x + 3x^2 + \cdots = 1/(1-x)^2$, see equation (2.13), and also the infinite series $1 + x + x^2 + \cdots = 1/(1-x)$ as,

$$S_2 = x \left(\frac{1 - x^N}{(1-x)^2} - \frac{Nx^N}{1-x} \right).$$

Therefore equation (6.7) reduces to $\langle R^2 \rangle = Nl^2 + 2Nl^2 S_1 - 2l^2 S_2$ and

$$\langle R^2 \rangle = Nl^2 + 2Nl^2 x \left(\frac{1 - x^N}{1 - x} \right) - 2l^2 x \left(\frac{1 - x^N}{(1 - x)^2} - \frac{Nx^N}{1 - x} \right), \quad (6.8)$$

where as stated above, $x = \cos \theta'$ and θ' is the supplement of the valence angle. At large values of N, equation (6.8) approaches,

$$\langle R^2 \rangle = Nl^2 \left(\frac{1 + x}{1 - x} \right). \quad (6.9)$$

As already noted, $\langle R^2 \rangle$ is proportional to N rather than N^2. The reduction in end-to-end distance over that of an extended repeated conformation caused by the introduction of bond rotation manifests itself by the dependence of $\langle R^2 \rangle^{1/2}$ on N being reduced to proportionality to $N^{1/2}$. This result is characteristic of random walk or diffusive processes. The progress of the walk is proportional to the square root of the number of steps or of the diffusing particle to the square root of time. In the current case, progress is the displacement of one end of the chain from the other and the number of steps is the number of bonds. This part of the result is independent of the assumption of free rotation and depends only on allowing the rotational states to be populated. The linear dependence of $\langle R^2 \rangle$ on N forms the basis for the definition of the characteristic ratio in equation (6.1).

The characteristic ratio, C_N, from equation (6.1) and equation (6.8) is

$$C_N = \langle R^2 \rangle_0 / Nl^2 = \left(\frac{1 + x}{1 - x} \right) - \frac{2x}{N} \left(\frac{1 - x^N}{(1 - x)^2} \right).$$

For the infinitely long freely rotating chain from equations (6.1) and (6.9), making a replacement of the supplement of the valence angle by the valence angle itself, $\cos \theta = -\cos \theta' = -x$, the characteristic ratio is therefore (Kuhn, 1934),

$$C_\infty = \langle R^2 \rangle_0 / Nl^2 = \left(\frac{1 - \cos \theta}{1 + \cos \theta} \right). \quad (6.10)$$

It is of some importance to see the degree to which the chain structure and conformational behavior of real chains are reflected in the characteristic ratio. Thus it is of interest to compare the results for simple models like the above to experimental values for actual chains. If equation (6.10) is applied to the polyethylene chain assuming tetrahedral bond angles (cos $\theta = -1/3$), $C_\infty = 2.00$ results. The experimental value is 6.7 and thus there is a large discrepancy. The disagreement is presumably due to the assumption of free rotation or symmetrically disposed equal rotational energy states. It would be of great interest therefore if the postulated rotational energy diagram (Figure 5.3) could be used to calculate the correct characteristic ratio.

6.1.3 *The characteristic ratio of the independent bond chain*

The next step then is to generalize equation (6.10) by not assuming that the component of a bond perpendicular to the preceding bond vanishes on the average. It will be assumed, however, that the rotational potentials for each bond are independent of the conformation of other bonds in the chain and therefore that this component has an average value that depends only on the rotational potential for the bond with which it is associated.

To accomplish this generalization it is helpful to set up coordinate systems for each bond so that the jth bond can be projected on the ith bond by successive projection of the components of the jth bond on the intervening bonds. This is most easily done by means of a transformation matrix first employed in molecular geometry problems by Eyring (1932). Its derivation is given in Appendix A6.1. In its local coordinate system as defined in the appendix, each bond can be written as a vector, $\mathbf{l}_j = (0, 0, 1)$. Let $\mathbf{l} = (0, 0, l)$, a generic local vector. In terms of the local coordinate system of the preceding bond, $j - 1$, $\mathbf{l}_j = \mathbf{t}(j - 1)\mathbf{l}$ where $\mathbf{t}(j - 1)$ is the Eyring transformation matrix that transforms bond j and contains the torsional angle, ϕ_{j-1}, formed by bonds $j - 2, j - 1, j$ and valence angle, θ formed by bonds $j - 1, j$. The dot product, $\mathbf{l}_{l-1} \cdot \mathbf{l}_j$ can then be written as $\mathbf{l}^T\mathbf{t}(j - 1)\mathbf{l}$ where $\mathbf{l}^T$ is the transpose of $\mathbf{l} = (0, 0, l)$. Proceeding back along the chain to the coordinate system of the ith bond, the vector $\mathbf{l}_j$ can be written as $\mathbf{l}_j = \mathbf{t}^{j-i}\mathbf{l}$ where $\mathbf{t}^{j-i}$ is a product of successive transformations, $= \mathbf{t}(i)\mathbf{t}(i + 1) \cdots \mathbf{t}(j - 2)\mathbf{t}(j - 1)$ and $\mathbf{l}_i \cdot \mathbf{l}_j = \mathbf{l}^T\mathbf{t}^{j-1}\mathbf{l}$. The average value, $\langle \mathbf{l}_i \cdot \mathbf{l}_j \rangle$, $= \mathbf{l}^T \langle \mathbf{t}^{j-i} \rangle \mathbf{l}$. Averaging brackets on matrices, $\langle \mathbf{t} \rangle$, for example, mean that each matrix element, t_{nm}, in $\mathbf{t}$ is to be statistically mechanically averaged, $\langle t_{nm} \rangle$. Since the bonds are independent and identical, then

$$\langle \mathbf{t}^{j-i} \rangle = \langle \mathbf{t}(i) \rangle \langle \mathbf{t}(i + 1) \rangle \cdots \langle \mathbf{t}(j - 2) \rangle \langle \mathbf{t}(j - 1) \rangle$$

and therefore equal to a typical $\mathbf{t}(i)$ averaged over ϕ and raised to the power $|j - i|$, or, $= \langle t \rangle^{j-i}$. Equation (6.6) can therefore be written

$$\langle R^2 \rangle = Nl^2 + 2(N - 1)\mathbf{l}^T \langle \mathbf{t} \rangle \mathbf{l}$$
$$+ 2(N - 2)\mathbf{l}^T \langle \mathbf{t}^2 \rangle \mathbf{l} + 2(N - 3)\mathbf{l}^T \langle \mathbf{t}^3 \rangle \mathbf{l} + \cdots. \tag{6.11}$$

In the limit of the large N, as was invoked in equation (6.8), it follows that,

$$\langle R^2 \rangle = Nl^2 + 2Nl^T \langle \mathbf{t} \rangle (1 + \langle \mathbf{t} \rangle + \langle \mathbf{t} \rangle^2 + \langle \mathbf{t} \rangle^3 + \cdots)\mathbf{l}, \tag{6.12}$$

where $\mathbf{1}$ is the identity matrix. The terms in the parentheses in equation (6.12) can be summed as a geometric series and there results,

$$\langle R^2 \rangle = Nl^2 + 2Nl^T \langle \mathbf{t} \rangle (1 - \langle \mathbf{t} \rangle)^{-1}\mathbf{l}. \tag{6.13}$$

For a rotational potential like that of polyethylene that is symmetrical about 180°, the $\sin \phi$ terms occurring in $\mathbf{t}$ average to zero and in this case the matrix below results,

$$\langle \mathbf{t} \rangle = \begin{pmatrix} \langle \cos \phi \rangle & 0 & 0 \\ 0 & \langle \cos \phi \rangle \cos \theta' & -\langle \cos \phi \rangle \sin \theta' \\ 0 & \sin \theta' & \cos \theta' \end{pmatrix}. \tag{6.14}$$

The matrix $(1 - \mathbf{t})^{-1}$ is found as an inverse by replacing each element of $(1 - \mathbf{t})$ by $-1^{(i+j)}M_{ij}$ where M_{ij} is the minor of the i, j element, then dividing each term of the resulting matrix by the determinant of $(1 - \mathbf{t})$ and finally taking the transpose of this matrix. Therefore,

$$(1 - \langle \mathbf{t} \rangle)^{-1} = \frac{1}{(1 + \langle \cos \phi \rangle)(1 - \cos \theta')}$$

$$\times \begin{pmatrix} (1 + \langle \cos \phi \rangle)(1 - \cos \theta') & 0 & 0 \\ 0 & 1 - \cos \theta' & -\langle \cos \phi \rangle \sin \theta' \\ 0 & \sin \theta' & (1 - \langle \cos \phi \rangle \cos \theta' \end{pmatrix}$$

$$\tag{6.15}$$

and placing this matrix in equation (6.13) gives

$$\langle R^2 \rangle = Nl^2 + 2Nl^2 \left[\frac{-\langle \cos \phi \rangle \sin^2 \theta' + (1 - \langle \cos \phi \rangle \cos \theta') \cos \theta'}{(1 + \langle \cos \phi \rangle)(1 - \cos \theta')} \right] \tag{6.16}$$

$$= Nl^2 \left(\frac{1 + \cos \theta'}{1 - \cos \theta'} \right) \left(\frac{1 - \langle \cos \phi \rangle}{1 + \langle \cos \phi \rangle} \right). \tag{6.17}$$

Replacing the supplement bond angle θ' by the bond angle, θ, the characteristic ratio is found to be (Oka, 1942; Taylor, 1947; Benoit, 1947; Kuhn, 1947)

$$C_\infty = \langle R^2 \rangle / Nl^2 = \left(\frac{1 - \cos \theta}{1 + \cos \theta} \right) \left(\frac{1 - \langle \cos \phi \rangle}{1 + \langle \cos \phi \rangle} \right). \qquad (6.18)$$

Evaluation of $\langle \cos \phi \rangle$ is accomplished by (cf. equation (5.1))

$$\langle \cos \phi \rangle = \frac{\int_0^{2\pi} \cos \phi \, e^{-U(\phi)/k_B T} \, d\phi}{\int_0^{2\pi} e^{-U(\phi)/k_B T} \, d\phi} \qquad (6.19)$$

where $U(\phi)$ is the rotational potential (see Figure 5.3). This evaluation is simplified if the rotational isomeric state model is invoked (Section 5.1.1) and it is assumed that the major contributions to the integral are in the neighborhood of the minima in $U(\phi)$, i.e. near the trans and gauche states. In that event, equation (6.19) becomes

$$\langle \cos \phi \rangle = \frac{\cos \phi_{G^+} \, e^{-E_G/k_B T} + \cos \phi_T \, e^{-E_T/k_B T} + \cos \phi_{G^-} \, e^{-E_G/k_B T}}{e^{-E_G/k_B T} + e^{-E_T/k_B T} + e^{-E_G/k_B T}}. \qquad (6.20)$$

Making this evaluation for polyethylene, using $\phi_{G^+} = 60°$, $\phi_{G^-} = 300°$, $\phi_T = 180°$, $E_G = E_{G^+} = E_{G^-} = 2.5 \text{ kJ/mol}$, $E_T = 0$, $\theta = 109.5°$, it is found

$$\langle \cos \phi \rangle = (e^{-2.5/RT} - 1)/(2 \, e^{-2.5/RT} + 1). \qquad (6.21)$$

Notice that the introduction of energy differences among the states has made the characteristic ratio temperature-dependent. This provides additional information for checking the agreement between calculated and experimental values of C_∞. At 413 K where the characteristic ratio of polyethylene has been measured, the calculated value above is $\langle \cos \phi \rangle = -0.261$ and C_∞ from equation (6.18) equals 3.42, Table 6.1. This is a substantial increase over the value of $C_\infty = 2.00$ for free rotation but still considerably short of the measured value of 6.7. The use of more realistic values of $\phi_{G^+} = 67.5°$, $\phi_{G^-} = 292.5°$ and $\theta = 112°$ gives a calculated value of $C_\infty = 4.25$, which is still smaller than the experimental one.

6.1.4 *The effect of interactions between neighboring bonds*

It is apparent that for the polyethylene example, although the preference of trans bonds to gauche causes a considerable expansion of the coiled molecule over that for free rotation, the expansion is still not sufficient to reproduce the experimental dimensions. There must be other conformational factors up to now neglected that operate to extend further the

Table 6.1 *Comparison of the characteristic ratio of polyethylene as calculated from various models and from experiment.*

model	C_∞
(1) *experiment*, 413 K	6.7
(2) *freely rotating chain*,	2.0
equation (6.10), tetrahedral valence angles	
(3) *independent bonds*,	
equation (6.18), gauche energy = 2500 J/mol	
(*a*) tetrahedral valence angles, gauche torsion = 60, 300°	3.42
(*b*) valence angle = 112°, gauche torsion = 67.5, 292.5°	4.25
(4) *interacting bond pairs*	
equations (7.53), (7.65), (7.73), Figure 7.2	
(*a*) G^+G^- excluded, other parameters as in 3(*b*) above	9.0
(*b*) G^+G^- disfavored by 5400 J/mol	7.5

average dimensions of the coiled molecule. Upon investigating the validity of the assumption of the previous section that the rotational potentials of each bond are independent it is readily concluded there are circumstances when it is obviously not true. To show this, one investigates neighboring *sequences* of bond conformations with the aid of models. There are no serious steric interactions between successive bond conformations of the types TT, TG^+, TG^-, G^+T, G^-T, or G^+G^+, G^-G^-. However, the sequences G^+G^- and G^-G^+ do have a steric interference, Figure 6.4. Because five carbon atoms are involved and pentane is the first member of the alkane series that can show this effect, it is sometimes called the 'pentane interference' (Taylor, 1948). The repulsion energy is so high as effectively to exclude completely G^+G^- or G^+G^- sequences

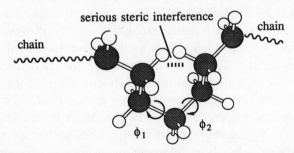

Fig. 6.4 'Pentane interference', steric repulsion in a G^+G^- sequence in polyethylene. Bond ϕ_1 is $-60°$ and bond ϕ_2 is $60°$, this brings two methylene units into steric interference.

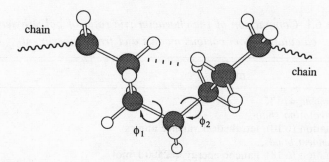

Fig. 6.5 Alleviation of steric strain in a G^+G^- sequence in polyethylene. The conformation is identical to that in Figure 6.4 except that the torsional angle ϕ_2 has been increased to 100° from 60°. The five-center methylene group distance has been greatly increased and the repulsion energy reduced.

for the sequences as drawn in Figure 6.4. However, the interaction can be greatly reduced by partial rotations about the G^+ or G^- bonds. This is accomplished at the expense of adding more energy to the rotated bond since it is no longer at the minimum energy position of the rotational potential. However, it is the total energy, the sum of the rotational barrier energies plus the steric interaction energy that determines the stability of the sequence. Calculations show that rotation of either G^+ or G^- towards trans (to about 100° or 260°) and leaving the other nearly undistorted is the best way of reducing the G^+G^- interference and minimizes the total energy, Figure 6.5. This energy is certainly higher than separated G^+, G^- bonds but is much less than the G^+G^- energy would be if they were locked in at 60°, 300° with the large steric repulsions. Therefore it might be expected that G^+G^- sequences are disfavored but not eliminated. Excluding or disfavoring G^+G^- sequences tends to disfavor the occurrence of G^+ and G^- bonds and biases in favor of trans bonds. This favoring of trans would result in a more expanded average dimension of the molecular than for the independent bond rotation model calculation of the preceding section. The G^+G^- sequences in the latter case were not discriminated against at all beyond the separate energies of two gauche bonds. There could be interactions between groups of atoms engendered by conformational sequences of three or even more bonds. For a simple chain like polyethylene, inspection of models leads to the expectation that for relatively nearby bonds such interactions will be so rare compared to the number of allowed sequences that their effect on the dimensions will be negligible. In very long chains the excluded volume effect from self-interactions comes into play but in the definition of the characteristic ratio, phantom or unperturbed conditions are invoked.

What is needed then is a method for generalizing the calculation of the characteristic ratio to take into account the interactions between pairs of local bond conformations. This can be accomplished but because its development is somewhat lengthy its treatment is deferred to a separate chapter (Chapter 7). However, for comparison with the independent bond model, the effect on the characteristic ratio of completely excluding and also disfavoring G^+G^- pairs as calculated by the methods of Chapter 7 is shown in Table 6.1. It may be seen that completely excluding G^+G^- pairs now overestimates C_∞, there are now too many trans bonds. However, an energy weighted disfavoring as suggested by Figure 6.5 and assigned a value of 5400 J/mol from conformational energy calculations leads to fairly reasonable agreement, see Section 7.6.1. An important conclusion is that the characteristic ratio can be very sensitive to the conformational features of a polymer chain and experimental values thus provide an important testing ground for models of the conformational behavior.

6.2 Long phantom chains, the Gaussian coil

Under the simplifying conditions of phantom chains or no excluded volume and in the limit of long chains not only can the average dimensions be formulated but the distribution of end-to-end distances can be found. This result forms the framework for much of the conceptualization about polymer solutions and melts and also rubber elasticity.

6.2.1 The end-to-end vector distribution function

The derivation of the distribution function for end-to-end separation of phantom chains proceeds most easily by finding its form in one dimension and then inferring what its three dimensional representation must be. To this end, a one dimensional random walk where self-intersection is allowed is considered. In Figure 6.6 is shown a walk of 17 total steps that consists of the sequence of $+, -$ steps as $+ + - + + + - - + - - - - + - - +$. Let n_+ be the number of $+$ steps and n_- be the number of $-$ ones and $N = n_+ + n_-$ be the total number of steps. The progress of the walk after N steps will be denoted as $m = n_+ - n_-$. If the $+, -$ steps are equally probable, then the probability of a *particular* sequence like the one above occurring will be $(1/2)^N$. However there are many different sequences

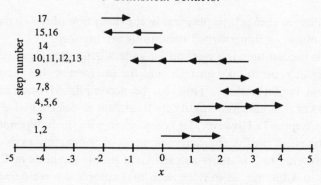

Fig. 6.6 A one dimensional walk of 17 total steps.

of walks that will have the same net progress value $= m$. In fact, the number of walks of N steps with a progress of m is the number of ways of mixing or ordering n_+ and n_- objects and is equal to $N!/n_+!n_-!$. Therefore the probability, $P(m, N)$ of a walk of $n_+ + n_- = N$ steps of progress $m = n_+ - n_-$ is

$$P(m, N) = (\tfrac{1}{2})^N \frac{N!}{n_+!n_-!} = (\tfrac{1}{2})^N \frac{N!}{\left(\dfrac{N+m}{2}\right)!\left(\dfrac{N-m}{2}\right)!}. \qquad (6.22)$$

It is easily verified that equation (6.22) is properly normalized by observing that normalization involves summation over all values of n_+, n_- that add to N and this sum can be found from the binomial theorem

$$(\tfrac{1}{2})^N \sum_{n_+ + n_- = N} \frac{N!}{n_+!n_-!} (a^{n_+} b^{n_-}) = (\tfrac{1}{2})^N (a + b)^N = 1 \text{ for } a, b = 1. \qquad (6.23)$$

In dealing with very long chains it is appropriate to invoke Stirling's approximation for the factorial terms,

$$n! \cong (2\pi)^{1/2} n^{n+1/2} e^{-n}. \qquad (6.24)$$

On introduction into equation (6.23), it is found that,

$$P(n, m) = \frac{(\tfrac{1}{2})^N (2\pi)^{\frac{1}{2}} N^{N+1/2} e^{-N}}{(2\pi)^{\frac{1}{2}} [(N+m)/2]^{[(N+m)/2+1/2]}} \qquad (6.25)$$

$$\times e^{-(N+m)/2} (2\pi)^{\frac{1}{2}} [(N+m)/2]^{[(N-m)/2+1/2]} e^{-(N-m)/2}$$

and on algebraic manipulation,

$$P(n, m) = \left(\frac{2}{\pi N}\right)^{\frac{1}{2}} \left[1 + \frac{m}{N}\right]^{-[(N+m)/2 + 1/2]} \left[1 - \frac{m}{N}\right]^{-[(N-m)/2 + 1/2]}$$

$$= \left(\frac{2}{\pi N}\right)^{\frac{1}{2}} \left[1 - \left(\frac{m}{N}\right)^2\right]^{-(N+1)/2} \left[1 - \frac{m}{N}\right]^{m/2} \left[1 + \frac{m}{N}\right]^{-m/2}. \quad (6.26)$$

At this point, the restriction that the progress, m, be small compared to the total number of steps is introduced, or, $m \ll N$. In that case, the terms in square brackets in equation (6.26) can be recognized as the leading terms of expansions of exponentials and approximated by the latter. Thus

$$P(m, N) \cong \left(\frac{2}{\pi N}\right)^{\frac{1}{2}} e^{(m/N)^2(N+1)/2} e^{-m^2/2N} e^{-m^2/2N} \quad (6.27)$$

and on neglecting a term $1/N$ compared to 1 it follows that

$$P(m, N) = \left(\frac{2}{\pi N}\right)^{\frac{1}{2}} e^{-m^2/2N}. \quad (6.28)$$

Equation (6.28) is the *Gaussian approximation* to the binomial distribution of equation (6.22). Obviously, the original walks were discrete ones, a unit step at a time. However, in the accumulated result as expressed by the Gaussian function, m is a small number compared to N and the function for the large values of N anticipated is a smooth function of m. It is now appropriate to recognize this and to replace the discrete progress index, m, by a continuous distance. This is easily done by defining a step length along the x axis in Figure 6.6 as l_x and setting the distance progress, x, equal to ml_x. There is a further subtlety however. In a 'bar graph' sense, the distribution of equation (6.22) or equation (6.28) actually spans two discrete steps. If the total steps, N, is odd, then m must also be odd and conversely if N is even, m is also even. Since N is fixed in a given example, m must be either even or odd and therefore change by two units along the bar graph. Let the continuous distribution be denoted as $P(x)$. Thus in establishing the differential length element in the continuous distribution, $P(x)\Delta x = P(m, N)$, the element is taken to be $\Delta x = 2l_x$. Therefore, the continuous distribution function for a one dimensional walk in terms of the step length is

$$P(x) = \frac{1}{2l_x} \left(\frac{2}{\pi N}\right)^{\frac{1}{2}} e^{-x^2/2l_x^2 N}. \quad (6.29)$$

The generalization to three dimensions is accomplished by observing that in a self-intersecting random walk, the paths must not be correlated with any specific direction. The progress along the x axis must be independent of y and z and vice-versa for an arbitrary coordinate system. This requires the distribution function to be the product of one dimensional functions as in equation (6.29), or $P(x, y, z) = P(x)P(y)P(z)$. The step length, l_x, along the x axis will no longer be the actual discrete step length but rather its projection on the axis. However, the effective projected values on the three axes must be the same or, $l_x = l_y = l_z$. Therefore,

$$P(x, y, z) = \left[\frac{1}{2l_x} \left(\frac{2}{\pi N} \right)^{\frac{1}{2}} \right]^3 e^{-R^2/2l_x^2 N}, \tag{6.30}$$

where $R^2 = x^2 + y^2 + z^2$.

It now remains to make the connection between the effective step along an axis, l_x, and the actual discrete step length. The latter is, of course, the bond length, l. On reflection, it is seen that the parameter, l_x, is actually what sets one type of polymer chain apart from others. It is the only parameter that depends on the structure of a particular chain example. Its value, especially relative to l, describes how expanded or contracted the coil is. The case has been made above (Section 6.1) that the characteristic ratio is a convenient measure of chain extension that can be directly calculated from theory if the conformational energetics are known and also is capable of experimental measurement. Thus it is appropriate to formulate the effective projected step length in terms of actual bond length through the characteristic ratio. This proceeds by a 'calibration procedure' by computing $\langle R^2 \rangle_0$ in equation (6.1) using equation (6.30). Since the Gaussian function is spherically isotropic, the volume element, $dx\, dy\, dz$ can be replaced by $4\pi R^2\, dR$ and,

$$\langle R^2 \rangle_0 = \int_0^\infty R^2 \left[\frac{1}{2l_x} \left(\frac{2}{\pi N} \right)^{\frac{1}{2}} \right]^3 e^{-R^2/2l_x^2 N} 4\pi R^2\, dR. \tag{6.31}$$

This is a standard integral and has a value of $3Nl_x^2$. From equation (6.1), $\langle R^2 \rangle_0 = C_\infty Nl^2$ and thus,

$$l_x = \left(\frac{C_\infty}{3} \right)^{\frac{1}{2}} l. \tag{6.32}$$

Finally, equation (6.30) can be rewritten as,

$$P(\mathbf{R}) = \left(\frac{B}{\pi} \right)^{3/2} e^{-BR^2}, \tag{6.33}$$

where

$$B = \frac{3}{2\langle R^2 \rangle_0} = \frac{3}{2C_\infty N l^2}$$

and C_∞ is the characteristic ratio of equation (6.1). It is appropriate to restate the assumptions that were made in arriving at the Gaussian function. The chains may self-intersect, they are long (Stirling's approximation for factorials of large numbers) and the function is not valid for extensions comparable to the extended chain length, i.e. equation (6.27) from equation (6.26).

6.2.2 *The radius of gyration of the Gaussian chain*

The radius of gyration of the Gaussian chain is of interest in several contexts. It is to be recalled that in Chapter 3 (Section 3.3.3) the rotational motion of random coils in solution was pertinent to the connection between solution viscosity and molecular length. Since for a spherically symmetric rigid body the rotational inertia that determines the angular velocity of rotation behaves as if the mass is concentrated at the radius of gyration, this, in turn, suggested that the latter was an appropriate measure of effective molecular size. Therefore it is of interest to establish the value of the radius of gyration for the Gaussian chain. In addition, it is found that the relation between the rms end-to-end distance and the radius of gyration is a sensitive test of whether the behavior of chains of finite length is actually Gaussian under a given circumstance. That is, it is possible to calculate numerically both quantities for various models. Whether or not their ratio has approached the Gaussian value is a monitor of the applicability of Gaussian statistics for the finite chains.

The radius of gyration, s, is defined as the mass weighted rms average of the magnitudes of the vectors leading from the center-of-mass to the mass points making up the rigid body, or,

$$s^2 = M^{-1} \sum m_i (\mathbf{r}_i - \mathbf{R}_0)^2, \tag{6.34}$$

where $\mathbf{r}_i$ and $\mathbf{R}_0$ are the mass point and center of mass position vectors as shown in Figure 6.7 and M is the total mass. Since the center of mass, $\mathbf{R}_0$, is determined as,

$$\mathbf{R}_0 = M^{-1} \sum m_i \mathbf{r}_i, \tag{6.35}$$

then

$$s^2 = M^{-1} \sum m_i (r_i^2 - R_0^2). \tag{6.36}$$

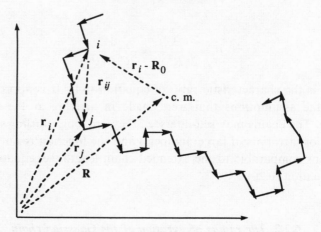

Fig. 6.7 The radius of gyration is defined in terms of the vectors from the center of mass (c, m) to each mass point, $r_i - R_0$.

Introduction of equation (6.35) into equation (6.36) gives,

$$s^2 = M^{-1} \sum m_i r_i^2 - M^{-2} \sum_i \sum_j m_i m_j \mathbf{r}_i \cdot \mathbf{r}_j. \tag{6.37}$$

The products $\mathbf{r}_i \cdot \mathbf{r}_j$ in equation (6.37) can be expressed in terms of the vectors $\mathbf{r}_{ij}$ spanning i to j (see Figure 6.7) by squaring $\mathbf{r}_j - \mathbf{r}_i$ and thus $\mathbf{r}_i \cdot \mathbf{r}_j = (r_{ij}^2 - r_i^2 - r_j^2)/2$. Therefore in terms of r_{ij} vectors equation (6.37) becomes,

$$s^2 = M^{-1} \sum m_i r_i^2 - M^{-2} \sum_i \sum_j m_i m_j (r_i^2 + r_j^2 - r_{ij}^2)/2$$

and it follows that,

$$s^2 = \tfrac{1}{2} M^{-2} \sum_i \sum_j m_i m_j r_{ij}^2. \tag{6.38}$$

If all the chain beads are the same and have the same mass, equation (6.38) when averaged over the configurations of the chain gives

$$\langle s^2 \rangle = \tfrac{1}{2} N^{-2} \sum_i \sum_j \langle r_{ij}^2 \rangle. \tag{6.39}$$

Equation (6.39) forms the basis upon which the average radius of gyration may be computed for the Gaussian chain. This proceeds by realizing that for such a chain, an individual term, $\langle r_{ij}^2 \rangle$, will have the value associated with a Gaussian chain of $|j - i|$ bonds. Thus $\langle r_{ij}^2 \rangle = |i - j| l^2 C_\infty$. Rewriting equation (6.39) in terms of a sum over $j \geq i$

and making the substitution $k = j - i$ gives,

$$\langle s^2 \rangle = C_\infty N^{-2} l^2 \sum_{i=1}^{N} \sum_{k=0}^{N-i} k^2. \tag{6.40}$$

The sum of k^2 over k may be verified to be $(N - i)(N - i + 1)/2$ which approaches $(N - i)^2/2$ at large N. The substitution, $q = N - i$ results in

$$\langle s^2 \rangle = C_\infty N^{-2} l^2 \sum_{q=0}^{N-1} \frac{q^2}{2}. \tag{6.41}$$

The sum may be verified by induction to be

$$\sum_{q=0}^{N-1} \frac{q^2}{2} = \frac{1}{2} \left[\frac{(N-1)^3}{3} + \frac{(N-1)^2}{2} - \frac{(N-1)}{6} \right] \tag{6.42}$$

and therefore at large N

$$\langle s^2 \rangle = C_\infty N l^2 / 6$$

or

$$\langle s^2 \rangle = \langle R^2 \rangle / 6. \tag{6.43}$$

Thus for the Gaussian chain it is seen that the mean-square radius of gyration is one-sixth of the mean-square end-to-end distance.

6.2.3 The approach of phantom chains to Gaussian statistics with increasing chain length

There are a number of measures of the effect of deviation from the long chain length assumption in the Gaussian distribution. Three are illustrated here. First, as suggested by the treatment of the freely rotating chain, equation (6.10), the average dimensions divided by the number of bonds, as described by the characteristic ratio, $\langle R^2 \rangle_0 / N l^2$, depend on chain length for shorter chains. Second, the radius of gyration, $\langle s^2 \rangle$, compared to $\langle R^2 \rangle_0$ will approach the $\frac{1}{6}$ value only at long chain lengths. Finally, the actual distribution function calculated numerically for a specific model can be compared to the long chain Gaussian limit. In this section these comparisons are made for such a specific chain model. The model is one for a polyethylene-like chain with exact trans (180°) and gauche (60°) states and has the valence angles fixed at the tetrahedral value. Such a chain fits on a diamond lattice, a circumstance that speeds the enumeration numerically of the chain configurations. The calculations are made under these circumstances using a Monte Carlo method that has been implemented for convenient computer use (Nairn, 1990). In Figure 6.8, the

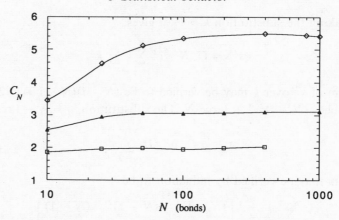

Fig. 6.8 The dependence of characteristic ratio on chain length. The chain is polyethylene-like with exact trans (180°) and gauche (60°) states and has the valence angles fixed at the tetrahedral value. Bottom curve (squares) is an independent bond chain with equal energy states (freely rotating chain); middle curve (triangles) is an independent bond chain (gauche energy = 0.5 kcal/mol = 2.1 kJ/mol); and upper curve (diamonds) an interacting bond model (GG′ interference energy = 2.0 kcal/mol = 8.4 kJ/mol, gauche energy = 0.5 kcal/mol = 2.1 kJ/mol). $T = 413$ K.

characteristic ratio is plotted as a function of chain length for three cases. The cases are a freely rotating chain, an independent chain and an interacting bond model. The C_N for the finite chain is smaller than C_∞ because of the truncation of positive terms in equation (6.3). The approach to the long chain limit depends on the characteristic ratio, the higher the latter, the slower the approach. This is simply a matter of persistence length. If the correlations in $\langle \mathbf{l}_i \cdot \mathbf{l}_j \rangle$ in equation (6.3) die away quickly as $|i - j|$ increases then the effect of the truncation in the sums in a finite chain will be lessened.

In Figure 6.9, a similar plot is shown for the ratio of the rms end-to-end distance to the rms radius of gyration. The approach to the long chain value of 6, equation (6.43), is seen to be sensitive again to the average dimensions. As observed above, the terms $\langle r_{ij}^2 \rangle$ in the sum in equation (6.39) each represent effectively an $\langle R^2 \rangle$ for a chain of length $|i - j|$ and therefore $\langle s^2 \rangle$ of a chain of N bonds is sensitive to $\langle R^2 \rangle$ of all lengths of chains shorter than N. Since the latter are smaller than that for the infinite chain, the ratio $\langle R^2 \rangle / \langle s^2 \rangle$, is larger than the asymptotic value of 6. The chain length must be relatively long for this effect to be attenuated.

Some comparisons of the distribution function calculated directly by numerical simulation with the Gaussian function are made in Figures

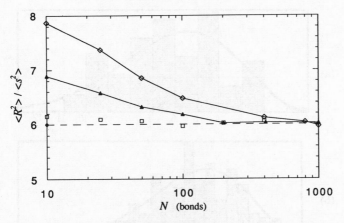

Fig. 6.9 Dependence on chain length of ratio of rms end-to-end distance to rms radius of gyration. Cases and symbols are the same as in Figure 6.8.

6.10 and 6.11. Figure 6.10 shows the functions for three chain lengths for the freely rotating chain and Figure 6.11 displays the same for the interacting bond chain with the same parameters as Figure 6.8. Two observations are warranted. First, the effect of the conformational states being restricted to exact discrete values shows up strongly in the distribution functions at the short chain lengths (10 bonds) and to some degree at the intermediate ones (25 bonds) as a noticeably irregular distribution. The irregularity is not due to finite sample size in the calculations but rather to the discrete nature of the chain states. The second observation is that for the more expanded, interacting bond chain of Figure 6.11 the distribution deviates noticeably from Gaussian at the shorter chain lengths.

6.3 The freely jointed chain: the distribution function for an exactly treatable model

As has been seen, the Gaussian function not only is based on phantom chain behavior, but also incorporates the approximations that the chain length is long and the chain extension is small compared to the contour length. There is a simple chain model that can be treated exactly within the phantom chain assumption and the above two approximations associated with Gaussian statistics removed. This is useful because it provides a distribution function that is valid up to chain extensions equal to the contour length, a region where the Gaussian function fails. The

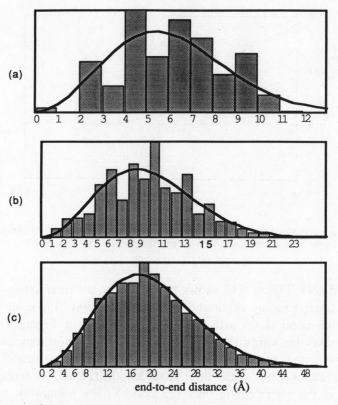

Fig. 6.10 End-to-end distance radial distribution function for the equal states chain of Figures 6.8 and 6.9. The bars are the results calculated by direct numerical simulation. The solid curves are the Gaussian function of equation (6.33) multiplied by $4\pi R^2$. $B = 3/(2\langle R^2 \rangle)$ where $\langle R^2 \rangle$ is taken as that calculated from the simulation. Curve (a) is for 10 bonds; (b) 25 bonds; (c) 100 bonds.

chain model is one where not only is there free rotation about the bond torsions, as in the freely rotating chain case, but the valence angles are also unconstrained and free to adjust without restriction. Only the bond lengths are fixed. Thus the model is known as the 'freely jointed' chain. As a model for polymer chain it was introduced by Kuhn (1936). The exact mathematical treatment as a problem in random walks was given by Rayleigh (1919). The development here follows that of Flory (1969, Chapter VIII) but see also Volkenstein (1963). Before taking up the derivation of the distribution function for the freely jointed chain some preliminaries are considered concerning the formulation of the end-to-end vector distribution function in general.

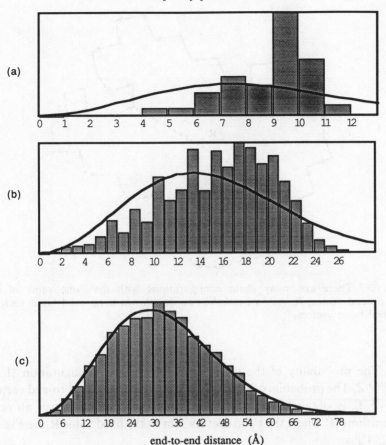

Fig. 6.11 End-to-end distance radial distribution function for the interacting bond chain of Figures 6.8 and 6.9. See Figure 6.10. Curve (a) is for 10 bonds; (b) 25 bonds; (c) 100 bonds.

6.3.1 *The end-to-end vector distribution function*

The classical configurational partition function of a chain of N connected bonds, $l_1, l_2, \ldots, l_N$ (see Figure 6.1) is

$$Z = \int \cdots \int e^{-\beta E(\mathbf{l})} \, d\{\mathbf{l}\}, \qquad (6.44)$$

where $\beta = (k_B T)^{-1}$ and $d\{\mathbf{l}\}$ means that integration over the chain configurations as represented by the chain bond vectors $\mathbf{l}_1, \mathbf{l}_2, \ldots, \mathbf{l}_N = \{\mathbf{l}\}$. The energy of the chain, $E(\mathbf{l})$, is, of course, a function of the configuration,

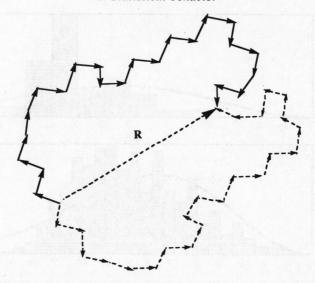

Fig. 6.12 There are many chain configurations with the same value of the end-to-end vector, **R**. As an example, two are shown here, solid bond vectors, dashed bond vectors.

{**l**}. The probability of the chain being in a single configuration {**l**} is $e^{-\beta E(\mathbf{l})}/Z$. The probability, $W(\mathbf{R})$, of the chain having an end-to-end vector, $\mathbf{R} = \sum \mathbf{l}_i$, is obtained by integrating the preceding factor over all configurations {**l**} where the bond vectors sum to the fixed value **R**, see Figure 6.12. Thus,

$$W(\mathbf{R}) = \int \cdots \int_{\sum \mathbf{l} = \mathbf{R}} e^{-\beta E(\mathbf{l})} \, d\{\mathbf{l}\}/Z, \tag{6.45}$$

where $\sum \mathbf{l} = \mathbf{R}$ means that the integration is subject to the constraint that the bond vectors sum to the fixed value = **R**. As an aside that will be useful later in the treatment of rubber elasticity a comment on the configurational free energy is made here. In general, the configurational contribution to the Helmholtz free energy, A, is related to the configurational partition function, Z, as

$$A = -k_\mathrm{B} T \ln Z. \tag{6.46}$$

The numerator in equation (6.45), constituting an integration over the Boltzmann factor, can also be regarded as a partition function. Thus, $Z(\mathbf{R})$

defined as

$$Z(\mathbf{R}) = \int \cdots \int_{\sum \mathbf{l} = \mathbf{R}} e^{-\beta E(\mathbf{l})} \, d\{\mathbf{l}\} \qquad (6.47)$$

can be regarded as the configurational partition function of a chain constrained to the end-to-end vector, $\mathbf{R}$, and the configurational Helmholtz free energy of such a constrained chain is

$$A(\mathbf{R}) = -k_B T \ln Z(\mathbf{R}). \qquad (6.48)$$

6.3.2 The distribution function for the freely jointed chain

Finding the distribution function proceeds by taking the Fourier transform of the distribution function and imposing the conditions of the freely jointed chain and then back transforming to the function itself. The Fourier transform of $W(\mathbf{R})$ of equation (6.45) is

$$F(\mathbf{s}) = \int_0^\infty W(\mathbf{R}) \, e^{\hat{\mathbf{i}} \mathbf{R} \cdot \mathbf{s}} \, d\mathbf{R}, \qquad (6.49)$$

where $\hat{\mathbf{i}} = \sqrt{(-1)}$. Since, according to equation (6.45), $W(\mathbf{R})$ represents an integration over all values of the coordinates with fixed $\mathbf{R}$, further integration over $d\mathbf{R}$ in equation (6.49) implies unrestricted integration over the coordinates and thus,

$$F(\mathbf{s}) = \int \cdots \int e^{-\beta E(\mathbf{l})} \, e^{\hat{\mathbf{i}} \mathbf{R} \cdot \mathbf{s}} \, d\{\mathbf{l}\}/Z. \qquad (6.50)$$

For the freely jointed chain, $E(\mathbf{l})$ is independent of configuration with respect to valence and torsional angles and subject only to the restriction of constant bond lengths, l_i. Therefore, using $\mathbf{R} = \sum \mathbf{l}_i$ it is seen that

$$F(\mathbf{s}) = \prod_i \int e^{\hat{\mathbf{i}} \mathbf{s} \cdot \mathbf{l}_i} \, d\mathbf{l}_i \Big/ \prod_i \int d\mathbf{l}_i, \qquad (6.51)$$

where the integrations are to be carried out over the orientations of the bond vectors, $\mathbf{l}_i$. Using $\mathbf{s} \cdot \mathbf{l}_i = ls \cos \alpha$, where all bonds are supposed of the same length, l, and $d\mathbf{l}_i = l^2 \sin \alpha \, d\alpha \, d\gamma \, dl$ where α is the angle between $\mathbf{s}, \mathbf{l}_i$ and γ completes the polar coordinate system for integrating over $d\mathbf{l}_i$, then at constant bond length (no integration over dl), equation (6.51) for a chain of N bonds reduces to,

$$F(\mathbf{s}) = \left(\frac{\sin ls}{ls} \right)^N. \qquad (6.52)$$

Back transformation of equation (6.52) using

$$W(\mathbf{R}) = [1/(2\pi)^3] \int F(\mathbf{s}) \, e^{-i\mathbf{s} \cdot \mathbf{R}} \, d\mathbf{s} \tag{6.53}$$

yields

$$W(\mathbf{R}) = \frac{1}{2\pi^2} \int_0^\infty \left(\frac{\sin ls}{ls}\right)^N (\sin Rs)(s/R) \, ds. \tag{6.54}$$

Equation (6.54) constitutes the exact distribution function for the freely jointed chain (Rayleigh, 1919). It can be integrated (Treloar, 1946, 1975; Flory, 1969) but the result is rather cumbersome and for large values of N, i.e. longer chains, numerical integration is just as convenient.

6.3.3 The inverse Langevin function approximation for the freely jointed chain

There is an alternative derivation of the distribution function for the freely jointed chain, one in which an approximation is introduced that leads to a somewhat simpler form (Kuhn and Gruen, 1942; James and Guth, 1943b). Although it is approximate it does preserve the property of the function properly vanishing for the fully extended chain, $\mathbf{R} = Nl$, rather than at $R = \infty$ as for the Gaussian function. Its derivation is given in Appendix A6.2. The result for the distribution function is

$$W(\mathbf{R}) = \frac{Cu}{2\pi Rl^2} \, (e^{-uR/l}) \left(\frac{\sinh u}{u}\right)^N, \tag{6.55}$$

where

$$u = L^{-1}(R/Nl)$$

and where L^{-1} is the inverse Langevin function as defined in equations (A6.13) and (A6.14).

6.3.4 Comparisons among distribution functions for the freely jointed chain

It is interesting to compare the exact distribution function with the inverse Langevin function approximation and also to compare these two with the Gaussian approximation. In doing so it is first appropriate to establish that the exact function for the freely jointed chain does, in fact, reduce to the Gaussian function under the conditions previously supposed for the validity of the latter. In inspecting equation (6.54), the exact function, it is to be noticed that for large values of N, the number of bonds, the factor

$[\sin(ls)/ls]^N$ will be very small except for values of the term in brackets near unity. From the series expansion, $\sin(ls)/ls = 1 - (ls)^2/3! + \cdots$, it is apparent that non-vanishing values will occur near small values of ls. Thus the approximation $[\sin(ls)/ls]^N \cong e^{-Nl^2s^2/6} = [1 - (ls)^2/6 + \cdots]^N$ becomes increasingly accurate at large N. In this circumstance equation (6.54) becomes

$$W(\mathbf{R}) = \frac{1}{2\pi^2} \int_0^\infty e^{-Nl^2s^s/6} \sin(Rs)(s/R) \, ds. \tag{6.56}$$

Integration by parts, $\int u \, dv = uv - \int v \, du$, with $dv = e^{-Nl^2s^s/6}s \, ds$, $u = \sin(Rs)$ leads to a standard definite integral with the value

$$W(\mathbf{R}) = \left(\frac{3}{2\pi Nl^2}\right)^{3/2} e^{-3R^2/2Nl^2}. \tag{6.57}$$

Thus the distribution reduces to Gaussian for large N. In comparing equation (6.57) with equation (6.33) it is apparent that $C_\infty = 1$ for the freely jointed chain. Referring to equation (6.3) or equation (6.5) it is seen that since $\langle \mathbf{l}_i \cdot \mathbf{l}_j \rangle = 0$ for $i \neq j$ in the present chain model, direct calculation of the characteristic ratio yields $C_N = 1$. This is true for all chain lengths, not just for $N \to \infty$.

In Figures 6.13 and 6.14 comparisons are made between the exact function, the inverse Langevin approximation and the Gaussian function for two chain lengths, $N = 5$ and 10. It is to be kept in mind that the

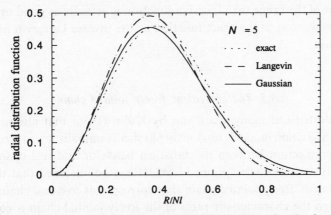

Fig. 6.13 Comparisons among three distribution functions for the freely jointed chain. The radial distribution functions $4\pi R^2 W(\mathbf{R})$ for the exact function, the inverse Langevin function approximation and the Gaussian approximation are shown for a chain of five bonds. (The bond length $l = 1$.)

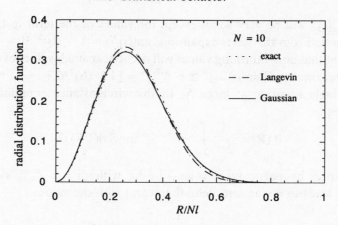

Fig. 6.14 Comparisons among three distribution functions for the freely jointed chain. See Figure 6.13. Results for a chain of ten bonds.

exact function and the inverse Langevin approximation both have the property that they properly go to zero exactly at the extended chain length and are not defined beyond this distance whereas the Gaussian approximation tails toward zero as $R \to \infty$ and the extended length plays no special role. However, it is only for extremely short chains that the differences in the approach to zero are apparent. In the $N = 5$ example the Gaussian function can be seen to lie above the exact function near $R = Nl$. However, by $N = 10$ all the functions are extremely small well before the extended length is reached. Even for $N = 5$, it may be seen that over most of the range the Gaussian function is actually as good or better an approximation to the exact function as the inverse Langevin function distribution.

6.3.5 *The equivalent freely jointed chain*

It was pointed out many years ago by Kuhn (1936) that although the freely jointed chain model is obviously physically unrealistic in a structural sense, a connection between its statistical behavior and real chains can be established at a coarse grained level. It can be imagined that there is an 'equivalent' freely jointed chain that corresponds to a real chain. That is, although the characteristic ratio of the freely jointed chain is equal to 1, and thus far from the values for a real chain, a correspondence can be established between both $\langle R^2 \rangle_0$ and the extended or contour length of the real and idealized chain by adjusting the number of bonds and their lengths in the latter. Thus if N_K and l_K are the number of bonds and their lengths (often called the number of Kuhn segments and the Kuhn segment length respectively) in the equivalent freely jointed chain and N, l are the values for the real chain, then, to establish correspondence,

$$C_\infty N l^2 = \langle R^2 \rangle_0 = N_K l_K^2 \qquad (6.58)$$

and

$$K_{geom} N l = \text{extended length} = N_K l_K,$$

where K_{geom} is a geometrical constant that expresses the projection of the bond length on the axis of the extended real chain. As an example, for polyethylene $C_\infty = 6.7$ and $K_{geom} = \sin(109.47°/2) = 0.816$ (tetrahedral valence angles). This results in $N_K/N \cong 1/10$ and $l_K/l \cong 8$ for the equivalent chain (Figure 6.15).

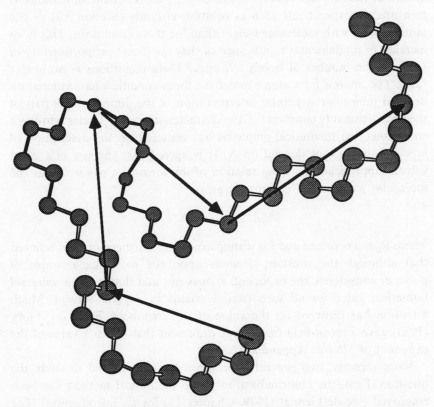

Fig. 6.15 An equivalent chain. A *freely jointed* chain (vectors) with $\sim 1/10$ of the bonds in a polyethylene chain and each of length ~ 8 times that of the polyethylene bond length has the same average dimensions, $\langle R^2 \rangle_0$, and extended length as polyethylene.

6.4 Excluded volume

In the foregoing discussions in this chapter, phantom or unperturbed chains have been the basis for development of statistical behavior. The interesting question thus arises as to the degree to which chains not subject to this condition differ in their statistical description. This has proven to be a difficult problem to attack. On the other hand it has attracted a great deal of attention from the theoretical point of view. As already touched upon in Section 3.3.3, it is useful to think of a chain immersed in a 'good' solvent, with polymer bead–solvent interactions more favorable than the polymer–polymer, solvent–solvent interactions, as expanded and thus occupying more volume in comparison with a phantom chain. This extra volume due to the effective mutual exclusion of the polymer beads is known as the *excluded volume*. It causes the effective chain dimensions as measured in experiments such as solution viscosity (Section 3.3) to rise more rapidly with molecular weight than for theta conditions. This is an increase of fundamental significance in that the direct proportionality of $\langle R^2 \rangle$ to the number of bonds, N, under theta conditions is no longer valid. The interest for systems under the theta condition has centered on detailed molecular structural interpretation of the dimensionless part of the proportionality constant (i.e. the characteristic ratio). Under non-theta conditions, the theoretical emphasis has centered on the description of how the dimensions depend on N. It is agreed that the size in a good solvent approaches a *scaling* relation of the form of a *power law* as the molecular weight or N becomes large, or

$$\langle R^2 \rangle^{1/2} \to R_0 N^\nu, \tag{6.59}$$

where R_0 is a constant and ν is scaling exponent. Furthermore it is believed that although the constant depends upon the particular example of polymer considered, the exponent, ν, does not and that it has a *universal* numerical value for all such flexible chains in a good solvent. Much attention has centered on the value of this exponent. Early on, Flory (1953) gave a mean-field free energy argument that led to a value of the exponent of 3/5 (see Appendix A6.3).

More recently, two general approaches have been used to study the question. First, the renormalization group theoretical method has been employed (see de Gennes (1979, Chapter 11) for an introduction). The basic idea is to consider the chain to be divided into many fine grained subunits ('monomers'). Then the subunits are grouped into larger units and the collection of larger units still partitions the chain. Relations are established between the properties of the original partitioned chain and the regrouped one. Then this process is repeated iteratively. When continued enough times the relations connecting the regrouped chain with the previous iteration approach fixed values. For a property such as the rms end-to-end distance it is possible to show that the approach to the fixed relations is only consistent with power law behavior. Further, the exponent is independent of the values of the coupling between iterations and therefore of the detailed structure of the chain. Hence it has a 'universal' value for the property considered. Thus power law behavior with universal exponent can be established. However, determining the value of the exponent is more difficult and requires elaborate calculation. For the case at hand, the most widely accepted value from renormalization group theory is $v = 0.588$ (le Guillou and Zinn-Justin, 1977). It is apparent that this value is only marginally different from the Flory mean field result.

The other approach to determination of the exponent has been employment of computer simulations. These usually involve generating a chain as a random walk on a lattice. The condition that multiple occupation of sites is forbidden is imposed in order to generate self-avoiding walks ('SAW's). For long walks, i.e. chains, *the attrition* problem arises. That is, the chances of completing a successful non-interfering walk drops exponentially with the number of bonds. It becomes difficult to create an adequate population of chains for statistical purposes. However, methods for overcoming this problem to the extent of being able to generate large numbers of quite long (hundreds of bonds) chains have been introduced (Alexandrowicz, 1969, 1983). Following the introduction of this method by Wall (Wall, Hiller and Wheeler, 1954; Wall, Widmer and Gans, 1963), a large number of such simulations have been carried out (see McKenzie (1976) and Dayantis and Palierne (1991) for summaries) and they confirm with apparent impressive accuracy that power law behavior is obeyed. An illustration, based on results of Dayantis and Palierne (1991) is shown in Figure 6.16. Simulations agree with the renormalization method that the exponent is very near the 3/5 value, but noticeably smaller. They also agree with the universality nature of the dependence in that the exponent appears to be independent of the lattice type used, e.g., simple cubic, BCC, FCC or diamond.

It is also possible to address the question of the distribution function for the end-to-end vector. Theory has provided the asymptotic form of the chain length dependence at both short and long chain lengths. The distribution function is expressed as a function of the reduced variable,

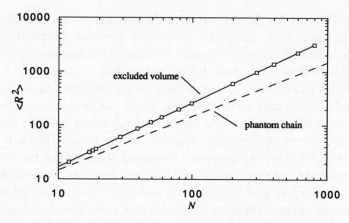

Fig. 6.16 The dependence of mean-square end-to-end distance on chain length for a chain with excluded volume. The points are from Monte Carlo simulation of self-avoiding walks (SAWs) on a cubic lattice (Dayantis and Palierne, 1991). The full curve is a least-square fit to the data and is of the form $\langle R^2 \rangle = AN^{2n}$, $A = 1.09$, $n = 0.595$. The dashed curve is drawn with slope $2n$, where $n = \frac{1}{2}$, the phantom chain value.

$$\mathbf{x} = \mathbf{R}/\langle R^2 \rangle^{1/2} \text{ where } \langle R^2 \rangle^{1/2} = R_0 M^\nu \text{ as}$$

$$P(R) = c(R_0 N^\nu)^{-3} f(x) \tag{6.60}$$

and where $f(x)$ is a *universal function* of the reduced chain length variable, x (c is a normalization constant). The function, $f(x)$ has been shown to approach e^{-x^δ} as $x \to \infty$ (Fisher and Sykes, 1959; Fisher and Hiley, 1961; Fisher, 1966) where δ is an exponent related to ν as $\delta = (1 - \nu)^{-1}$ ($= \frac{5}{2}$ for $\nu = \frac{3}{5}$). McKenzie and Moore (1971) proposed the form

$$f(x) = x^{g'} e^{-x^\delta}; \; x \to \text{large}, \tag{6.61}$$

where g' is a exponent related to other scaling exponents and in three dimensions has a value of $\sim \frac{1}{3}$. For short chains $f(x)$ has the form

$$f(x) = x^{g''}; \; x \to \text{small}, \tag{6.62}$$

where g'' is related to the exponent γ that scales the total number of configurations available to the chain as function of chain length, as $g'' = (\gamma - 1)/\nu$ (des Cloiseaux, 1974, 1980). The latter exponent has the value $\gamma = \frac{7}{6}$ in three dimensions so that $g'' = \frac{5}{18}$ for $\nu = \frac{3}{5}$. Since the values of exponents g' and g'' are not greatly different ($\sim \frac{1}{3}$ vs $\frac{5}{18}$) it has been suggested (des Cloiseaux and Jannik, 1987) that a function that spans the entire range of x that is based on combining them, and thus is of the form,

$$f(x) = x^g e^{-\beta x^\delta} \tag{6.63}$$

but where an extra parameter β is introduced, can be constructed that is of acceptable accuracy. The constants are given as $g = g'' = 0.275$, $\delta = 2.427$ and $\beta = 0.318$. The first two are slightly different from the $g'' = \frac{5}{18}$, $\delta = \frac{5}{2}$ values since v is usually taken as slightly smaller than exactly $\frac{3}{5}$. Dayantis and Palierne (1991) from computer simulation found optimized values for the constants that were very similar to the scaling based values above.

Computer simulations have also been carried out without the lattice simplification, a condition usually designated as *off-lattice*. Bishop, Clarke, Rey and Freire (1991) investigated the end-to-end vector distribution function off-lattice for chains of $N = 50$ and 100 and for the conditions of (1) no excluded volume (phantom chains), (2) the bead interaction potential set to mimic theta conditions and (3) the bead interaction potential set to cause chain expansion. They found that, as expected, the distribution function was Gaussian for condition (1) and nearly so for condition (2). For the excluded volume case (3), the distribution function fitted the des Cloiseaux (1974) scaling form well.

The scaling exponent $2v$ for $\langle R^2 \rangle$ vs N has been investigated as a function of the strength of the excluded volume via off-lattice simulations (Kranbuehl and Verdier, 1992). The hard sphere diameter, d, of the beads was varied from 0 to 1 times the bond length, l. It was found that the parameter $2v$ varied smoothly from very close to one, for $d = 0$, to values slightly *above the mean-field theory results of* $\frac{6}{5} = 1.2$, for values of $d > 0.63l$. As emphasized above both the lattice simulations and renormalization group theory indicate limiting values of $2v$ slightly *less* than $\frac{6}{5}$. It is not yet clear what the significance of this result is but it could be that the off-lattice conditions require longer chain lengths to reach the infinite chain limiting values.

6.5 Further reading

The Flory (1969) book *Statistical Mechanics of Chain Molecules* is a good source on the freely jointed chain statistics as well as characteristic ratio matters. Volkenstein's (1963) book, *Configuration Statistics of Polymeric Chains*, is a classic. On modern methods involving scaling, de Gennes' book (1979) *Scaling Concepts in Polymer Physics* and Freed's *Renormalization Group Theory of Macromolecules* are recommended.

Appendix A6.1 The Eyring transformation matrix

A local coordinate system is set up for each bond as depicted in Figure 6.17. The components (x, y, z) of a vector in the coordinate system of the jth bond can be transformed to the coordinate system (x'', y'', z'') of the $j - 1$ bond in a two-step operation. First, the y, z coordinates are rotated about the x axis through the angle θ' so that the new z' axis coincides with the z axis (z'') of the $j - 1$ bond, see Figure 6.18. The new coordinates are given by

$$\left.\begin{array}{l} x' = x \\ y' = y \cos \theta' - z \sin \theta' \\ z' = y \sin \theta' + z \cos \theta'. \end{array}\right\} \tag{A6.1}$$

This rotation is followed by a rotation through an angle, ϕ, about the new z axis (z') to superimpose the x axis on the x axis (x'') of the preceding bond (see Figure 6.19). This rotation results in

$$\left.\begin{array}{l} x'' = x' \cos \phi - y' \sin \phi \\ y'' = x' \sin \phi + y' \cos \phi \\ z'' = z'. \end{array}\right\} \tag{A6.2}$$

The first transformation can be written in matrix form as

$$\begin{bmatrix} x' \\ y' \\ z' \end{bmatrix} = \begin{bmatrix} 1 & 0 & 0 \\ 0 & \cos \theta' & -\sin \theta' \\ 0 & \sin \theta' & \cos \theta' \end{bmatrix} \begin{bmatrix} x \\ y \\ z \end{bmatrix} \tag{A6.3}$$

and the second one as

$$\begin{bmatrix} x'' \\ y'' \\ z'' \end{bmatrix} = \begin{bmatrix} \cos \phi & -\sin \phi & 0 \\ \sin \phi & \cos \phi & 0 \\ 0 & 0 & 1 \end{bmatrix} \begin{bmatrix} x' \\ y' \\ z' \end{bmatrix}. \tag{A6.4}$$

The combined effect of both transformations is given by the product, **t**, of the

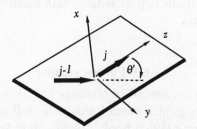

Fig. 6.17 Coordinate system for bond, j. The x axis is perpendicular to the plane of the bonds $j - 1, j$ and its direction determined by $\mathbf{l}_{j-1} \times \mathbf{l}_j$, a right-handed rotation of $j - 1$ into j. The z axis lies along $\mathbf{l}_j$. The y axis is in the $j - 1, j$ plane in a right-handed sense. θ' is the supplement of the bond angle.

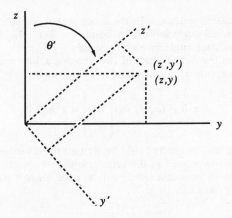

Fig. 6.18 Rotation about x axis to superimpose z axis on z axis of preceding bond.

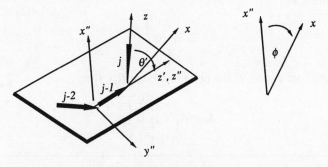

Fig. 6.19 Coordinate systems of adjacent bonds. (x, y, z for bond j; x'', y'', z'' for bond $j - 1$.) x', y', z' is an intermediate coordinate system for j that has its z' axis along the z'' axis of bond $j - 1$, but the same x axis, $x' = x$. The torsional angle, ϕ, is the angle between x'' and x, positive for clockwise rotation of x away from x'' looking down bond $j - 1$ from x'' to x.

two transformation matrices above

$$\mathbf{t} = \begin{bmatrix} \cos \phi & -\sin \phi \cos \theta' & \sin \phi \sin \theta' \\ \sin \phi & \cos \phi \cos \theta' & -\cos \phi \sin \theta' \\ 0 & \sin \theta' & \cos \theta' \end{bmatrix} \quad (A6.5)$$

as,

$$\begin{bmatrix} x'' \\ y'' \\ z'' \end{bmatrix} = \mathbf{t} \begin{bmatrix} x \\ y \\ z \end{bmatrix}. \quad (A6.6)$$

The above equation states that the components (x, y, z) of a vector in its own coordinate system based on the bond, j, it is attached to can be transformed to its components based on the coordinate system of the preceding bond, $j - 1$, by use of the transformation matrix, **t**. In the latter matrix, equation (A6.5), the angle, ϕ, is the torsional angle formed by bonds $j - 2$, $j - 1$, j and θ' is the supplement of the valence angle formed by bonds $j - 1$, j.

In matrix notation the dot product of two vectors, **a**, **b** considered as column matrices with components a_1, a_2, a_3 and b_1, b_2, b_3, respectively, is given by

$$\mathbf{a} \cdot \mathbf{b} = (a_1 \ a_2 \ a_3) \begin{bmatrix} b_1 \\ b_2 \\ b_3 \end{bmatrix} = \mathbf{a}^{\mathrm{T}} \mathbf{b}. \tag{A6.7}$$

In equation (6.6) the dot products could be written in this form, if both vectors had their components expressed in the same coordinate system. The transformation given by equation (A6.5) can be used for this purpose. For example in the dot product

$$\mathbf{l}_1 \cdot \mathbf{l}_2$$

$\mathbf{l}_2$ in terms of its own coordinate system is the bond itself (along the z axis) and can be written as

$$\mathbf{l}_2 = \begin{bmatrix} 0 \\ 0 \\ l \end{bmatrix}. \tag{A6.8}$$

In the coordinate system of the preceding bond ($\mathbf{l}_1$), $\mathbf{l}_2$ is

$$\mathbf{l}_2 = \mathbf{t}\mathbf{l}_2 = \mathbf{t} \begin{bmatrix} 0 \\ 0 \\ l \end{bmatrix}. \tag{A6.9}$$

Therefore

$$\mathbf{l}_1 \cdot \mathbf{l}_2 = \mathbf{l}_1^{\mathrm{T}} \mathbf{l}_2 = (0, 0, l)\mathbf{t} \begin{bmatrix} 0 \\ 0 \\ l \end{bmatrix} \tag{A6.10}$$

Explicit evaluation using equation (A6.5) for **t** gives $\mathbf{l}_1 \cdot \mathbf{l}_2 = l^2 \cos \theta'$ as already concluded in evaluating equation (6.6).

Appendix A6.2 The inverse Langevin function approximation for the freely jointed chain

The derivation (Flory, 1969, Chapter VIII) proceeds by computing the work required to maintain the average extension of the chain at a given value. The distribution function is then computed from the Boltzmann distribution appropriate for the work of extension. Let the extension be produced in the presence of a contactile force of magnitude, f, in response to a force applied at the end, R, and along a direction taken to be the x axis. If the end-to-end vector, **R**, has components X, Y, Z, the average value of the X component, in the direction of the force, is given by,

$$\langle X \rangle = \frac{\int \cdots \int X \, \mathrm{e}^{-\beta[E(\mathbf{l}) - fX]} \, \mathrm{d}\{\mathbf{l}\}}{\int \cdots \int \mathrm{e}^{-\beta[E(\mathbf{l}) - fX]} \, \mathrm{d}\{\mathbf{l}\}}, \tag{A6.11}$$

where the term, fX, in the integral arises from the work of extension in the presence of the force. Using $X = \sum_i x_i$ where x_i is the component of an individual bond on the x axis and the fact that for the freely jointed chain $E(\mathbf{l})$ is independent of orientation, equation (A6.11) reduces to

$$\langle X \rangle = \sum_i \langle x_i \rangle = N\langle x_i \rangle,$$

where

$$\langle x_i \rangle = \frac{\int_{-l}^{+l} x_i \, e^{\beta f x_i} \, dx_i}{\int_{-l}^{+l} e^{\beta f x_i} \, dx_i} \tag{A6.12}$$

and where the integration is over the possible limits of the projection on the x axis of an individual bond of $-l$ to $+l$. This integrates to

$$\langle x_i \rangle = l[\coth u - 1/u] = lL(u) \tag{A6.13}$$

where $u = fl/k_B T$ and $L(u) = \coth u - 1/u$ is known as the *Langevin function*. The constant force, f, that produces the average extension, $\langle X \rangle$, is found by solving equation (A6.13), in terms of the *inverse Langevin function*, L^{-1}, or,

$$u = L^{-1}(\langle x_i \rangle/l) = L^{-1}(\langle X \rangle/Nl) \tag{A6.14}$$

and therefore,

$$f = (k_B T/l)u. \tag{A6.15}$$

At this point the approximation is introduced. It is assumed that the average force that arises along the x axis as a response to an applied extension, X, is the same as the constant force giving rise to $\langle X \rangle$. On this basis the work of extension to X is given by

$$w = \frac{k_B T}{l} \int_0^X u' \, dX', \tag{A6.16}$$

where the force is approximated by equation (A6.15) and u is found from equation (A6.14). The integration above may be carried out by using equation (A6.13) to find

$$dX' = Nl[-\operatorname{cosech}^2 u' + 1/u'^2] \, du'. \tag{A6.17}$$

Using this relation equation (A6.16) integrates to

$$w = k_B T[uX/l - N \ln(\sinh(u)/u)], \tag{A6.18}$$

where u is to be found from the upper limit, X, of the integration by using equation (A6.14), with $X = \langle X \rangle$. Using the work represented in equation (A6.18) in a Boltzmann distribution, $\exp(-\beta w)$, results in

$$W(X) = (C/l) \, e^{-uX/l}[\sinh u/u]^N, \tag{A6.19}$$

where

$$u = L^{-1}(X/Nl)$$

and C is a dimensionless normalization constant chosen such that $\int W \, dX = 1$; $W(X)$ is the distribution function that expresses the probability of finding a component equal to X of the end-to-end vector $\mathbf{R}$ along an axis, x. It remains to find the distribution function $W(\mathbf{R})$ in terms of $W(X)$. This is accomplished in the following manner. The vectors $\mathbf{R}$ will be distributed uniformly over any sphere of radius R. In three dimensions, the projections of the $\mathbf{R}$ for a given radius R onto an axis such as x will be distributed uniformly over the interval

$-R$ to $+R$. The uniform density of points along x contributed from the vectors distributed over the sphere at $R + dR$ will be $W(R)4\pi R^2 \, dR/2R$. Since all spheres of radius equal to or larger than X contribute to $W(X)$, the latter function can be constructed from $W(\mathbf{R})$ by the relation,

$$W(X) = \int_{R=X}^{R_{max}} W(R)2\pi R \, dR \tag{A6.20}$$

or

$$\left. \frac{dW(X)}{dX} \right|_{X=R} = -2\pi R W(R). \tag{A6.21}$$

From

$$\frac{dW(X)}{dX} = \frac{\partial W}{\partial u}\frac{\partial u}{\partial X} + \frac{\partial W}{\partial X} \tag{A6.22}$$

and from the fact that $\partial W/\partial X = 0$ when found from equation (A6.19), equation (A6.21) leads to

$$W(\mathbf{R}) = \frac{Cu}{2\pi R l^2} (e^{-uR/l})\left(\frac{\sinh u}{u}\right)^N \tag{A6.23}$$

where

$$u = L^{-1}(R/Nl).$$

This constitutes the inverse Langevin function approximation to the freely jointed chain distribution function. It differs from the original one of Kuhn and Gruen (1942), equation (A6.19), in the introduction of the distinction between X and R in leading from equation (A6.19) to (A6.23), see Flory (1969, Chapter VIII).

Appendix A6.3 Mean-field derivation of the excluded volume expansion of R

This much simplified version of the Flory (1953) derivation follows that of de Gennes (1979).

In a good solvent, the expansion of the coil over that of the phantom chain case is occasioned by two opposing effects. One of these is the effective repulsive interaction between polymer beads that arises because the beads are on average more attracted to solvent molecules than to themselves. This causes an increase in the free energy. The latter is ameliorated by expansion of the chain since the beads become on average more distant. However the expansion also changes the free energy of the chain through perturbation of the internal chain statistics of the coil. That is, if the mean-square chain end-to-end distance increases this will be accompanied by an increase of the free energy that arises statistically. This may be regarded as an elastic effect since it is similar to the free energy increase that arises on stretching of chains in rubber elasticity (Chapter 8). Thus the two effects are opposing. The bead–bead repulsion free energy decreases with increasing coil size while that due to the elastic effect increases.

The repulsive energy of one polymer bead will be proportional to the number *concentration* of other beads and to a parameter that characterizes the strength of the interaction. The average (or mean-field) concentration of beads is proportional to N/R^3 where N is the number of beads and R is the end-to-end distance. Thus the repulsive free energy of a chain may be written as N times the repulsive energy of one bead as,

$$A_{rep} = k_B T N b(N/R^3), \tag{A6.24}$$

where b is the excluded volume strength parameter (with the dimensions of a volume). Borrowing some results from Chapter 9, the elastic free energy may be written as follows. According to equation (9.67), the potential of mean force or free energy of a single Kuhn segment is $U_{el} = \frac{1}{2}kR^2$ where R is the end-to-end distance of the segment and $k = 3k_B T/l_K^2$. If the Kuhn segment is taken to be the entire chain, $l_K^2 = Nl^2$ where l is the bond length (or bead size). Thus the elastic free energy is

$$A_{el} = \frac{3}{2}[k_B T/(Nl^2)]R^2. \tag{A6.25}$$

The total free energy,

$$A_{tot} = k_B T N b(N/R^3) + \frac{3}{2}[k_B T/(Nl^2)]R^2, \tag{A6.26}$$

may be differentiated with respect to R to find the minimum at

$$R = (bl^2)^{1/5}N^{3/5}. \tag{A6.27}$$

Thus the exponent v in equation (6.59) is given by $v = \frac{3}{5}$.

Nomenclature

A = Helmholtz free energy

C_N = characteristic ratio, equation (6.1), of a chain with N bonds

C_∞ = limiting value of the characteristic ratio for very long chains

$f(x)$ = reduced distribution function for chains with excluded volume, equation (6.60)

g, g', g'' = scaling exponent in the end-to-end vector distribution function of a self avoiding chain

E_G = energy associated with a gauche state

E_T = energy associated with a trans state

G = gauche conformation

G^+, G^- = $\sim +60°$ or $\sim -60°$ gauche torsional angle. A plus or minus *in parentheses* indicates that there is a significant distortion of the angle from its standard value but it still in the range occupied by gauche, e.g. $G^+(+)$, means there is a significant distortion away from $\sim +60°$ in the + direction but that the value is still less than $120°$ $G^+(+) = 100°$ would be an example

l = length of a chain bond

l_K = length of a Kuhn segment in an equivalent freely jointed chain

L_P = the persistence length in a polymer chain, equation (6.4)

m_i = the mass of a polymer chain bead, used in radius of gyration calculation

N = the number of bonds in a chain

N_K = the number of Kuhn segments in an equivalent freely jointed chain

$P(R)$ = Gaussian chain probability distribution function, equation (6.33)

$\mathbf{r}_i$ = position vector, vector from an origin to a chain bead

$R, \mathbf{R}$ = end-to-end distance, end-to-end vector of a polymer chain

$\langle R^2 \rangle, \langle R^2 \rangle_0$ = mean-square end-to-end distance, mean-square end-to-end distance under unperturbed theta conditions

$s, \langle s^2 \rangle$ = radius of gyration, equation (6.34), and its mean-square value

T = trans bond conformation, $\sim 180°$, $T(+)$, $T(-)$ means that the torsional angle is significantly distorted away from the $180°$ value in the direction indicated but that the value is still in the trans range, between $-120°$, $+120°$

$W(R)$ = probability distribution function for the freely jointed chain, either the exact function, equation (6.54), or the inverse Langevin function approximation, equation (6.55)

$\mathbf{x}$ = reduced variable for end-to-end vector of chains with excluded volume = $\mathbf{R}/\langle R^2 \rangle^{1/2}$, $\langle R^2 \rangle^{1/2} = R_0 N^\nu$

$Z, Z(\mathbf{R})$ = configurational partition function of a polymer chain, configurational partition function for the constraint of a fixed value of $\mathbf{R}$

β = parameter in the end-to-end vector distribution function of a self-avoiding chain

δ = parameter in the exponential dependence of end-to-end vector distribution function of a self avoiding chain, $= 1/(1 - \nu)$

ϕ = chain torsional angle

ν = scaling exponent in the dependence of rms end-to-end distance on chain length

θ, θ' = chain valence angle, supplement of the chain valence angle

Problems

6.1   Calculate the characteristic ratio of POM (at $T = 453\ K$) assuming the independent bond rotation model and the rotational isomeric state approximation. Assume the trans conformation lies 7 kJ/mol above the two gauche states and that the ϕs are $60°$, $180°$, and $300°$ and that $\theta = 109.47°$.

6.2   Calculate the fraction of bonds in the T state for POM for the conditions of problem 6.1.

6.3 Just as **R** is the vector sum of the bond vectors $\mathbf{l}_i$, equation
(6.2), the dipole moment, **μ**, of a chain is the sum of the individual
bond moments $\mathbf{m}_i$. In POM each C—O bond may be considered to
have a moment $\mathbf{m}_i$ directed from C to O,

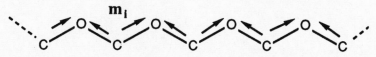

so that the vectors *alternate* down the chain rather than all pointing
in the same direction as in $\mathbf{R} = \sum \mathbf{l}_i$. Modify equation (6.6) and
equation (6.10) to represent $\mu^2 = \mathbf{\mu} \cdot \mathbf{\mu} = \sum m_i \cdot \sum m_j$ and $\langle \mu^2 \rangle$,
respectively for the square of the dipole moment of a freely rotating
POM chain. Evaluate for $\theta = 109.470°$, $\langle \mu^2 \rangle$, in terms of μ_0 (the
magnitude of each μ_i).

6.4 Follow up on problem 6.3 by modifying equations (6.12) and (6.17)
to represent the average square dipole moment of a POM chain in
the independent bond rotation model. Calculate $\langle \mu^2 \rangle$ in terms of μ_0
for $\theta = 109.47°$ and $E_T = 0$ and $E_G = E_{G'} = -7 \text{ kJ/mol}$ at 453 K.

7

The interacting bond model for the average properties of coiling chains

In Chapter 6 it was pointed out that comparison of the mean unperturbed solution dimensions, i.e., the characteristic ratio, derived from experiment and from calculations based on conformational models could form an important testing ground for the structural models. It was concluded, however (Section 6.1.4), that in order for realistic comparisons to be made, the computation of characteristic ratio would have to allow for interactions between nearby bonds. The steric interference between a pair of adjacent gauche bonds of opposite sense in a three-state chain was given as an important example of such interactions between nearby bonds. It is possible to formulate a treatment of chains where adjacent bond pairs interact. This is accomplished by means of a representation of the statistical mechanical partition function of a one-dimensional chain using the methods of matrix algebra (Kramers and Wannier, 1941) and adapted to the polymer chain conformation problem (Birshtein and Ptitsyn, 1959, 1966; Birshtein, 1959; Nagai, 1959; Lifson, 1959). This development and some applications to real chains are described in this chapter.

7.1 The partition function and scalar averages for a one dimensional chain with interacting bonds

In Section 6.3.1 the classical configurational partition function was written in equation (6.44) as an integration over all chain configurations. In the present development, attention will be focused on a discrete number of local bond states, those judged as important in the rotational isomeric state context (see Chapter 5, especially Section 5.1.1). In the discrete state approximation, the partition function of a system that may exist in energy states $E_1, E_2, \ldots, E_n, \ldots$, is

$$Z = \sum_n e^{-E_n/k_B T}. \tag{7.1}$$

The thermodynamic properties may be calculated from the relation

$$A = -k_B T \ln Z, \tag{7.2}$$

where A is the Helmholtz free energy. The average value of a property, f_n, of a system that depends on the energy state, n, is found from the population of each state, N_n, as,

$$\langle f \rangle = \sum_n \frac{f_n N_n}{\sum N_n} = \sum_n \frac{f_n e^{-E_n/k_B T}}{Z}. \tag{7.3}$$

Since the system represents a polymer chain, each state of the system is a particular conformational sequence of bonds along the chain. To find averages using equation (7.3), the energy of the chain must be expressed as a function of the conformational sequences and then summed over all of the possible sequences. For the independent bond rotational model this is an easy task, the total energy is just the sum of individual bond rotational state energies. For example, if the sequence of bonds for the nth state of an eight-bond chain were

$$\begin{array}{cccccccc} T & T & G^+ & T & G^- & G^+ & G^+ & T \\ 1 & 2 & 3 & 4 & 5 & 6 & 7 & 8 \end{array}$$

the energy in that state would be

$$E_n = E_1^T + E_2^T + E_3^{G^+} + E_4^T + E_5^{G^-} + E_6^{G^-} + E_7^{G^+} + E_8^T \tag{7.4}$$

where the subscript refers to the number of the bond and the superscript to the energy state of that bond. The partition function is evaluated by summing over each energy state of each bond $\alpha, \beta, \ldots, \omega = T, G^+, G^-$,

$$Z = \sum_\alpha \sum_\beta \cdots \sum_\omega e^{-(E_1^\alpha + E_2^\beta + \cdots + E_8^\omega)/k_B T} \tag{7.5}$$

or since the energies are independent,

$$Z = \left(\sum_\alpha e^{-(E^\alpha/k_B T)} \right)^8$$

$$= (e^{-E^T/k_B T} + e^{-E^{G^+}/k_B T} + e^{-E^{G^-}/k_B T})^8. \tag{7.6}$$

If the average value of $\cos \phi$ for the fifth bond is desired then

$$\cos \langle \phi_5 \rangle = \sum_n e^{-E_n/k_B T} \cos \phi_5 / Z$$

$$= \left(\sum_\alpha e^{-E^\alpha/k_B T} \right)^4 \left(\sum_\beta \cos \phi_5^\beta e^{-E_5^\beta/k_B T} \right) \left(\sum_\gamma e^{-E^\gamma/k_B T} \right)^3 \Big/ \left(\sum_\alpha e^{-E_\alpha/k_B T} \right)^8$$

$$= \frac{\cos \phi_5^T e^{-E^T/k_B T} + \cos \phi_5^{G^+} e^{-E^{G^+}/k_B T} + \cos \phi_5^{G^+} e^{-E^{G^+}/k_B T}}{e^{-E^T/k_B T} + e^{-E^{G^+}/k_B T} + e^{-E^{G^+}/k_B T}} \tag{7.7}$$

in agreement with equation (6.20). Thus in the independent bond model, the partition function of the chain is just that of one bond raised to the power of the number of bonds. The average value of a bond property of one bond (i.e., $\cos \phi_5$) is, of course, independent of the other bonds in the chain.

Equation (7.4) is generalized to include the effect of interactions between *adjacent pairs* of the above $TTG^+TG^-G^+G^+T$ sequence by writing

$$E_n = E_1^T + E_2^{TT} + E_3^{TG^+} + E_4^{G^+T} + E_5^{TG^-} + E_6^{G^-G^+} + E_7^{G^+G^+} + E_8^{G^+T}. \quad (7.8)$$

The first bond not being influenced by a preceding bond is simply denoted as a T bond. But E_2^{TT} denotes the energy of the second bond when in the T state but as influenced by being preceded by a T bond in the first position. The term E_3^{TG} specifies the energy of the third bond in the G state in the presence of bond 2 in the T state, etc. For the special example of polyethylene, it was postulated that the only case where the energy of a bond is influenced by its neighbor is the G^-G^+ or G^+G^- sequence. Thus for polyethylene the following terms in equation (7.8) can be evaluated as $E_1^T = E_T, E_2^{TT} = E_T, E_3^{TG^+} = E_{G^+}, E_4^{G^+T} = E_T, E_5^{TG^-} = E_{G^-}, E_7^{G^+G^+} = E_{G^+},$ $E_8^{G^+T} = E_T$. However, $E_6^{G^-G^+}$ would have to be given a special value $E_{G^-G^+}$ characteristic of a G^+ bond preceded by a G^- bond that includes the effect of the 'pentane interference.' It is evident that the factoring of the sums in the independent bond model is illustrated by equation (7.6) and equation (7.7) is no longer possible.

Introducing the notation

$$u = e^{-E/k_BR}; \quad u_2^{TT} = e^{-E_2^{TT}/k_BT}; \quad u_6^{G^+G^-} = e^{-E_6^{G^+G^-}/k_BT}; \quad \text{etc.} \quad (7.9)$$

the partition function becomes in general,

$$Z = \sum_\alpha \sum_\beta \sum_\gamma \cdots \sum_\omega u_1^\alpha u_1^{\alpha\beta} u_3^{\beta\gamma} \cdots u_N^{\psi\omega}, \quad (7.10)$$

where $\alpha, \beta, \gamma, \ldots, \psi, \omega$ are summed over the local bond conformational states such as T, G^+, G^-. Thus the evaluation of the partition function involves writing the sum of the products of the Boltzmann weighting factors, u, for every possible conformational sequence of the chain. It is possible to write this sum in a very compact way using the notation of matrix algebra and the properties of matrix multiplication. To demonstrate this, consider for simplicity a chain of five bonds, each of which can exhibit just two conformational states which are designated A and B. The first and last bonds in the chain do not have defined rotational states so that they can be omitted from consideration. Explicit enumeration of the partition function results in

$$Z = u_2^A u_3^{AA} u_4^{AA} + u_2^A u_3^{AA} u_4^{AB} + u_2^A u_3^{AB} u_4^{BA} + u_2^B u_3^{BA} u_4^{AA} + u_2^A u_3^{AB} u_4^{BB}$$
$$+ u_2^B u_3^{BA} u_4^{AB} + u_2^B u_3^{BB} u_4^{BA} + u_2^B u_3^{BB} u_4^{BB}. \tag{7.11}$$

Now define *statistical weight matrices* such that

$$\mathbf{u}_2 = \begin{bmatrix} u_2^A & 0 \\ 0 & u_2^B \end{bmatrix} \tag{7.12}$$

for the initial bond for which rotational states are defined and

$$\mathbf{u}_i = \begin{bmatrix} u_i^{AA} & u_i^{AB} \\ u_i^{BA} & u_i^{BB} \end{bmatrix} \tag{7.13}$$

for the succeeding ones. The product of the statistical weight matrices for each bond results in

$$\mathbf{u}_2 \mathbf{u}_3 \mathbf{u}_4 = \begin{bmatrix} u_2^A u_3^{AA} u_4^{AA} + u_2^A u_3^{AB} u_4^{BA}, & u_2^A u_3^{AA} u_4^{AB} + u_2^A u_3^{AB} u_4^{BB} \\ u_2^B u_3^{BA} u_4^{AA} + u_2^B u_3^{BB} u_4^{BA}, & u_2^B u_3^{BA} u_4^{AB} + u_2^B u_3^{BB} u_4^{BB} \end{bmatrix}. \tag{7.14}$$

Notice that the above matrix contains all of the terms involved in the partition function, equation (7.11). It now remains to collect the elements of this array into a single sum. This can be accomplished by defining a column matrix $\mathbf{b} = \begin{bmatrix} 1 \\ 1 \end{bmatrix}$ and a row matrix $\mathbf{b}^T = [1, 1]$. These are used to post- and pre-multiply the product of equation (7.14). The result is that the partition function is reproduced by

$$Z = \mathbf{b}^T (\mathbf{u}_2 \, \mathbf{u}_3 \, \mathbf{u}_4) \mathbf{b}. \tag{7.15}$$

In general for a polymer chain whose bonds each have r rotational conformational states A, B, C, ..., R

$$\mathbf{u}_2 = \begin{bmatrix} u_A & 0 & 0 & \cdots \\ 0 & u_B & 0 & \cdots \\ 0 & 0 & u_C & \cdots \\ \cdots & & & \end{bmatrix} \tag{7.16}$$

and for $1 < i < N$,

$$\mathbf{u}_i = \begin{bmatrix} u_i^{AA} & u_i^{AB} & u_i^{AC} & \cdots \\ u_i^{BA} & u_i^{BB} & u_i^{BC} & \cdots \\ u_i^{CA} & u_i^{CB} & u_i^{CC} & \cdots \\ \cdots & & & \end{bmatrix} \tag{7.17}$$

and *if the first and last bonds are included* but represented by $r \times r$ *identity matrices* $= \mathbf{e}_r$, or, $\mathbf{u}_1 = \mathbf{u}_N = \mathbf{e}_r$, then

$$Z = \mathbf{b}^T \mathbf{u}_1 \mathbf{u}_2 \mathbf{u}_3 \cdots \mathbf{u}_N \mathbf{b}, \qquad (7.18)$$

where $\mathbf{b}^T = (1, 1, 1, \ldots)$ and $\mathbf{b}$ is the column vector transpose to $\mathbf{b}^T$.

Now the average value of a quantity that depends on the state of one of the bonds is found. Suppose it is wished to find the average value of a quantity, f, that depends on the state of one bond ($\cos \phi$ for example) for the third bond of the five-bond, two-state example treated above. This quantity will have two values f_3^A and f_3^B for the two states of the third bond. Explicit evaluation of equation (7.3) results in

$$\langle f_3 \rangle = \begin{bmatrix} u_2^A u_3^{AA} f_3^A u_4^{AA} + u_2^A u_3^{AA} f_3^A u_4^{AB} \\ + u_2^A u_3^{AB} f_3^B u_4^{BA} + u_2^B u_3^{BA} f_3^A u_4^{AA} \\ + u_2^A u_3^{AB} f_3^B u_4^{BB} + u_2^B u_3^{BA} f_3^A u_4^{AB} \\ + u_2^B u_3^{BB} f_3^B u_4^{BA} + u_2^B u_3^{BB} f_3^B u_4^{BB} \end{bmatrix} / Z, \qquad (7.19)$$

where Z is given by equation (7.11). This result can be reproduced by matrix multiplication by defining the matrix

$$\mathbf{f}_3 = \begin{bmatrix} f_3^A & 0 \\ 0 & f_3^B \end{bmatrix} \qquad (7.20)$$

and replacing $\mathbf{u}_3$ in equation (7.14) by

$$\mathbf{u}_3 \mathbf{f}_3 = \begin{bmatrix} u_3^{AA} f_3^A & u_3^{AB} f_3^B \\ u_3^{BA} f_3^A & u_3^{BB} f_3^B \end{bmatrix}$$

and dividing by Z,

$$\langle f_3 \rangle = \mathbf{b}^T \mathbf{u}_2 \mathbf{u}_3 \mathbf{f}_3 \mathbf{u}_4 \mathbf{b} / \mathbf{b}^T \mathbf{u}_2 \mathbf{u}_3 \mathbf{u}_4 \mathbf{b}. \qquad (7.21)$$

In general, with the convention of including the first and last bonds, 1 and N, by means of identity matrices for $\mathbf{u}$, then,

$$\langle f_i g_j h_k \rangle = \mathbf{b}^T \mathbf{u}_1 \mathbf{u}_2 \cdots \mathbf{u}_i \mathbf{f}_i \mathbf{u}_{i+1} \cdots \mathbf{u}_j \mathbf{g}_j \mathbf{u}_{j+1} \cdots \mathbf{u}_k \mathbf{h}_k \mathbf{u}_{k+1} \cdots \mathbf{u}_N \mathbf{b} / Z, \quad (7.22)$$

where

$$\mathbf{f}_i = \begin{pmatrix} f_i^A & 0 & 0 \\ 0 & f_i^B & \\ 0 & & f_i^C \cdots \end{pmatrix}, \qquad \mathbf{g}_j = \begin{pmatrix} g_j^A & 0 & 0 \\ 0 & g_j^B & \\ 0 & & g_j^C \cdots \end{pmatrix}, \quad \text{etc.}$$

and Z is given by equation (7.18). As is manifest from the coupled nature of matrix multiplication, and in contrast to the independent bond model as exemplified by equation (7.7), the average value of a property of a specified bond depends on the averaging over the states of *all* of the bonds in the chain.

Equations (7.18) and (7.22) may be used directly to find the properties of finite chains by direct multiplication. For a chain whose bonds are chemically identical (e.g., polyethylene), the **u** matrices (after the second) are identical and equation (7.18) reduces to

$$Z = \mathbf{b}^T \mathbf{u}_1 \mathbf{u}^{N-1} \mathbf{b}, \tag{7.23}$$

where for convenience in what follows, N now refers to the number of bonds with *defined* rotational states and $\mathbf{u}_1$ is for the first such bond. It is an easy matter for a computer to evaluate equation (7.23) for quite large values of N by successively squaring the matrix **u**.

For an infinite uniform chain the partition function can be evaluated exactly in closed form (Kramers and Wannier, 1941). Define a matrix **a** that diagonalizes the statistical weight matrix by a similarity transformation,

$$\mathbf{a}^{-1}\mathbf{u}\mathbf{a} = \lambda, \tag{7.24}$$

$$\lambda = \begin{pmatrix} \lambda_1 & 0 & 0 \\ 0 & \lambda_2 & \\ 0 & & \cdots \end{pmatrix}.$$

Since

$$\mathbf{u} = \mathbf{a}\lambda\mathbf{a}^{-1},$$

then

$$\mathbf{u}^{N-1} = \mathbf{a}\lambda^{N-1}\mathbf{a}^{-1}$$

and

$$Z = \mathbf{b}^T \mathbf{u}_1 \mathbf{a}\lambda^{N-1}\mathbf{a}^{-1}\mathbf{b}. \tag{7.25}$$

The partition function is a scalar quantity (1×1 matrix) and hence equal to its trace when considered a matrix, therefore,

$$Z = \text{Trace}(\mathbf{b}^T \mathbf{u}_1 \mathbf{a}\lambda^{N-1}\mathbf{a}^{-1}\mathbf{b}). \tag{7.26}$$

The trace of a product of matrices does not change under cyclic permutation of the product and thus

$$Z = \text{Trace}(\lambda^{N-1}\mathbf{a}^{-1}\mathbf{b}\mathbf{b}^T \mathbf{u}_1 \mathbf{a}). \tag{7.27}$$

Denote the product $\mathbf{a}^{-1}\mathbf{b}\mathbf{b}^T \mathbf{u}_1 \mathbf{a}$ by **c**. In that case equation (7.27) can be written in terms of the elements of λ and **c** as

$$Z = \lambda_1^{N-1}c_{11} + \lambda_2^{N-1}c_{22} + \lambda_3^{N-1}c_{33} + \cdots. \tag{7.28}$$

Suppose that the eigenvalues $\lambda_1, \lambda_2, \ldots$ have been ordered so that $\lambda_1 > \lambda_2 > \lambda_3 \ldots$. In that case as N becomes very large $(\lambda_2/\lambda_1)^{N-1}$, $(\lambda_3/\lambda_1)^{N-1} \cdots$ approach zero and

$$Z \to c_{11}\lambda_1^{N-1}. \tag{7.29}$$

In evaluating thermodynamic functions $\ln Z$ is required and $\ln c_{11}/\lambda_1$ may be neglected compared to $N \ln \lambda_1$. The log of the partition function thus may be taken for indefinitely long chains as

$$\ln Z = N \ln \lambda_1,$$
$$\lambda_1 = \text{dominant eigenvalue of } \mathbf{u}. \tag{7.30}$$

A similar reduction of the expression for the average of the product of properties for adjacent bonds in a uniform chain may be effected. The expression

$$\langle f_i g_{i+1} h_{i+2} \rangle = \mathbf{b}^T \mathbf{u}_1 \mathbf{u}^{i-2}(\mathbf{fuguh})\mathbf{u}^{N-i-1}\mathbf{b}/Z \tag{7.31}$$

may be written with the aid of the same similarity transformation as above as

$$\langle f_2 g_{i+1} h_{i+2} \rangle = \mathbf{b}^T \mathbf{u}_1 \mathbf{a}\lambda^{i-2}\mathbf{a}^{-1}(\mathbf{fuguh})\mathbf{a}\lambda^{N-i-1}\mathbf{a}^{-1}\mathbf{b}/Z$$

$$= \text{Trace}[(\mathbf{fuguh})\mathbf{a}(\lambda^{N-i-1}\mathbf{c}\lambda^{i-2})\mathbf{a}^{-1}]/Z, \tag{7.32}$$

where $\mathbf{c} = \mathbf{a}^{-1}\mathbf{bb}^T\mathbf{u}_1\mathbf{a}$ as above. The product involving the λ matrices may be expanded as

$$\mathbf{P} = \lambda^{N-i-1}\mathbf{c}\lambda^{i-2} = \begin{pmatrix} \lambda_1^{N-3}c_{11}, & \lambda_1^{N-i-1}\lambda_2^{i-2}c_{12}, \ldots \\ \lambda_2^{N-i-1}\lambda_1^{i-2}c_{21}, & \lambda_2^{N-3}c_{22}, \ldots \\ \vdots & \end{pmatrix}. \tag{7.33}$$

If λ_1 is the largest eigenvalue, then as $N \to \infty$ the above matrix approaches

$$\mathbf{P} = \begin{bmatrix} \lambda_1^{N-3}c_{11} & 0 & 0 & \ldots \\ 0 & 0 & 0 & \ldots \\ 0 & 0 & 0 & \ldots \end{bmatrix}. \tag{7.34}$$

It then follows that the product $\mathbf{aPa}^{-1}$ reduces to

$$\mathbf{aPa}^{-1} = \lambda_1^{N-3}c_{11}\mathbf{a}_1\mathbf{a}_1^*, \tag{7.35}$$

where

$$\mathbf{a}_1 = \begin{pmatrix} a_{11} \\ a_{21} \\ a_{31} \\ \vdots \end{pmatrix}, \quad \text{the first column of } \mathbf{a} \text{ (the eigenvector of } \mathbf{u} \text{ corresponding to } \lambda_1, \text{ the largest eigenvalue)} \quad (7.36)$$

and

$\mathbf{a}_1^* = (a_{11}^*, a_{12}^*, \ldots)$, the row of $\mathbf{a}^{-1}$ corresponding to the column of $\mathbf{a}$ belonging to λ_1.

Using equation (7.35), the desired average in equation (7.32) can be written as,

$$\langle f_i g_{i+1} h_{i+2} \rangle = \text{Trace}(c_{11}\lambda_1^{N-3}\mathbf{a}_1^*\mathbf{fuguha}_1)/Z \quad (7.37)$$

and using equation (7.29) for Z,

$$\langle f_i g_{i+1} h_{i+2} \rangle = \mathbf{a}_1^*\mathbf{fuguha}_1/\lambda^2. \quad (7.38)$$

The above equation forms the basis for computation of average products of properties associated with finite serial sequences of bonds embedded in long chains. The generalization to sequences of lengths other than the three bonds illustrated is apparent. The application to actual chains is taken up later after the subject of setting up the statistical weight matrices, equation (7.17), is considered.

7.2 Averaging of a matrix and matrix products

In Section 6.1.3, the characteristic ratio was formulated in terms of transformation matrices, $\mathbf{t}$. In the independent bond model, the averages of products of the transformation matrices were easily found in terms of averages of the individual matrices. However, in the present case in order to evaluate the characteristic ratio averages of matrices and matrix products $\langle \mathbf{t} \rangle$, $\langle \mathbf{t}_1\mathbf{t}_2 \rangle$, $\langle \mathbf{t}_1\mathbf{t}_2\mathbf{t}_3 \rangle$, etc. will be needed. The meaning of $\langle \mathbf{t}_1\mathbf{t}_2\cdots\mathbf{t}_n \rangle$ being that each element of the final product is a statistical mechanical average, $\langle (\mathbf{t}_1\mathbf{t}_2\cdots\mathbf{t}_n)_{ij} \rangle$, computed according to equation (7.31) or equation (7.38). Since the elements are complicated sums of products obtained on multiplying $\mathbf{t}_1\mathbf{t}_2\mathbf{t}_3\cdots\mathbf{t}_n$, it would be extremely tedious to apply equation (7.31) or equation (7.38) to the averaging of the elements. Therefore a procedure for directly averaging matrix products such as $\mathbf{t}_1\mathbf{t}_2\cdots\mathbf{t}_n$ is needed. This can be accomplished by a method closely analogous to that developed above for averages of products of scalars. The formalism of Lifson (1959) is employed below.

It resulted above that averaging of a scalar involved setting up appropriate matrices whose elements represent the quantities to be averaged. Similarly, matrices are averaged by setting up 'supermatrices' whose elements are matrices. Define a statistical weight 'supermatrix', $\mathbf{U}$,

$$\mathbf{U} = \begin{bmatrix} \mathbf{u} & \mathbf{0} & \mathbf{0} \\ \mathbf{0} & \mathbf{u} & \mathbf{0} \\ \mathbf{0} & \mathbf{0} & \mathbf{u} \end{bmatrix}, \qquad (7.39)$$

where $\mathbf{u}$ is the statistical weight matrix as defined in equation (7.17). If there are r rotational states $(A, B, C \cdots R)$ $\mathbf{u}$ is an $r \times r$ matrix and $\mathbf{U}$ is a $3r \times 3r$ matrix. Each of the $\mathbf{0}$ elements represents an $r \times r$ array of zeros. The 3×3 transformation matrix $\mathbf{t}$ has elements t_{11}, t_{12} etc. defined in Appendix A6.1, equation (A6.5). Set up, in analogy to equation (7.20), a diagonal $r \times r$ matrix, $\mathbf{T}_{ij}$, for each of the elements of $\mathbf{t}$,

$$\mathbf{T}_{11} = \begin{bmatrix} t_{11}(A) & 0 & 0 & \cdots \\ 0 & t_{11}(B) & \cdot & \\ 0 & & \cdot & \\ \cdots & & & \cdot \end{bmatrix},$$

$$\mathbf{T}_{12} = \begin{bmatrix} t_{12}(A) & 0 & 0 & \cdots \\ 0 & t_{12}(B) & \cdot & \\ 0 & & \cdot & \\ \cdots & & & \cdot \end{bmatrix}, \qquad (7.40)$$

etc.

where $t_{11}(A) = \cos \phi_A$, $t_{11}(B) = \cos \phi_B$, $t_{12}(A) = -\sin \phi_A \cos \theta'$, $t_{12}(B) = -\sin \phi_B \cos \theta'$ etc., see equation (A6.5), and $\phi_A, \phi_B, \ldots$ are the values of ϕ in the conformational states $A, B, \ldots$. Now define a $3r \times 3r$ supermatrix,

$$\mathbf{T} = \begin{bmatrix} \mathbf{T}_{11} & \mathbf{T}_{12} & \mathbf{T}_{13} \\ \mathbf{T}_{21} & \mathbf{T}_{22} & \mathbf{T}_{23} \\ \mathbf{T}_{31} & \mathbf{T}_{32} & \mathbf{T}_{33} \end{bmatrix}, \qquad (7.41)$$

whose elements are the diagonal matrices defined in equation (7.40). Properly conformable matrices whose elements are matrices follow the same multiplication rules as matrices whose elements are scalars. Therefore, an analogy to equations (7.20) and (7.21) can be set up, where $\mathbf{T}$ is quantity averaged,

$$\mathbf{UT} = \begin{bmatrix} \mathbf{uT}_{11} & \mathbf{uT}_{12} & \mathbf{uT}_{13} \\ \mathbf{uT}_{21} & \mathbf{uT}_{22} & \mathbf{uT}_{23} \\ \mathbf{uT}_{31} & \mathbf{uT}_{32} & \mathbf{uT}_{33} \end{bmatrix}. \tag{7.42}$$

Pre- and post-multiplying $\mathbf{UT}$ by $\mathbf{U}$ for the bonds flanking that bond, now designated n, for which $\mathbf{T}$ was set up gives a 3×3 supermatrix ($3r \times 3r$ matrix)

$$\mathbf{U}_{n-1}\mathbf{UT}_n\mathbf{U}_{n+1} = \begin{bmatrix} \mathbf{u}_{n-1}(\mathbf{uT}_{11})_n\mathbf{u}_{n+1}, \ \mathbf{u}_{n-1}(\mathbf{uT}_{12})_n\mathbf{u}_{n+1}, \ \mathbf{u}_{n-1}(\mathbf{uT}_{13})_n\mathbf{u}_{n+1} \\ \mathbf{u}_{n-1}(\mathbf{uT}_{21})_n\mathbf{u}_{n+1}, \ \mathbf{u}_{n-1}(\mathbf{uT}_{22})_n\mathbf{u}_{n+1}, \dots \\ \mathbf{u}_{n-1}(\mathbf{uT}_{31})_n\mathbf{u}_{n+1}, \dots, \dots \end{bmatrix}. \tag{7.43}$$

Repeated pre- and post-multiplication by $\mathbf{U}$ generates a 3×3 supermatrix ($3r \times 3r$ matrix),

$$\mathbf{U}_1\mathbf{U}_3 \cdots \mathbf{U}_{n-1}(\mathbf{UT})_n\mathbf{u}_{n+1} \cdots \mathbf{U}_N,$$

whose elements are the $r \times r$ matrices,

$$\mathbf{u}_1\mathbf{u}_2 \cdots \mathbf{u}_{n-1}(\mathbf{uT}_{11})_n\mathbf{u}_{n+1} \cdots \mathbf{u}_N,$$

$$\mathbf{u}_1\mathbf{u}_2 \cdots \mathbf{u}_{n-1}(\mathbf{uT}_{12})_n\mathbf{u}_{n+1} \cdots \mathbf{u}_N, \text{ etc.}$$

Further, define a 3×3 supermatrix ($3r \times 3$ matrix) from $\mathbf{b}$, equation (7.18),

$$\mathbf{B} = \begin{bmatrix} \mathbf{b} & 0 & 0 \\ 0 & \mathbf{b} & 0 \\ 0 & 0 & \mathbf{b} \end{bmatrix} \tag{7.44}$$

and also the transpose, a 3×3 supermatrix ($3 \times 3r$ matrix),

$$\mathbf{B}^{\mathrm{T}} = \begin{bmatrix} \mathbf{b}^{\mathrm{T}} & 0 & 0 \\ 0 & \mathbf{b}^{\mathrm{T}} & 0 \\ 0 & 0 & \mathbf{b}^{\mathrm{T}} \end{bmatrix}.$$

The product

$$\mathbf{B}^{\mathrm{T}}\mathbf{U}_1\mathbf{U}_2 \cdots \mathbf{U}_{n-1}(\mathbf{U}_n\mathbf{T}_n)\mathbf{U}_{n+1}\mathbf{U}_{n+2} \cdots \mathbf{U}_{N-1}\mathbf{U}_N\mathbf{B}$$

produces a 3×3 matrix whose elements are the scalars

$$\mathbf{b}^{\mathrm{T}}\mathbf{u}_1\mathbf{u}_2 \cdots (\mathbf{u}\mathbf{T}_{11})_n \cdots \mathbf{u}_N\mathbf{b}$$

$$\mathbf{b}^{\mathrm{T}}\mathbf{u}_1\mathbf{u}_2 \cdots (\mathbf{u}\mathbf{T}_{12})_n \cdots \mathbf{u}_N\mathbf{b} \text{ etc.}$$

It is clear from equation (7.22) and the definition of $\mathbf{T}_{11}$, $\mathbf{T}_{12}$ etc., equation (7.40), that dividing the above elements by Z gives the average value of those elements

$$\langle t_{11} \rangle = \mathbf{b}^{\mathrm{T}}\mathbf{u}_1\mathbf{u}_2 \cdots (\mathbf{u}\mathbf{T}_{11})_n \cdots \mathbf{u}_N\mathbf{b}/Z$$

$$\langle t_{12} \rangle = \mathbf{b}^{\mathrm{T}}\mathbf{u}_1\mathbf{u}_2 \cdots (\mathbf{u}\mathbf{T}_{12})_n \cdots \mathbf{u}_N\mathbf{b}/Z \text{ etc.}$$

Therefore,

$$\langle \mathbf{t} \rangle = \begin{bmatrix} \langle t_{11} \rangle & \langle t_{12} \rangle & \langle t_{13} \rangle \\ \langle t_{21} \rangle & \langle t_{22} \rangle & \langle t_{23} \rangle \\ \langle t_{31} \rangle & \langle t_{32} \rangle & \langle t_{33} \rangle \end{bmatrix}$$

$$= \mathbf{B}^{\mathrm{T}}\mathbf{U}_1\mathbf{U}_2 \cdots \mathbf{U}_{n-1}(\mathbf{U}_n\mathbf{T}_n)\mathbf{U}_{n+1}\mathbf{U}_{n+2} \cdots \mathbf{U}_{N-1}\mathbf{U}_N\mathbf{B}/Z. \quad (7.45)$$

Products of $\mathbf{t}$ matrices are averaged by setting up the appropriate products of supermatrices; for $\mathbf{t}_n\mathbf{t}_{n+1}$, the product is set up as

$$\mathbf{U}_n\mathbf{T}_n\mathbf{U}_{n+1}\mathbf{T}_{n+1}.$$

From equation (7.42) it is seen that the above multiplication leads to a supermatrix whose elements combine in a manner exactly analogous to the combination of elements from $\mathbf{t}_n$ and $\mathbf{t}_{n+1}$ in the desired product $\mathbf{t}_n\mathbf{t}_{n+1}$. These elements, however, are matrices formed in the proper manner for finding an average and exactly analogous to the product **ufug** in averaging scalars. Continuing to post- and pre-multiply by **U** results in

$$\langle \mathbf{t}_n\mathbf{t}_{n+1} \rangle = \mathbf{B}^{\mathrm{T}}\mathbf{U}_1\mathbf{U}_2 \cdots \mathbf{U}_{n-1}(\mathbf{U}_n\mathbf{T}_n)(\mathbf{U}_{n+1}\mathbf{T}_{n+1})\mathbf{U}_{n+2} \cdots \mathbf{U}_N\mathbf{B}/Z. \quad (7.46)$$

The above equation establishes the procedure for finding the average of a product of matrices. The generalization to products for other numbers of bonds is apparent.

Recall that the partition function, equation (7.18), and the expression for the average of a product of scalar properties of adjacent bonds, equation (7.31), could be reduced for an indefinitely long chain of uniform bonds to simple expressions involving the largest eigenvalue and its eigenvector, equations (7.30) and (7.38), of the statistical weight matrix. Equation (7.46) can also be reduced to simpler forms under these circumstances. If the following supermatrices are defined from the eigenvectors $\mathbf{a}_1$ ($r \times 1$ column matrix) and $\mathbf{a}_1^*$ ($1 \times r$ row matrix) in equation (7.38),

$$\mathbf{A} = \begin{bmatrix} \mathbf{a}_1 & \mathbf{0} & \mathbf{0} \\ \mathbf{0} & \mathbf{a}_1 & \mathbf{0} \\ \mathbf{0} & \mathbf{0} & \mathbf{a}_1 \end{bmatrix}, \quad 3r \times 3 \text{ matrix}$$

$$\mathbf{A}^* = \begin{bmatrix} \mathbf{a}_1^* & \mathbf{0} & \mathbf{0} \\ \mathbf{0} & \mathbf{a}_1^* & \mathbf{0} \\ \mathbf{0} & \mathbf{0} & \mathbf{a}_1^* \end{bmatrix}, \quad 3 \times 3r \text{ matrix}.$$

(7.47)

An argument paralleling that leading to equation (7.38) gives for the matrix product averages in long chains the results,

$$\langle \mathbf{t}_n \mathbf{t}_{n+1} \rangle = \mathbf{A}^* \mathbf{TUTA}/\lambda,$$
$$\langle \mathbf{t}_n \mathbf{t}_{n+1} \mathbf{t}_{n+2} \rangle = \mathbf{A}^* \mathbf{TUTUTA}/\lambda^2, \text{ etc.}$$

(7.48)

7.3 The characteristic ratio of long uniform chains

Following the development of Sections 6.1.2 and 6.1.3, equation (6.3) can be written as,

$$\langle R^2 \rangle = \sum l_i^2 + 2 \sum_{i=1}^{N-1} \sum_{j=i+1}^{N} \langle \mathbf{l}_i \cdot \mathbf{l}_j \rangle \tag{7.49}$$

Using the transformation matrices $\mathbf{t}$ of the previous section, the above equation can be written in matrix form

$$\langle R^2 \rangle = \sum l_i^2 + 2 \sum_{i=1}^{N-1} \sum_{j=i+1}^{N} \mathbf{l}_i \langle \mathbf{t}^{j-i} \rangle \mathbf{l}_j, \tag{7.50}$$

where, as in Section 6.1.3, $\mathbf{t}^{j-i}$ is the serial product of matrices, $\mathbf{t}(i)\mathbf{t}(i+1)\cdots\mathbf{t}(j-2)\mathbf{t}(j-1)$ and $\mathbf{l}_i$, $\mathbf{l}_j$, the bond vectors, are expressed in the local bond coordinate system of each. Equation (7.50) is valid for all chain lengths and bond compositions. However, if the chain is of uniform composition all of the statistical weight matrices will be identical and if the chain is indefinitely long ($N \to \infty$) the averages, $\langle \mathbf{t}^{j-i} \rangle$, will become independent of the value of i and depend only on the magnitude $|j - i| = m$ at large N. Replacing the sum over i by multiplying by N, equation (7.50) becomes

$$\langle R^2 \rangle = Nl^2 + 2N(\mathbf{l}^{\mathrm{T}}\langle \mathbf{t} \rangle \mathbf{l} + \mathbf{l}^{\mathrm{T}}\langle \mathbf{t}^2 \rangle \mathbf{l} + \mathbf{l}^{\mathrm{T}}\langle \mathbf{t}^3 \rangle \mathbf{l} + \cdots). \tag{7.51}$$

Using equation (7.48) for the long chain limit values for the averages $\langle t^m \rangle$ gives

$$\langle R^2 \rangle = Nl^2 + 2Nl^T A^* \left(T + \frac{TUT}{\lambda} + \frac{TUTUT}{\lambda^2} + \cdots \right) Al. \quad (7.52)$$

The series in the brackets is summed as a geometric series and there results (Lifson, 1959),

$$\langle R^2 \rangle = Nl^2 + 2Nl^2 l^T A^* T (1 - UT/\lambda)^{-1} Al, \quad (7.53)$$

where **1** is the $3r \times 3r$ identity matrix.

7.4 The characteristic ratio of finite chains

As observed, equation (7.50) is generally valid. Using equation (7.46) for $\langle t^{j-i} \rangle$ and substituting into equation (7.50), $\langle R^2 \rangle$ for a general chain can be expressed as

$$\langle R^2 \rangle = \sum l_i^2 + 2 \sum_{i=1}^{N-1} \sum_{j=i+1}^{N} l_i^T B^T U_1 U_2 \cdots U_{i-1} (UT)^{j-i} U_j \cdots U_N Bl_j / Z, \quad (7.54)$$

where

$$(UT)^{j-i} = (U_i T_i)(U_{i+1} T_{i+1}) \cdots (U_{j-2} T_{j-2})(U_{j-1} T_{j-1})$$

and

$$Z = b^T u_1 u_2 \cdots u_N b.$$

Direct evaluation of the double sum would be tedious. However, just as the evaluation of the partition function itself and the averages of scalars and matrices was shown above to be amenable to computation by matrix multiplication, so can the double sum over the averaged matrix products be collected by further matrix operations as shown by Flory (Flory, 1964; Flory and Jernigan, 1965; Flory, 1969) whose formalism is followed below but with the Lifson (1959) definitions of the **U** and **T** supermatrices. In accomplishing this, first some alterations to equation (7.54) are made. First it is observed that the serial products of statistical weight supermatrices, $U_1 \cdots U_{i-1}$ and $U_j \cdots U_N$ can be reduced to the supermatrix of the product of the statistical weight matrices, **u**, themselves. To demonstrate this it is convenient to use an operator for creating the supermatrices. This can be accomplished through the *direct product* operation,

$$c = a \otimes b, \quad (7.55)$$

which has the meaning that each element of $\mathbf{a}$, a_{ij} is replaced by $a_{ij}\mathbf{b}$. Thus the $\mathbf{U}$ supermatrix of equation (7.39) can be created by the operation, $\mathbf{e}_3 \otimes \mathbf{u}$ where $\mathbf{e}_3$ is the 3×3 identity matrix. Products of direct products have the property $(\mathbf{a} \otimes \mathbf{b})(\mathbf{c} \otimes \mathbf{d}) = (\mathbf{ac}) \otimes (\mathbf{bd})$ when the matrices are properly conformable. Therefore the product of two supermatrices, $\mathbf{U}_1\mathbf{U}_2$, for example, is

$$\mathbf{U}_1\mathbf{U}_2 = (\mathbf{e}_3 \otimes \mathbf{u}_1)(\mathbf{e}_3 \otimes \mathbf{u}_1) = (\mathbf{e}_3\mathbf{e}_3) \otimes (\mathbf{u}_1\mathbf{u}_2) = \mathbf{e}_3 \otimes (\mathbf{u}_1\mathbf{u}_2).$$

Iteratively it results that

$$\mathbf{U}_1 \cdots \mathbf{U}_{i-1} = \mathbf{e}_3 \otimes (\mathbf{u}_1 \cdots \mathbf{u}_{i-1})$$
$$\mathbf{U}_j \cdots \mathbf{U}_N = \mathbf{e}_3 \otimes (\mathbf{u}_j \cdots \mathbf{u}_N).$$

Continuing

$$\mathbf{B}^\mathrm{T}\mathbf{U}_1 \cdots \mathbf{U}_{i-1} = \mathbf{e}_3 \otimes (\mathbf{b}^\mathrm{T}\mathbf{u}_1 \cdots \mathbf{u}_{i-1}) \tag{7.56}$$

and

$$\mathbf{U}_j \cdots \mathbf{U}_N\mathbf{B}^\mathrm{T} = \mathbf{e}_3 \otimes (\mathbf{u}_j \cdots \mathbf{u}_N\mathbf{b}). \tag{7.57}$$

Next, the positions of the vectors $\mathbf{l}_i^\mathrm{T}, \mathbf{l}_j$ are brought from their terminal positions in equation (7.54) to positions flanking the $(\mathbf{UT})^{j-i}$ product. It can be verified by direct expansion that

$$\mathbf{l}_i^\mathrm{T}(\mathbf{e}_3 \otimes (\mathbf{b}^\mathrm{T}\mathbf{u}_1 \cdots \mathbf{u}_{i-1})) = (\mathbf{b}^\mathrm{T}\mathbf{u}_1 \cdots \mathbf{u}_{i-1})(\mathbf{l}_i^\mathrm{T} \otimes \mathbf{e}_r) \tag{7.58}$$

and

$$\mathbf{e}_3 \otimes (\mathbf{u}_j \cdots \mathbf{u}_N\mathbf{b}^\mathrm{T})\mathbf{l}_j = (\mathbf{l}_j \otimes \mathbf{e}_r)(\mathbf{u}_j \cdots \mathbf{u}_N\mathbf{b}^\mathrm{T}), \tag{7.59}$$

where $\mathbf{e}_r$ is the $r \times r$ identity matrix (r rotational states). The matrices $\mathbf{l}_i^\mathrm{T} \otimes \mathbf{e}_r$, $\mathbf{l}_j \otimes \mathbf{e}_r$ are supermatrices with a correspondence to $\mathbf{l}_i^\mathrm{T}$ and $\mathbf{l}_j$ similar to that with $\mathbf{b}$, $\mathbf{B}$ and also $\mathbf{b}^\mathrm{T}$, $\mathbf{B}^\mathrm{T}$. It is convenient then to define two supermatrices, $\mathbf{L} = \mathbf{l} \otimes \mathbf{e}_r$, $\mathbf{L}^\mathrm{T} = \mathbf{l}^\mathrm{T} \otimes \mathbf{e}_r$, such that

$$\mathbf{L} = \begin{bmatrix} \mathbf{l}_1 \\ \mathbf{l}_2 \\ \mathbf{l}_3 \end{bmatrix}, \qquad \mathbf{L}^\mathrm{T} = [\mathbf{l}_1 \ \ \mathbf{l}_2 \ \ \mathbf{l}_3], \tag{7.60}$$

where $\mathbf{l}_1, \mathbf{l}_2, \mathbf{l}_3$ are $r \times r$ diagonal matrices with the designated components l_1, l_2, l_3 of $\mathbf{l}$ repeated on the diagonal, e.g.,

$$\mathbf{l} = \begin{bmatrix} l_1 & 0 & 0 & \cdots \\ 0 & l_1 & & \\ 0 & & l_1 & \\ 0 & \cdots & & l_1 \end{bmatrix} \quad \text{etc.}$$

It is appropriate to comment on the inclusion of components l_1, l_2 in equation (7.60) since according to the local coordinate system convention, Appendix A6.1, the local bond vector is written as $(0, 0, l)$. The complete components are included in anticipation that the resulting formalism for $\langle R^2 \rangle$, where $\mathbf{R} = \sum \mathbf{l}_i$, may be applied to the mean-square value of a sum of any vectors associated with local groups. The prominent example is that of the mean-square dipole moment of a molecule arising from the sum of local moments. The local dipole moments may not be directed along bonds but are nevertheless associated with the local coordinate system. The pendant C—Cl bonds in polyvinyl chloride would be a typical example.

With the above definitions of $\mathbf{L}$, $\mathbf{L}^T$ placed into equations (7.58) and (7.59) which are, in turn, inserted into equation (7.54) via equations (7.56) and (7.57), $\langle R^2 \rangle$ can now be expressed as

$$\langle R^2 \rangle = \sum l_i^2 + 2 \sum_{i=1}^{N-1} \sum_{j=i+1}^{N} \mathbf{b}^T \mathbf{u}_1 \mathbf{u}_2 \cdots \mathbf{u}_{i-1} \mathbf{L}_i^T (\mathbf{UT})^{j-i} \mathbf{L}_j \mathbf{u}_j \mathbf{u}_{j+1} \cdots \mathbf{u}_N \mathbf{b}/Z.$$

(7.61)

The evaluation of $\langle R^2 \rangle$ via equation (7.61) is found by means of a *generator matrix* that on repeated multiplication creates the double sum. For the generator matrix write,

$$\mathbf{G}_i = \begin{bmatrix} \mathbf{u}_i & \mathbf{L}_i^T(\mathbf{UT})_i & 0 \\ 0 & (\mathbf{UT})_i & \mathbf{L}_i \mathbf{u}_i \\ 0 & 0 & \mathbf{u}_i \end{bmatrix}.$$

(7.62)

The $\mathbf{0}$ elements are matrices of orders appropriate to maintain conformability and thus $\mathbf{G}_i$ is a $5r \times 5r$ square matrix. Set up the row $[\mathbf{1}\ \mathbf{0}\ \mathbf{0}]$ where $\mathbf{1}$ is the $r \times r$ $\mathbf{e}_r$ identity matrix and perform the operation

$$[\mathbf{1}\ \ \mathbf{0}\ \ \mathbf{0}]\mathbf{G}_1$$

$$= [\mathbf{1}\ \ \mathbf{0}\ \ \mathbf{0}] \begin{bmatrix} \mathbf{u}_1 & \mathbf{L}_1^T(\mathbf{UT})_1 & 0 \\ 0 & (\mathbf{UT})_1 & \mathbf{L}_1 \mathbf{u}_1 \\ 0 & 0 & \mathbf{u}_1 \end{bmatrix}$$

$$= [\mathbf{u}_1,\ \mathbf{L}_1^T(\mathbf{UT})_1, 0].$$

Follow this by multiplying from the right by $\mathbf{G}_2$ or

$$[\mathbf{1}\ \ \mathbf{0}\ \ \mathbf{0}]\mathbf{G}_1\mathbf{G}_2 = [\mathbf{u}_1\mathbf{u}_2,\ \mathbf{u}_1\mathbf{L}_2^T(\mathbf{UT})_2 + \mathbf{L}_1^T(\mathbf{UT})_1(\mathbf{UT})_2,\ \mathbf{L}_1^T(\mathbf{UT})_1\mathbf{L}_2\mathbf{u}_2]$$

and continuing,

$$[1 \quad 0 \quad 0]G_1G_2G_3 = [\mathbf{u}_1\mathbf{u}_2\mathbf{u}_3, \ \mathbf{u}_1\mathbf{u}_2\mathbf{L}_3^T(\mathbf{UT})_3 + \mathbf{u}_1\mathbf{L}_2^T(\mathbf{UT})_2(\mathbf{UT})_3$$
$$+ \ \mathbf{L}_1^T(\mathbf{UT})_1(\mathbf{UT})_2(\mathbf{UT})_3, \ \mathbf{u}_1\mathbf{L}_2^T(\mathbf{UT})_2\mathbf{L}_3\mathbf{u}_3$$
$$+ \ \mathbf{L}_1^T(\mathbf{UT})_1(\mathbf{UT})_2\mathbf{L}_3\mathbf{u}_3 + \mathbf{L}_1^T(\mathbf{UT})_1\mathbf{L}_2\mathbf{u}_2\mathbf{u}_3].$$

It may be seen that the desired sum accumulates in the third element of the product. Thus on completion of the requisite number of multiplications of $\mathbf{G}_i$, up to $i = N$, the sum can be collected by multiplication from the right by $[0 \ 0 \ 1]^T$. The final expression for $\langle R^2 \rangle$ from equation (7.61) then becomes (Flory, 1969)

$$\langle R^2 \rangle = \sum l_i^2 + 2\mathbf{b}^T[1 \quad 0 \quad 0]G_1G_2 \cdots G_{N-1}G_N[0 \quad 0 \quad 1]^T\mathbf{b}/Z. \quad (7.63)$$

The above equation forms the basis for the calculation of average moments of finite chains. It is also to be emphasized that, since each $\mathbf{G}_i$ and the matrices contained therein can be defined separately with different conformational parameters in each, chains of non-uniform chemical structure may be treated. In its use the remarks accompanying equations (7.16) and (7.18) concerning the bonds at the ends of the chain are to be kept in mind.

A summary of the matrices introduced is presented in Table 7.1.

7.5 Illustrative calculations, the three-state chain

As already developed in Chapters 5 and 6, the three state chain, containing trans bonds and the two gauche conformations of opposite sense is an important prototype for flexible polymers. The application of the interacting bond model to chains of this type is demonstrated here by way of illustration.

7.5.1 Statistical weight matrices

The appropriate matrices are set up here for the evaluation of averages for a chain with three rotational states designated as T, G^+ and G^- that has interactions between neighboring bonds. The statistical weight matrix, equation (7.17) in this instance is,

$$\mathbf{u} = \begin{bmatrix} u^{TT} & u^{TG^+} & u^{TG^-} \\ u^{G^+T} & u^{G^+G^+} & u^{G^+G^-} \\ u^{G^-T} & u^{G^-G^+} & u^{G^-G^-} \end{bmatrix}. \quad (7.64)$$

Table 7.1 *Summary of matrices used in calculation of* $\langle R^2 \rangle$.

$\mathbf{b} = \begin{bmatrix} 1 \\ 1 \\ \vdots \\ 1 \end{bmatrix}, \quad \mathbf{b}^{\mathrm{T}} = (1\ 1 \cdots 1),$	$r \times 1$ and $1 \times r$ matrices used to collect elements of a product of statistical weight ($r \times r$) matrices into a single element (partition function).
$\mathbf{B} = \begin{bmatrix} \mathbf{b} & 0 & 0 \\ 0 & \mathbf{b} & 0 \\ 0 & 0 & \mathbf{b} \end{bmatrix},$	3×3 supermatrix ($3r \times 3$ matrix) made up of $\mathbf{b}$ matrices.
$\mathbf{B}^{\mathrm{T}} = \begin{bmatrix} \mathbf{b}^{\mathrm{T}} & 0 & 0 \\ 0 & \mathbf{b}^{\mathrm{T}} & 0 \\ 0 & 0 & \mathbf{b}^{\mathrm{T}} \end{bmatrix},$	3×3 supermatrix ($3 \times 3r$ matrix) made up of $\mathbf{b}^{\mathrm{T}}$ matrices.
$\mathbf{u} = \begin{bmatrix} u_{\mathrm{AA}} & u_{\mathrm{AB}} & \cdots & u_{\mathrm{Ar}} \\ u_{\mathrm{BA}} & u_{\mathrm{BB}} & \cdots & u_{\mathrm{Br}} \\ \vdots & & & \\ u_{r\mathrm{A}} & \cdots & & u_{rr} \end{bmatrix},$	$r \times r$ statistical weight matrix for a bond that has r rotational states. $u_{\mathrm{AB}} = \mathrm{e}^{-E_{\mathrm{AB}}/kT}$, $E_{\mathrm{AB}} = $ energy of a bond in its B rotational state when the preceding bond is in its A rotational state.
$\mathbf{U} = \begin{bmatrix} \mathbf{u} & 0 & 0 \\ 0 & \mathbf{u} & 0 \\ 0 & 0 & \mathbf{u} \end{bmatrix},$	3×3 supermatrix ($3r \times 3r$ matrix) made up of $\mathbf{u}$ matrices on diagonal.
$\mathbf{t},$	3×3 transformation matrix (Appendix A6.1).
$\mathbf{T}_{ij} = \begin{bmatrix} t_{ij}(\mathrm{A}) & 0 & 0 & 0 \\ 0 & t_{ij}(\mathrm{B}) & & \\ & & \ddots & \\ 0 & 0 & 0 & t_{ij}(r) \end{bmatrix},$	$r \times r$ diagonal matrix made up of values of one of the elements of $\mathbf{t}$, one for each rotational state. $t_{11}(\mathrm{A}) = \cos \phi_{\mathrm{A}},$ $t_{11}(\mathrm{B}) = \cos \phi_{\mathrm{B}}$, etc. $t_{12}(\mathrm{A}) = -\sin \phi_{\mathrm{A}} \cos \theta',$ $t_{12}(\mathrm{B}) = -\sin \phi_{\mathrm{B}} \cos \theta'$, etc. according to equation (A6.5).
$\mathbf{T} = \begin{bmatrix} \mathbf{T}_{11} & \mathbf{T}_{12} & \mathbf{T}_{13} \\ \mathbf{T}_{21} & \mathbf{T}_{22} & \mathbf{T}_{23} \\ \mathbf{T}_{31} & \mathbf{T}_{32} & \mathbf{T}_{33} \end{bmatrix},$	3×3 supermatrix ($3r \times 3r$ matrix) made up of $\mathbf{T}_{ij}$ matrices.
$\mathbf{a},$	$r \times r$ matrix that diagonalizes statistical weight matrix in a similarity transformation, $\mathbf{a}^{-1}\mathbf{u}\mathbf{a} = \boldsymbol{\lambda}$.

Table 7.1 (*continued*)

$\lambda = \begin{bmatrix} \lambda_1 & 0 & & \\ 0 & \lambda_2 & & \\ & & \ddots & \\ 0 & & & \lambda_r \end{bmatrix}$,	eigenvalues of **u** ordered such that λ_1 is largest.
$\mathbf{a}_1 = \begin{bmatrix} a_{11} \\ a_{21} \\ \vdots \\ a_{r1} \end{bmatrix}$,	eigenvector of the largest eigenvalue, λ_1, of the statistical weight matrix, **u**; used to find average of scalar property of infinite uniform chain.
$\mathbf{a}_1^* = (a_{11}^*, a_{12}^* \cdots a_{1r}^*)$,	row of $\mathbf{a}^{-1}$ corresponding to $\mathbf{a}_1$ used to find average of scalar property for infinite uniform chain.
$\mathbf{A} = \begin{bmatrix} \mathbf{a}_1 & \mathbf{0} & \mathbf{0} \\ \mathbf{0} & \mathbf{a}_1 & \mathbf{0} \\ \mathbf{0} & \mathbf{0} & \mathbf{a}_1 \end{bmatrix}$,	3×3 supermatrix ($3r \times 3$ matrix) made up of $\mathbf{a}_1$ on diagonal; used to find averages of matrix products for infinite chain.
$\mathbf{A}^* = \begin{bmatrix} \mathbf{a}_1^* & \mathbf{0} & \mathbf{0} \\ \mathbf{0} & \mathbf{a}_1^* & \mathbf{0} \\ \mathbf{0} & \mathbf{0} & \mathbf{a}_1^* \end{bmatrix}$,	3×3 supermatrix ($3 \times 3r$ matrix) made up of $\mathbf{a}_1^*$ on diagonal; used to find averages of matrix products for infinite chain.

Finite chain matrices

$\mathbf{G}_i$,	generator matrix ($5r \times 5r$) for summing the terms in the expansion of $\langle R^2 \rangle = \sum_i \sum_j \mathbf{l}_i \cdot \mathbf{l}_j$; defined in equation (7.62).
$\mathbf{L} = \begin{bmatrix} \mathbf{l}_1 \\ \mathbf{l}_2 \\ \mathbf{l}_3 \end{bmatrix}$, $\mathbf{L}^T = [\mathbf{l}_1 \ \mathbf{l}_2 \ \mathbf{l}_3]$,	supermatrices, $3r \times r$, $r \times 3r$ matrices, made up of matrices representing the components of a bond vector, **l**.
$\mathbf{l}_1 = \begin{bmatrix} l_1 & 0 & 0 & \cdots \\ 0 & l_1 & & \\ 0 & & l_1 & \\ & & \cdots & \\ 0 & \cdots & & l_1 \end{bmatrix}$ etc.,	$r \times r$ diagonal matrix with one of the components of a bond vector repeated on the diagonal.

The conformational energies are measured relative to the TT state of $E_{TT} = 0$, $u_{TT} = 1$. The energy of a gauche bond is considered to be that of the bond when preceded by a trans bond. It is designated as E_{TG} and thus $u_{TG^+} = u_{TG^-} = e^{-E_{TG}/k_B T}$. It is often found for three state chains that except for the 'pentane interference' effect (Pitzer, 1940) occurring for G^+G^- pairs, Section 6.1.4, the conformational energy of a gauche bond is nearly independent of its neighboring bond environment. For present purposes, the approximation that $E_{TG} = E_{G^+G^+} = E_{G^-G^-} = E_G$ will be invoked and the weight for all taken as $g = e^{-E_G/k_B T}$. The case of the steric interference in $G^-G^+ = G^+G^-$ pairs is now parameterized. In order to preserve the concept that the conformational states of bonds considered above have a basic utility and are only *influenced* by their neighbors, $u_{G^+G^-} = u_{G^-G^+}$ are written as the product of two factors, one equal to g (for TG $= G^+G^+ = G^-G^-$) and the other, $\omega = e^{-E_\omega/k_B T}$, based on E_ω, the 'excess pentane interference energy', that corrects for the influence of a preceding neighbor that is gauche of the opposite sense. Therefore, $u_{G^+G^-} = u_{G^-G^+} = g\omega = (e^{-E_G/k_B T})(e^{-E_\omega/k_B T})$. In this scheme, the statistical weight matrix becomes (Flory, 1969)

$$\mathbf{u} = \begin{bmatrix} 1 & g & g \\ 1 & g & g\omega \\ 1 & g\omega & g \end{bmatrix}. \tag{7.65}$$

In applying the infinite chain results of Section 7.1 and 7.2, the largest eigenvalue and its associated eigenvector of $\mathbf{u}$ are needed. The characteristic equation of $\mathbf{u}$,

$$|\mathbf{u} - \lambda| = 0$$

is

$$\begin{vmatrix} 1 - \lambda & g & g \\ 1 & g - \lambda & g\omega \\ 1 & g\omega & g - \lambda \end{vmatrix} = 0$$

or

$$(1 - \lambda)[(g - \lambda)^2 - g^2\omega^2] - 2g(g - \lambda) + 2g^2\omega = 0. \tag{7.66}$$

By inspection a solution

$$\lambda_3 = g(1 - \omega)$$

is found or readily verified by substitution. Using this solution to factor equation (7.66) the other two roots are found to be,

$$\lambda_{1,2} = \tfrac{1}{2}(1 + g(\omega + 1) \pm \{[1 - g(1 + \omega)]^2 + 8g\}^{\tfrac{1}{2}}). \qquad (7.67)$$

The largest root is the one with the plus sign and is numbered λ_1. The eigenvectors are found to be

$$
\begin{array}{ccc}
\mathbf{a}_1 & \mathbf{a}_2 & \mathbf{a}_3
\end{array}
$$

$$
\mathbf{a} = \begin{bmatrix}
\lambda_1 - g(1 + \omega) & \lambda_2 - g(1 + \omega) & 0 \\
1 & 1 & -1 \\
1 & 1 & 1
\end{bmatrix} \qquad (7.68)
$$

The first column contains the vector associated with the largest eigenvalue, λ_1. In order to find the eigenvector $\mathbf{a}_1^*$ of $\mathbf{a}^{-1}$ it is not necessary to invert $\mathbf{a}$ formally. Observe that from equation (7.24)

$$(\mathbf{a}^{-1}\mathbf{u}\mathbf{a})^{\mathrm{T}} = \lambda^{\mathrm{T}}$$

$$\mathbf{a}^{\mathrm{T}}\mathbf{u}^{\mathrm{T}}\mathbf{a}^{-1\mathrm{T}} = \lambda$$

Thus the matrix, $\boldsymbol{\alpha}$, that diagonalizes $\mathbf{u}^{\mathrm{T}}$ as $\boldsymbol{\alpha}^{-1}\mathbf{u}^{\mathrm{T}}\boldsymbol{\alpha}$ is the same as $\mathbf{a}^{-1\mathrm{T}}$ and therefore $\mathbf{a}^{-1} = \boldsymbol{\alpha}^{\mathrm{T}}$. Further, the latter matrix has the same eigenvalues as $\mathbf{u}$. The eigenvector of $\mathbf{u}^{\mathrm{T}}$ corresponding to λ_1 is

$$a_1^* = -\frac{1}{2(\lambda_1 - \lambda_2)}(2, \lambda_1 - 1, \lambda_1 - 1). \qquad (7.69)$$

7.5.2 Bond conformational populations

As an application of the averaging methods developed above, it is of interest to calculate the fractions of bonds in the three-state chain that are *trans* and *gauche*. In a finite chain the probability that a given bond is in one of the specific conformational states regardless of the conformation of other bonds will depend on the position of the bond in the chain. In the infinite chain all bonds are equivalent. The fraction of bonds for the entire chain in a given conformational state is the same as the probability of a given bond being in one of the states. To find the probability of a bond being in one of its states, the matrices of equations (7.20) or (7.22) are utilized,

$$
\mathbf{f} = \begin{bmatrix}
f_1 & 0 & \\
0 & f_2 & \\
& & \ddots
\end{bmatrix},
$$

Table 7.2 *Calculated fraction of bonds in the trans state in polyethylene at 137°C.*

model	fraction
(a) 'free rotation' ($\omega = 1$, $E_G = 0$, $g = 1$)	0.333
(b) independent bonds, gauche/trans energy difference	0.510
$\quad \omega = 1$, $E_G = 600\,\text{cal/mol}$ (2500 J/mol)	
(c) G^+G^- sequences suppressed	0.606
$\quad \omega = 0.202$, $E_\omega = 1300\,\text{cal/mol}$ (5449 J/mol); same E_G as (b)	
(d) G^+G^- sequences forbidden	0.628
$\quad \omega = 0$, same E_G as (b)	

where $f_1, f_2 \cdots$ is a property of the bond in each of its conformational states $1, 2, \ldots$. Designate $f_1 = $ a property in T state, f_2 in G^+ and f_3 in G^-. If the property to be represented is whether or not the bond is in the *trans* state, then it is appropriate to take $f_1 = 1$, $f_2 = f_3 = 0$, and thus

$$\mathbf{f} = \begin{bmatrix} 1 \\ 0 \\ 0 \end{bmatrix}. \tag{7.70}$$

The probability that a bond is in the *trans* state regardless of the conformations of other bonds is found by averaging over all of the states of all of the bonds, or from equation (7.38) for the infinite chain,

$$\text{fraction trans bonds} = \langle f_T \rangle = \mathbf{a}_1^* \mathbf{f} \mathbf{a}_1. \tag{7.71}$$

Using equations (7.68) and (7.69) it is found that,

$$\langle f_T \rangle = (1 - \lambda_2)/(\lambda_1 - \lambda_2). \tag{7.72}$$

The factors λ_1, λ_2 are evaluated from equation (7.67) using the values for the statistical weight matrix parameters $g = \text{e}^{-E_G/k_B T}$ and $\omega = \text{e}^{-E_\omega/k_B T}$. Using polyethylene as the example, Table 7.2 shows values of the fraction of bonds in *trans* state for a reasonable choice of E_g and two choices of ω. It may be seen that for this example the energy difference between *gauche* and *trans* states causes the obvious enhancement of population of *trans* states over that for equal energy states. The suppression or exclusion of G^+G^- sequences further reduces the probability of *gauche* states and this results in a further although not dramatic enhancement of the *trans* population.

7.5.3 *The characteristic ratio of the three-state chain*

The characteristic ratio of the infinite chain may be calculated from equation (7.53). The matrices $\mathbf{a}_1$, $\mathbf{a}_1^*$ in equations (7.68) and (7.69) are used to set up their supermatrix counterparts, $\mathbf{A}$, $\mathbf{A}^*$, equation (7.47). The statistical weight supermatrix, $\mathbf{U}$, is set up by entering equation (7.65) for $\mathbf{u}$ along the diagonals. The elements of the transformation supermatrix, $\mathbf{T}$, $\mathbf{T}_{ij}$, equation (7.41), are filled in with the elements of the transformation matrix, equation (A6.5), as

$$
\mathbf{T}_{11} = \begin{bmatrix} \cos(\phi_T) & 0 & 0 \\ 0 & \cos(\phi_{G^+}) & 0 \\ 0 & 0 & \cos(\phi_{G^-}) \end{bmatrix},
$$

$$
\mathbf{T}_{12} = \begin{bmatrix} -\sin(\phi_T)\cos(\theta_T') & 0 & 0 \\ 0 & -\sin(\phi_{G^-})\cos(\theta_{G^+}') & 0 \\ 0 & 0 & -\sin(\phi_{G^-})\cos(\theta_{G^-}') \end{bmatrix}
$$

$$
\text{etc.} \quad (7.73)
$$

where ϕ_T, ϕ_{G^+}, ϕ_{G^-}; θ_T', θ_{G^+}', θ_{G^-}' are the values of the torsional angle, ϕ, and the complement of the valence angle, θ' for the three states of the chain.

Figure 7.1 shows the characteristic ratio plotted against the *gauche* energy E_G divided by RT for various values of the excess G^+G^- energy E_ω/RT. For increasingly positive values of E_G (gauche less stable than trans) the coil expands towards the planar zig-zag limit. It may be observed that the suppression and finally near total exclusion of G^+G^- sequences through E_ω has a decided influence on C_∞. At a given value of E_G the suppression of G^+G^- sequences causes an expansion through depopulation of *gauche* states. For negative values of E_G (gauche more stable than trans) the size of the coil is especially sensitive to G^+G^- suppression. Without this effect a random mixture of G^+ and G^- without trans bonds (large negative E_G) becomes very small due to the ring-like nature of G^+G^- sequences. The suppression of G^+G^- pairs forces the chain into long sequences of $G^+G^+G^+$ or $G^-G^-G^-$ and it approaches the all G 2_1 helix at large negative E_G and $E_\omega = \infty$.

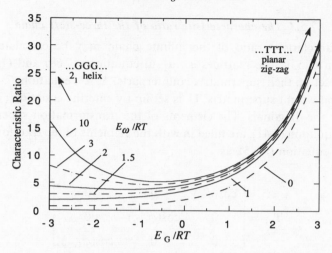

Fig. 7.1 Characteristic ratio of three-state chain plotted vs gauche energy (divided by RT). Different curves are for different values of the excess G^+G^- interference energy, E_ω (divided by RT). The values $\phi_T = 180°$, $\phi_{G^+} = 60°$, $\phi_{G^-} = 300°$ are used for the torsional angles and the tetrahedral value of $109.47°$ for the valence angle of all three states.

7.6 Some examples of calculations on actual chains

The ability of the statistical methods to reproduce experimentally measured theta solvent dimensions is illustrated by a few examples. In doing so, some methodology needed for calculations on chains containing chiral centers will also be introduced.

7.6.1 The characteristic ratio of polyethylene

Polyethylene is an important test for the methods developed above for calculating the average size of a coiled macromolecule. This is because of the simplicity of its conformational features and because considerable information is available from lower molecular weight hydrocarbons concerning the values of the appropriate conformational energy parameters and, of course, because the characteristic ratio has been experimentally determined. In addition to selecting numerical values of the two energy parameters, E_G and E_ω, there are some refinements to the three-state chain calculations carried out in the previous section (Figure 7.1) that are appropriate. These concern the numerical values of the torsional angles and also the valence angles. 'Ideal' values of these were used in Figure 7.1.

The actual values are sufficiently different to warrant adjustment. The valence angle can be inferred from electron diffraction experiments on homologs where a value of $112.6 \pm 0.3°$ has been derived (Bartell and Kohl, 1963). It may also be obtained from the 'c' dimension of the polyethylene unit cell if a value for the C—C bond length is known. Numerous studies show the latter varies but little in various alkanes and may be taken to be 1.533 ± 0.005 Å (Bartell and Kohl, 1963). This leads to $\theta = 111.5 \pm 0.5°$ (for $c = 2.534$ Å, Miller (1989)). The value $\theta = 112°$ can be adopted as consistent with both methods. The torsional angle of the *gauche* states ϕ_G is not known precisely experimentally but estimates can be made from conformational energy calculations using empirical energy functions. Calculations using different sets of energy functions place its value at 67.5° (Abe, Jernigan and Flory, 1966; Boyd and Breitling, 1972b). The characteristic ratio is fairly sensitive to this parameter but the value is probably reliable to $\pm 2°$. Turning to the energy parameters, the *gauche* energy has been determined from the intensities of temperature sensitive Raman vibrational bonds in several homologs (Szasz *et al.*, 1948; Sheppard and Szasz, 1949; Verma *et al.*, 1974). The values 3200 ± 300 J/mol (butane), 1900 ± 250 J/mol (pentane), and 2100 ± 300 J/mol (hexane) have been found. Electron diffraction of alkanes gives the value 2550 J/mol (Bartell and Kohl, 1963). Conformational energy calculations give values in the neighborhood of 2500 J/mol (Abe *et al.*, 1966; Boyd and Breitling, 1972b). Thus this parameter may be regarded as lying in the range ~ 1700–3000 J/mol. The excess G^+G^- energy must be inferred from conformational energy calculations. A value in the vicinity of $E_\omega = 5400$ J/mol is found (Boyd and Breitling, 1972b). In Figure 7.2 a plot is shown that is similar to that of Figure 7.1 but greatly blown up in the region of appropriate energy parameters and with the torsional and valence angle values selected above. It may be seen that a value of E_G at the lower end of the supposed range, 1700 J/mol, along with $E_\omega = 5400$ J/mol, gives good agreement with experiment. A mid-range value of 2500 J/mol for E_G is not far off the mark either. Thus it may be concluded that for this important example the experiments and the calculations are in accord.

7.6.2 *Conformational statistics for vinyl chains*

In this section appropriate statistical weight matrices are developed for chains containing asymmetric centers and exhibiting various tactic forms. Polypropylene will be invoked as the archetypal example of such chains

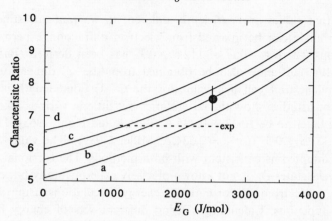

Fig. 7.2 The calculated characteristic ratio of polyethylene at 413 K (solid curves) in the region of reasonable conformational parameters (and $\theta = 112°$, $\phi_G = 67.5°$). Curve a $E_\omega = 4200$ J/mol, b 5400, c 6300, d 8400. The vertical bar for $E_G = 2500$ J/mol, $E_\omega = 5400$ J/mol shows the effect of variation of the gauche rotational angle minimum by $\pm 2.5°$ from $67.5°$. The experimental value of 6.7 is shown as the dashed line.

(Flory, Mark and Abe, 1966; Flory, 1969; Boyd and Breitling, 1972b). In Sections 5.2.2–5.2.4 the concept of the steric interactions *within dyads* controlling the conformational preferences of a vinyl chain was developed. This concept is now introduced into the statistical weight scheme. It is apparent, see **7.1**, that in accounting for the interactions between bonds

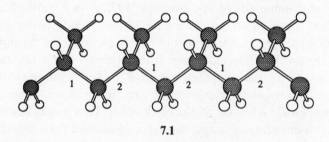

7.1

by considering the energy of a bond in the presence of the preceding one, bonds 1 and 2 are not equivalent in their statistical weight matrices. That is, bond 2, preceded by bond 1, contains the *within* dyad interactions whereas bond 1, preceded by bond 2, contains the *across* dyad interactions. Furthermore, the matrices will depend on the stereochemical configuration of the asymmetric centers.

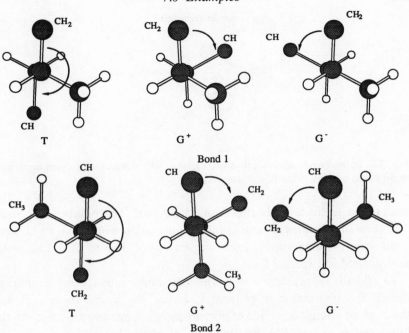

Fig. 7.3 Bond conformations about bonds 1 and 2 in **7.1**.

First, consider an isotactic all 'l' chain (l is chosen according to the convention presented in Section 5.2.1) as in **7.1**. The various conformations of bonds 1 and 2 above are shown in Figure 7.3. In the polyethylene chain the gauche vs trans energy difference arises due to the perturbation of the rotational energy of the C—C bond by the presence of carbon substituents at the ends of the bond. In polypropylene the skeletal C—C bond has one carbon with two carbon substituents. Here accounting for the conformation energy of the chain is attempted by assuming that each of the gauche-like interactions has an additive effect. However, the interaction of the main chain carbons $CH \cdots CH_2$ (gauche) and the gauche-like interaction of the substituent $—CH_3$ with the main chain CH groups will be distinguished, Figure 7.4. The term *skew methyl* interaction is adopted for this latter situation and assigned energy, E_{SK}, and statistical weight, $sk = e^{-E_{SK}/k_B T}$. It would be expected that the gauche and skew methyl interactions would be quantitatively similar to the gauche interaction in polyethylene. This, in fact, has been shown (Chang *et al.*, 1970; Schleyer, Williams and Blanchard, 1970) to be the case in low molecular weight homologs where the value of the skew methyl interaction E_{SK} has been

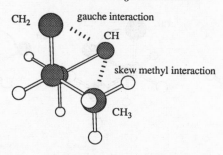

Fig. 7.4 Three bond skew methyl interaction, SK, defined and compared with main chain gauche three-bond interaction.

established in the neighborhood of 2500 J/mol. The number of gauche and skew interactions arising in the various conformations in an 1,1 meso dyad can be determined from inspection of Figure 7.3. These are summarized in Figure 7.5.

As already appreciated, it is the four-bond interactions similar to the G^+G^- interaction in polyethylene that are crucial to determining the conformational behavior of polypropylene. Focusing attention on the interferences occasioned by the pair of chain bonds *interior* to the asymmetric substituted carbon, it is found (Chapter 5) that there is a four-bond interaction involving the pendant methyl groups (Figures 5.21 or 5.26) and also one involving a methyl group and the chain CH_2 group. Inspection of models, or comparison with Figures 6.4 and 6.5 demonstrates the close similarity of these interactions with the G^+G^- interference in polyethylene. In all these cases, they involve repulsions between pairs of hydrogens brought into the same geometry of interference by the chain conformation. They differ only in the third substituent which is pointed away from the site of interference. For the present, the $CH_3 \cdots CH_3$ and $CH_3 \cdots CH_2$ interactions will be formally differentiated from the poly-ethylene four bond ω interaction by labeling then ω' and ω'' respectively. Just as in polyethylene, these interferences are greatly alleviated and a minimum energy established by rotation of one of the bonds away from the minimum normally associated with the bonds (cf. Figures 5.21 and 5.26). Conformational energy calculations (Boyd and Breitling, 1972b) indicate that a rotation of one of the bonds of $\sim 40°$ minimizes the energy. The establishment of a minimum energy in such interferences at a new value of the rotational angle of one of the bonds will be denoted by a + or − in parentheses indicating the direction of rotation to the new minimum. Thus TT(−) indicates the second bond in a TT sequence has rotated some $40°$ to the position $140°$ to establish the minimum. In not

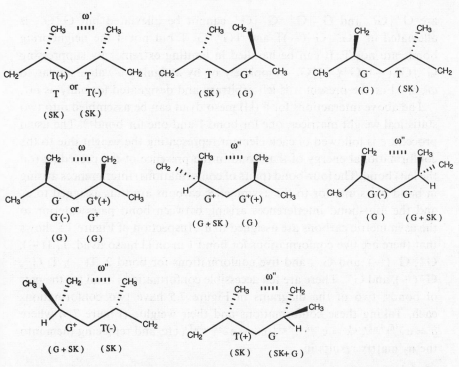

Fig. 7.5 Three- and four-bond interactions in the accessible conformations of a (l,l) meso dyad in polypropylene. The conformstions of the two bonds in the dyad are shown just below the bonds. A + or − in parentheses included with the conformations, T(+), T(−), G⁺(+), G⁻(−), means there is a significant torsional adjustment, about 40°, in that direction to relieve strain. In some cases there are alternative adjustments, i.e., T(+)T or TT(−). In other cases only one adjustment is possible. The four-bond interactions are indicated by vertical dashes (||||) and labeled ω, ω' or ω" according to the types of center brought into interference. The three-bond interactions are listed in parentheses just below the bond conformation.

all interference situations is it possible to reduce the energy by such rotations and they will be effectively ruled out by their high energy. The allowed or accessible conformations in a (l,l) meso dyad are displayed in Figure 7.5.

Turning to the interferences determined by the conformations of the two chain bonds exterior to the asymmetric carbons, there is a four-bond CH···CH G⁺G⁻ interference. However, in contrast to polyethylene, this interference may or may not be enhanced by the pendant methyl groups depending on the conformations of the neighboring bonds. Thus, the bond conformation pair interaction scheme is insufficient in this instance to account for the conformational interactions. The four-bond interferences are $G^+|G^-$ and $G^-|G^+$. $G^-|G^+$ cannot be alleviated but G^+G^- is alleviated in $TG^+|G^-(-)T$ and $TG^+|G^-T$ but not if the neighboring bonds are not T. It can be handled in limiting extremes by suppressing $G^+|G^-(-)$, $G^+(+)|G^-$ completely or by assigning a value similar to ω', ω''. For the present it is left arbitrary and designated formally as ω'''.

The above interactions for a (l,l) meso dyad can be assembled into two statistical weight matrices, one for bond 1 and one for bond 2. The usual procedure is followed of each element representing the weight due to the conformational energy of that bond in the presence of the preceding (on the left) bond. The four-bond (pairs of conformations) interferences arising in bond pairs interior to the asymmetric carbons are thus assigned to u_2 and the four-bond interferences arising between bond pairs exterior to the asymmetric carbons are assigned to u_1. Inspection of Figure 7.5 shows that there are five conformations for bond 1 in an l,l meso dyad, T, T($+$), G^+, $G^-(-)$, and G^-, and five conformations for bond 2, T($-$), T, G^+, $G^+(+)$, and G^-. There are ten accessible conformations in all for the pair of bonds, two of the diagrams in Figure 7.5 have two conformations each. Taking these conformations and their weights (Figure 7.5) where $g = e^{-(E_g/k_B T)}$, $sk = e^{-(E_{SK}/k_B T)}$, $\omega' = e^{-(E_{\omega'}/k_B T)}$, etc. and replacing them into the u_2 matrix results in,

$$
u_2^{ll} =
\begin{array}{c}
\\
T \\
T(+) \\
G^+ \\
G^-(-) \\
G^-
\end{array}
\begin{array}{c}
\begin{array}{ccccc}
T(-) & T & G^+ & G^+(+) & G^-
\end{array} \\
\left[
\begin{array}{ccccc}
sk\,\omega' & 0 & g & 0 & 0 \\
0 & sk\,\omega' & 0 & 0 & g\,sk\,\omega'' \\
sk\,\omega'' & 0 & 0 & g\,\omega'' & 0 \\
0 & 0 & g\omega & 0 & g\,sk\,\omega'' \\
0 & sk & 0 & g\,\omega & 0
\end{array}
\right]
\end{array}
. \quad (7.74)
$$

Incorporating the comments above about the across dyad four-bond interaction, ω''', and taking the three-bond interactions from Figure 7.5 gives for u_1,

$$
u_1^l =
\begin{array}{c}
\\
T(-) \\
T \\
G^+ \\
G^+(+) \\
G^-
\end{array}
\begin{array}{c}
\begin{array}{ccccc}
T & T(+) & G^+ & G^-(-) & G^-
\end{array} \\
\left[
\begin{array}{ccccc}
sk & sk & g\,sk & g & g \\
sk & sk & g\,sk & g & g \\
sk & sk & g\,sk & g\,\omega''' & 0 \\
sk & sk & g\,sk & 0 & g\,\omega''' \\
sk & sk & 0 & g & g
\end{array}
\right]
\end{array}
. \quad (7.75)
$$

Since the matrix for bond 2 depends on the preceding bond, 1, and this pair is interior to asymmetric carbon atoms the matrix, $\mathbf{u}_2$, depends on the stereochemical configuration of both asymmetric centers. Thus, the matrix requires a double superscript 'l,l' to denote this. However, the matrix for bond 1 depends on a pair exterior to one asymmetric center and depends only on its configuration. Thus, only a single superscript 'l' is required for this matrix.

The accessible conformations for a (l,d) racemic dyad are shown in Figure 7.6. Both bond 1 and bond 2 can occur in the conformations, T, $T(+), G^+, G^-(-), G^-$. Again there are ten conformational pairs allowed. These are assembled into the matrix, $\mathbf{u}_2^{ld}$,

$$
\mathbf{u}_2^{ld} =
\begin{array}{c}
\\
T \\
T(+) \\
G^+ \\
G^-(-) \\
G^-
\end{array}
\begin{array}{c}
\begin{array}{ccccc}
T & T(+) & G^+ & G^-(-) & G^-
\end{array} \\
\left[
\begin{array}{ccccc}
sk & 0 & 0 & g\,\omega'' & 0 \\
0 & 0 & sk\,g\,\omega' & 0 & g\,\omega'' \\
0 & sk\,\omega' & 0 & g\,\omega & 0 \\
sk\,\omega'' & 0 & sk\,g\,\omega & 0 & 0 \\
0 & sk\,\omega'' & 0 & 0 & g
\end{array}
\right]
\end{array}. \qquad (7.76)
$$

The matrix for bond 1, since it depends only on the chirality of the one center, l, is given by equation (7.75). The remaining matrices, $\mathbf{u}_2^{d,d}$, $\mathbf{u}_2^{l,d}$ and $\mathbf{u}_1^{d}$ may be obtained from those above by changing $+$ to $-$ (both as superscripts and in parentheses), d to l and vice versa, and ordering the rows and columns to make them conformable with the other matrices.

The characteristic ratio of a molecule of arbitrary stereosequence can be found using equation (7.63) and the matrices above for the various possible stereodyads occurring in such a molecule. For example, the sequence depicted in **5.16** would use generator matrices, $\mathbf{G}_i$, equation (7.62), based on the sequence of statistical weight matrices

$$\mathbf{u}_1^l \quad \mathbf{u}_2^{ld} \quad \mathbf{u}_3^d \quad \mathbf{u}_4^{dl} \quad \mathbf{u}_5^l \quad \mathbf{u}_6^{ll} \quad \mathbf{u}_7^{ll} \quad \mathbf{u}_8^{ld} \quad \mathbf{u}_9^d \quad \mathbf{u}_{10}^{dl} \quad \mathbf{u}_{11}^l \quad \mathbf{u}_{12}^{ld} \quad \mathbf{u}_{13}^d \quad \mathbf{u}_{14}^{dd}$$

The average behavior of molecules of arbitrary tacticity can be simulated by generating sequences of d and l centers with fixed values of replication probability (see Section 5.2.3). A Monte Carlo method may be used (Flory *et al.*, 1966; Flory, 1969) where a series of random numbers lying between 0 and 1 is generated. Each number is compared with the fixed desired replication probability, W (lying between 0 and 1). If the random number is less than W, the stereosense of the added unit is maintained. If it is greater than W, it is reversed. Thus, $W = 1.0$ generates isotactic polymer,

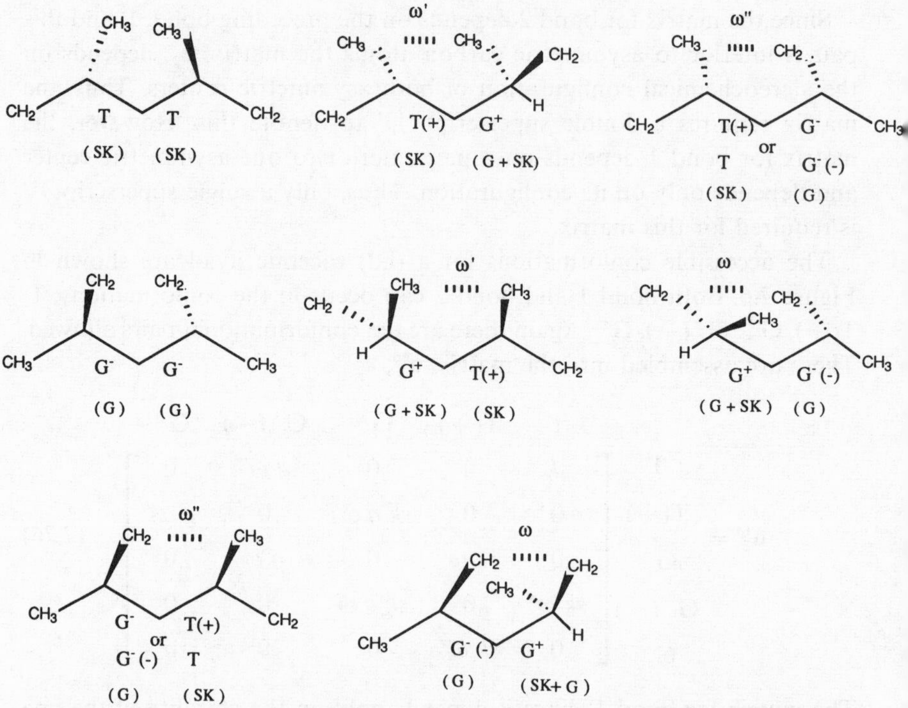

Fig. 7.6 Three- and four-bond interactions in the accessible conformations of a (l,d) racemic dyad in polypropylene. See Figure 7.5 for explanation.

$W = 0.0$ generates syndiotactic and $W = 0.5$ generates 'perfectly' atactic chains. The sequence of bonds can be made long enough for the characteristic ratio from equation (7.63) to depend on the stereosequence generated but not on 'end effects' from the finite chain size. That is, the chains are to be considerably longer than a typical persistence length. A large number of such sequences is generated, equation (7.63) applied to each and the average characteristic ratio for the population of sequences determined.

7.6.3 Characteristic ratios of polypropylenes

The statistical weight matrices developed in the preceding section can now be applied to the calculation of characteristic ratios of polypropylenes of varying tacticity. Since there are experimental data concerning the characteristic ratios of all three classes of polypropylenes, i.e., isotactic, syndiotactic and atactic, the results of the calculations are of considerable interest in establishing that realistic conformational models can be constructed that faithfully represent the behavior of real chains.

Conformational energy calculations (Boyd and Breitling, 1972b) indicate that the energy distinction between the gauche and skew methyl interactions allowed for in the statistical weight matrices, although not large, is, in fact, in order. Values of $E_G = 1700$ J/mol and $E_{SK} = 2500$ J/mol respectively give the best fit of the total calculated conformational energies of various conformations. Both of these values are close to values typically chosen for E_G in polyethylene. The calculations indicate no appreciable difference between the various four-bond ω interactions allowed for in polypropylene. The calculations indicate that $E_\omega = E_{\omega'} = E_{\omega''} = 5400$ J/mol, the same values as for E_ω in polyethylene. In addition, the conformational energy calculations result in the following values for the geometric parameters, $\phi_{T(-)} = 135°$, $\phi_T = 180°$, $\phi_{T(+)} = 225°$, $\phi_{G^+} = 60°$, $\phi_{G^+(+)} = 100°$, $\phi_{G^-(-)} = 260°$, $\phi_{G^-} = 300°$, θ (valence angle) $= 112°$ to be used in the transformation supermatrix, **T**. Figure 7.7 shows a comparison of calculated and experimental characteristic ratios using these values (and $\omega''' = 0$). It is apparent that there is excellent agreement for all three tactic forms.

In summary, the agreement between the calculated and experimental characteristic ratios of polyethylene and polypropylenes may be regarded

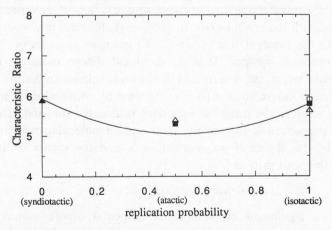

Fig. 7.7 Comparison of calculated (curve) and experimental (points) characteristic ratios for various tactic forms of polypropylene (Boyd and Breitling, 1972b; Boyd and Kesner, 1981a; see these for sources of experimental data). The curve is one calculated for the statistical weight model of Section 7.6.2 and using values of the energy and geometric parameters derived from conformational energy calculations.

as striking confirmation of the premise that the average conformational behavior of macromolecules in solution can be deduced from appropriate statistical mechanical averaging methods and from *a priori* determined conformational parameters. The latter may arise either from experiments on homologs or from conformational energy calculations.

7.6.4 Results for some other polymers

There has by now accumulated a large literature on applications of rotational isomeric state models and statistical averaging to a wide variety of polymers. No attempt is made here to summarize these efforts. However, some comments concerning the limitations of methodology for deriving a conformational model that will represent the statistical behavior in solution are in order.

Before undertaking the above it is appropriate to point out that the solution dimensions are not the only property experimentally measurable and amenable to the statistical treatment. Polymer chains possessing polar bonds can develop a molecular dipole moment that depends on the local bond conformations. In polyvinyl chloride, for example, the polar C—Cl bond moments add to form a resultant molecular moment that depends on the main chain bond conformations. As already commented in Section 7.4, the formalism developed for the mean-square end-to-end distance is immediately transferable to the mean-square value of any vector sum associated with the chain bonds. In poly(vinyl chloride) this would entail expressing the pendant side group C—Cl moment in terms of the local bond coordinate system. That is, the local dipole moment, **m**, has components, m_x, m_y, m_z determined by the local valence angles in contrast to the local bond vector $\mathbf{l} = [0 \; 0 \; l]$ directed by convention along the z axis. The result of the statistical averaging is usually expressed in analogy with the characteristic ratio as the mean-square molecular moment, $\langle \mu^2 \rangle$ divided by the degree of polymerization, x, and the square of the local moment, m^2, and thus as

$$\text{dipole moment ratio} = \langle \mu^2 \rangle / x m^2. \qquad (7.77)$$

There is a significant amount of experimental dipole moment data available for comparison with conformational models. Results for poly-(vinyl chloride) are tabulated in Table 7.3 along with the characteristic ratio results. Both types of data are in general agreement with the model derived from conformational energy calculations.

A major problem with conformational energy calculations is that they

Table 7.3 *Some calculated and experimental characteristic ratios and dipole moment ratios.*

polymer	temperature °C	$\langle R^2 \rangle_0 / Nl^2$ calculated	$\langle R^2 \rangle_0 / Nl^2$ experimental	$\langle \mu^2 \rangle / xm^2$ calculated	$\langle \mu^2 \rangle / xm^2$ experimental
polyethylene see text, Figure 7.2	140	7.5	6.7[a]		
polypropylene see text, Figure 7.7	145				
poly(vinyl chloride)	25	11.0 ± 0.4[b]	13.0[c], 8 ± 1[d]	0.72 ± 0.10[b]	0.70[e], 0.59[f], 0.67[g]
	155	8.7 ± 0.2[b]	9.8[h]		
polystyrene atactic ($W = 0.33$)[i]	25	10.0 ± 0.4[i]	10.0 ± 0.2[j]		
isotactic	25	11.0[i]	10.7 ± 0.5[k]		
poly(vinyl acetate)	30	9.3[l]	9.4[m]	0.72[l]	0.70[l], 0.89–0.94[n], 0.75–0.80[o]
poly(methyl acrylate)	66	8.6[l]	8.4[m]	0.67[l]	0.67[l], 0.67[q]
polyisobutylene	27	9.1[l]	8.4 ± 0.5[p]		
poly(vinylidene chloride)	25	6.0[r,s]	6.6[a]	0.89[s]	0.8[u]
polyoxymethylene	175	7.7[s]	8 ± 1[t]	0.176[u]	0.18[u]
	250			0.226	0.25

[a] From a tabulation by Flory, 1969; [b] Boyd and Kesner, 1981a; [c] Nakajima and Kato, 1966; [d] selected by Mark, 1972a; [e] Imamura, 1955; [f] Kotera, Shima, Fujisaki and Kobayahsi, 1962; [g] LeFevre and Sundaram, 1962; [h] Sato, Koshiishi and Asahina, 1963; [i] Stegen and Boyd, 1978; calculated for replication probability = 0.33; [j] selected by Mark, 1972b; [k] selected by Boyd and Stegen, 1978; [l] Smith and Boyd, 1991; [m] Matsumoto and Ohyanagi, 1961; [n] LeFevre, LeFevre and Parkins, 1960; [o] Takeda, Imamura, Okamura and Higashimura, 1960; [p] selected by Yoon, Suter, Sundararajan and Flory, 1975; [q] Tarazona and Saiz, 1983; [r] Boyd and Breitling, 1972a; [s] Boyd and Kesner, 1981b; [t] Matsuo and Stockmayer, 1975; [u] Porter and Boyd, 1971.

are usually carried out on isolated molecules and not in the condensed phase environment of the solution where the measurements take place. The theta solvent condition suppresses the effect of *long-range* interactions but solvent effects can nevertheless stabilize *local* conformations differentially over the vapor phase condition. In other words, in solution the effective conformational energy parameters can be different from those derived from isolated chain calculations. For the polyethylene and polypropylene examples above this effect appears to be minor. A case where it is not is polystyrene (Stegen and Boyd, 1978). The phenyl groups have large surface area and the amount exposed to solvent is highly dependent on the local bond conformations. If empirical corrections based on conformer equilibria in homologs are made for this, good agreement for the characteristic ratios of isotactic and atactic polymers can be attained (Table 7.3).

Sometimes the conformational calculations are not of high enough reliability *a priori* to represent accurately the experimental results but are nevertheless useful in rationalizing the experimental data. Poly(vinylidene chloride) is a crowded molecule sterically and is very polar as well. The characteristic ratio and dipole moment ratios are quite sensitive to the gauche vs trans energy parameter. The predicted value from conformational energy calculations does not accurately reproduce the data but adjustment of this one parameter allows both ratios to be fit (Boyd and Kesner, 1981b).

Another issue concerns assignments of statistical weights based on conformational energy differences between states in the rotational isomer state approximation. The partition function is approximated by a discrete sum based on the identified local minima in the conformational energy, i.e., the conformers. Each conformer is assigned an energy and the Boltzmann weighting is based on this. However, if the shapes or curvatures of the energy surface in the vicinity of the minima differ drastically among the conformers, the simple energy weighting of the discrete sum approximation to the complete phase space integration may be sensibly inadequate. It is possible to make corrections for this and still retain the averaging formalism based on statistical weight matrices. This is done by regarding the weights as derived from conformational *free energies* with an *entropic* component as well as the energetic one. The entropic corrections can be derived by integration over the energy surface near the minima, much in the spirit of the calculation concerning equation (5.1) and Figure 5.4. When done at several temperatures, the results can be fit with both an energetic and entropic term (Suter and Flory, 1975). Another procedure

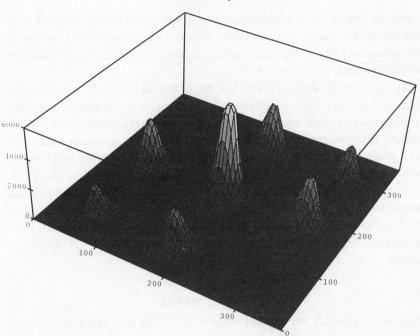

Fig. 7.8 Use of molecular dynamics in generating statistical weight matrices. In a long molecular dynamics run or *trajectory* statistics are accumulated on how much time is spent in various bond conformations. Here the relative probability (z axis) is shown of two neighboring bonds, ϕ_1, ϕ_2, having the values plotted as x, y. The TT, TG$^+$, TG$^-$, G$^+$G$^+$, G$^-$G$^-$ states are clearly seen. Higher magnification is required to see the less probable G$^+$G$^-$, G$^-$G$^+$ states. From Baher *et al.* (1991 with permission from *Macromolecules* © 1991, American Chemical Society).

which embraces the effects of all of the intramolecular degrees of freedom is to carry out a vibrational analysis on oligomer model compounds and compute free energy functions from the resulting vibrational frequencies (Boyd, 1968; Boyd *et al.*, 1973). The free energies can be parameterized with energetic and entropic components (Stegen and Boyd, 1978; Boyd and Kesner, 1981a,b). In most cases these corrections are minor. An example where they are significant is polystyrene (Stegen and Boyd, 1978). The phenyl group has a small inherent barrier to rotation about the attaching main chain carbon–phenyl group bond and therefore tends to undergo large vibrational excursions. The latter, being of low frequency, have a relatively large vibrational entropy. However, in some main chain conformations these are hindered by steric crowding. Thus there are fairly large differential entropy effects between conformations. These can be corrected for by the vibrational analysis scheme. Another method is to introduce the energy surface more directly in the statistical averaging procedure by Fourier expansion of the torsional energy (Allegra, Calagaris, Randaccio and Moraglio, 1973).

An interesting development has been the application of molecular dynamics simulations to isolated oligomer polymer chains. If run long enough and at high enough temperature, time trajectories that include visiting the more important, lower energy conformations can be generated. At a given temperature, conformational populations of given neighboring bond pairs can be established by accumulating statistics along the trajectory on the time spent in the various pairs. Thus the statistical weights can be generated directly without recourse to conformational analysis in terms of a model for which kinds of steric interactions are important (Zuniger, Dodge and Mattice, 1991). Figure 7.8 illustrates results.

7.7 Further reading

The basic reference on the application of matrix averaging methods to polymer molecules is Flory's *Statistical Mechanics of Chain Molecules* (Flory, 1969). Volkenstein's book (1963) is also a classic.

Nomenclature

See Table 7.1 for the definitions of the various matrices used in the averaging processes

A = Helmholtz free energy

E_G = energy assigned to a skew methyl interaction

E_ω = energy assigned to a ω four-bond 'pentane interference'-like steric interaction

G = gauche conformation

$G^+, G^- = \sim +60°$ or $\sim -60°$ gauche torsional angle. A plus or minus *in parentheses* indicates that there is a significant distortion of the angle from its standard value but it is still in the range occupied by gauche, e.g., $G^+(+)$, means there is a significant distortion away from $\sim +60°$ in the $+$ direction but that the value is still less than $120°$. $G^+(+) = 100°$ would be an example

$g = e^{-E_G/k_B T}$; statistical weight of gauche conformation

SK = a 'skew methyl' steric interaction, a pendant methyl group that has a conformation with its torsional angle $\sim \pm 60°$ with respect to the main chain

$sk = e^{-E_{SK}/k_B T}$; statistical weight of skew methyl interaction

T = trans bond conformation, $\sim 180°$, T(+), T(−) means that the torsional angle is significantly distorted away from the $180°$ value in the direction indicated but that the value is still in the trans range, between $-120°$, $+120°$

ω = a four-bond 'pentane interference'-like steric interaction. See Figures 7.5 and 7.6 for further refinement as to type, ω', ω'', ω''', in vinyl polymers

$\omega = e^{-(E_\omega/k_B T)}$, statistical weight of a four-bond 'pentane interference'-like steric interaction

Problems

7.1 Calculate the population of *trans* states in very high molecular weight polyoxymethylene at 453 K for $E_T = 0$, $E_{G^-} = E_{G^+} = -1700$ cal for:

    (*a*) the independent bond rotation (and rotational isomeric state) model;

    (*b*) for all G^+G^- sequences excluded.

7.2 Express the pendant C—Cl moment for polyvinyl chloride in the local bond coordinate system (Appendix A6.1). Assume the valence angles are tetrahedral.

8

Rubber elasticity

> If one end of a slip of Caoutchouc be fastened to a rod of metal
> or wood, and a weight be fixed to the other extremity, in order
> to keep it in a vertical position; the thong will be found to
> become shorter with heat and longer with cold.
>
> *Gough, 1805*

A material that can be deformed quickly to several hundred per cent strain, recovers rapidly and completely upon removal of the stress and is capable of having the process repeated numerous times is described as rubbery or elastomeric. The possibility of this behavior is due to the flexible long chain nature of polymer molecules and presents a type of response to mechanical deformation that is fundamentally quite different from the response given by rigid materials such as metals, ceramics and glassy or semi-crystalline polymers. This behavior is often called 'entropy elasticity' in contrast to the 'energy elasticity' of more familiar materials. The resistance to deformation is due largely to an entropy decrease rather than an energy increase. Entropy elasticity demonstrates itself in easily observed thermodynamic behavior such as the contraction of a stretched rubber band with increasing temperature as described by Gough (1805) in the quotation above. In this chapter, the contrast of energy vs entropy elasticity in thermodynamic behavior is developed. Then the molecular theory of rubber elasticity is discussed and its applicability to real elastomers is considered. Rubber elasticity is intimately connected with the presence of a network. Methods, both chemical and physical, used to establish networks are discussed. Since the elasticity response in rubber elasticity is to a large degree independent of the detailed chemical structure of the chain, the choice of polymer structures in applications as elastomers is largely based on ancillary properties. Some of these factors are also discussed.

8.1 Thermodynamics

In illustrating the contrast between energy and entropy elasticity it is useful to set down some thermodynamic relations appropriate for a very

266

simple situation, the stretching of a sample along one dimension (uniaxial stretch). The first law for the energy change, dE, for a body subjected to a transfer of heat, dQ, and change in volume, dV, under the pressure, P, and a uniaxial distortion is

$$dE = dQ - P\,dV + f\,dl, \tag{8.1}$$

where f is the retractive force and dl the change in length, l, of the sample. For a reversibly applied distortion where the entropy change $T\,dS = dQ$,

$$dE = T\,dS - P\,dV + f\,dl. \tag{8.2}$$

The Helmholtz free energy, $A = E - TS$, change is

$$dA = -S\,dT - P\,dV + f\,dl. \tag{8.3}$$

From this it follows that

$$f = (\partial A/\partial l)_{T,V} \tag{8.4}$$

and that

$$f = (\partial E/\partial l)_{T,V} - T(\partial S/\partial l)_{T,V}. \tag{8.5}$$

From the enthalpy, $H = E + PV$, the Gibbs free energy, $G = H - TS$, change is,

$$dG = -S\,dT + V\,dP + f\,dl. \tag{8.6}$$

Equations (8.4) and (8.5) thus have counterparts in terms of T and P as variables,

$$f = (\partial G/\partial l)_{T,P}, \tag{8.7}$$

$$f = (\partial H/\partial l)_{T,P} - T(\partial S/\partial l)_{T,P}. \tag{8.8}$$

Equation (8.8) shows that, in general, at constant T and P, the retractive force arises as the result of both enthalpy and entropy changes.

Since A and G are state functions, cross derivatives in equations (8.3) and (8.6) must be equal or

$$(\partial S/\partial l)_{T,V} = -(\partial f/\partial T)_{l,V} \tag{8.9}$$

and

$$(\partial S/\partial l)_{T,P} = -(\partial f/\partial T)_{l,P}. \tag{8.10}$$

The above two equations will be useful in employing the effect of temperature as a diagnostic in determining the relative importance of energy versus entropy as the source of the retractive force.

8.1.1 *Energy elasticity*

When a material body is deformed there is generally a force resisting that deformation and it requires work to accomplish this. One can imagine two extreme situations with respect to what this work accomplishes on the atomic or molecular scale. These extremes are energy vs entropy elasticity. In illustration of the energy case a crystal can be considered. A crystal is stable because the forces between the atoms or molecules comprising it are balanced. Each atom or molecule is in a state of mechanical equilibrium. Furthermore the crystal is stable with respect to small displacements of each of the atoms or molecules. The energy increases for all displacements of the atoms. If a crystal is deformed, neglecting any changes in the vibrational energy, the work of deformation would go into the energy required to displace the atoms or molecules from the positions in the undeformed lattice. The work of deformation would be equated with the increase in the total energy which is comprised of the interatomic interaction energies of the crystal. The work of deformation (in the absence of a surrounding atmosphere) is thus entirely expended in increasing the energy of the crystal (see Figure 8.1). In the presence of a surrounding atmosphere there will be '$P–V$' work due to the volume change on deformation. Therefore in an energy-elastic body like the crystal the deformation work plus the $P–V$ work is

$$f\,\mathrm{d}l = (\partial E/\partial l)_{T,P}\,\mathrm{d}l + P(\partial V/\partial l)_{T,P}\,\mathrm{d}l \qquad (8.11)$$

and therefore

$$f\,\mathrm{d}l = (\partial H/\partial l)_{T,P}\,\mathrm{d}l. \qquad (8.12)$$

Thus a material for which equation (8.8) reduces to

$$f \cong (\partial H/\partial l)_{T,P} \qquad (8.13)$$

and

$$(\partial S/\partial l)_{T,P} \cong 0 \qquad (8.14)$$

with the enthalpy dominating over the entropy term is consistent with the microscopic interpretation that in such a material the deformations are due to atom displacements against restraining interatomic forces. In such a material equation (8.14) compared with equation (8.10) leads to the important conclusion that

$$(\partial f/\partial T)_{l,P} \cong 0 \qquad (8.15)$$

and that the retractive force is temperature insensitive. For small deformations, the retractive force per unit strain is the tensile modulus.

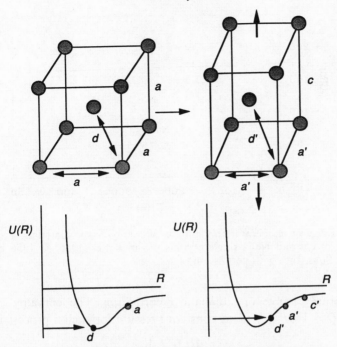

Fig. 8.1 An energy-elastic material. On the left is an undeformed BCC lattice unit cell of size, **a**. The one on the right has undergone a tensile distortion so the dimensions are now **c** in the tensile direction and **a′** in the transverse direction. The nearest neighbor distance in the undeformed material is **d**. The pair energy associated with **d** is near the minimum in the potential energy curve, lower left. On deformation, **d** increases to **d′** and the pair energy increases, lower right. The distance **c** is larger than **a** but **a′** is smaller (Poisson contraction). The net effect is for the total energy to increase. The total energy stored in the microscopic displacements equals the macroscopic work required to deform the specimen.

Thus the tensile stiffness is temperature insensitive. It is confirmed experimentally that the elastic constants of crystals tend to be relatively insensitive to temperature, see Figure 8.2. The small change with temperature is in the direction of decrease with increasing T. This can be rationalized as due to the secondary effect of thermal expansion causing increasing interatomic spacings. The same is true of disordered solids as well. In glasses, as long as the temperature is well below the vitrification point, the stiffness is relatively temperature insensitive. This is due on a microscale to the deformation process being the same as in ordered solids, atom displacements against interatomic forces.

Another distinguishing feature of energy elasticity is that there is no

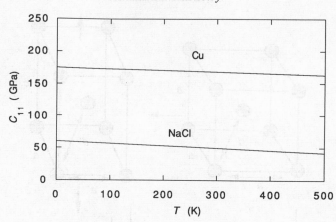

Fig. 8.2 Effect of temperature on stiffness of energy-elastic materials. C_{11} elastic constant of Cu and NaCl single crystals. From Kittel (1956, © 1956 by John Wiley & Sons, Inc., with permission).

temperature change on adiabatic deformation. The enthalpy change, $dH = dQ = V\,dP + f\,dl$, at constant pressure for such a process is

$$dH = f\,dl. \tag{8.16}$$

In terms of the variables T, P and l, at constant pressure the enthalpy change is

$$dH = C_P\,dT + (\partial H/\partial l)_{T,P}\,dl. \tag{8.17}$$

Thus for an energy-elastic material it follows from equation (8.13) and the above equation that

$$dH = C_P\,dT + f\,dl. \tag{8.18}$$

Comparison with equation (8.16) shows that

$$C_P\,dT = 0. \tag{8.19}$$

and therefore there is no temperature change.

8.1.2 Entropy elasticity

It is possible to imagine, in principle, a material which behaves in a manner opposite to the crystal considered in the previous section, i.e., one in which there is no enthalpy change on deformation (at constant temperature and pressure) and the refractive force is entirely due to an entropy change. It is useful to examine how this might be true for a collection of randomly coiled flexible polymer molecules. If the barrier to bond rotation were zero (free rotation), or if the rotational energy minima were all of equal energy, then the chain would have a large number of conformations available to it, all of equal energy. The ends of such a chain could undergo displacement relative to each other by means of bond rotations without any change of energy between initial and final states. Furthermore, it might be expected that the packing energy of such random coils would be insensitive to the precise conformation of each chain. Thus, it should be possible to deform a collection of randomly coiled chains without increasing their internal energy. That is, no increase in energy would accompany deformation. As has been seen, it is the change in *free* energy (equations (8.4) and (8.7)) that determines the retractive force. If there is an entropy decrease on stretching a retractive force can arise nevertheless. It will be shown in the next section that altering the extensions of the coiling chains from their undeformed states indeed causes an entropy decrease.

The assumption that the packing energy of the chains is independent of the sample length during deformation implies that there can be no change in volume on deformation. If there were a change in volume on stretching, it would indicate a change in the efficiency of packing and hence imply a change in energy. Thus, the conditions for an *ideal entropy-elastic* material are

$$(\partial E/\partial l)_{T,P} = 0 \tag{8.20}$$

and

$$(\partial V/\partial l)_{T,P} = 0. \tag{8.21}$$

These also imply

$$(\partial E/\partial l)_{T,V} = (\partial H/\partial l)_{T,P} = 0. \tag{8.22}$$

It then follows, equation (8.5) or (8.8), that

$$f = -T(\partial S/\partial l)_{T,P} \tag{8.23}$$

or

$$f = -T(\partial S/\partial l)_{T,V}. \tag{8.24}$$

It follows from equation (8.9) or equation (8.10) that the relations

$$f = T(\partial f/\partial T)_{l,P} \tag{8.25}$$

or

$$f = T(\partial f/\partial T)_{l,V} \tag{8.26}$$

hold for an ideal rubber. Thus, comparing with equation (8.15), it is seen that the dependence of retractive force on temperature is quite different

from that for an energy-elastic material and is therefore a sensitive
criterion for distinguishing entropy from energy elasticity.

Equation (8.25) fully specifies the form of the temperature dependence
of the retractive force at fixed length. In order to satisfy it, f must be
linear in temperature and therefore given by

$$f = C(l, P)T, \qquad (8.27)$$

where C is a function only of the length at constant P. This equation
forms the basis for the early observation by Gough quoted at the
beginning of this chapter. In order to sustain a constant force the stretched
length must decrease as temperature increases. Within limits, the linear
increase of f with absolute temperature and the implied vanishing of the
retractive force at $T = 0$ K is subject to experimental verification. The
temperature region over which rubbery behavior is exhibited does not
extend to indefinitely low temperatures, vitrification intervenes. However,
extrapolation can be carried out. Figure 8.3 shows data for natural rubber
over a limited temperature range near room temperature. Linear extrapolations to low temperature are also shown. It would appear that in this
case equation (8.27) is to some degree, but not exactly, satisfied. The linear
extrapolations show an intercept at $T = 0$ K. However, the value of the
intercept is a relatively small fraction of the total retractive force at
$T = 300$ K. It is to be emphasized that equation (8.27) is based on an
ideal entropy-elastic material, one for which equation (8.20) holds exactly.

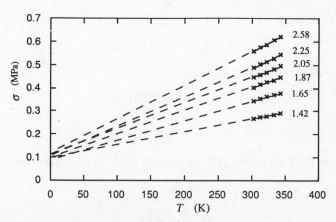

Fig. 8.3 Retractive force (expressed as engineering stress) vs absolute temperature
for natural rubber at the various fixed extension ratios indicated. In contrast to
crystals, the stiffness (retractive force at constant length) increases with temperature
and is proportional to absolute temperature (after Ciferri, 1961).

Table 8.1 *Energetic contribution to the retractive force from force vs temperature measurements.*

polymer	$T\,°C$	f_E/f
polyethylene	137–200	-0.5 ± 0.05
cis-poly(isoprene) (natural rubber)	20–70	0.18 ± 0.03
polyisobutylene (butyl rubber)	20–100	-0.05 ± 0.03
poly(butadiene/styrene) copolymer (SBR)	-35–100	-0.12 ± 0.05
polystyrene	120–175	0.16 ± 0.03
polyoxyethylene	30–100	0.10 ± 0.05
poly(dimethylsiloxane)	25–100	0.20 ± 0.05
poly(perfluoropropylene/vinylidene fluoride) copolymer	30	0.05

From Mark (1973) and Aklonis, MacKnight and Shen (1972).

This is by no means necessary in order to have rubbery behavior. Plots such as Figure 8.3 suggest expressing the degree to which a material is energy vs entropy elastic through using equation (8.8) to define the total force as the sum of an energy force, f_E and an entropy force f_S

$$f = f_E + f_S, \tag{8.28}$$

where

$$f_E = (\partial H/\partial l)_{T,P}$$

and

$$f_S = -T(\partial S/\partial l)_{T,P} = T(\partial f/\partial T)_{l,P}.$$

It follows, in general, from the above that

$$\frac{f_E}{f} = -T\frac{\partial \ln(f/T)}{\partial T}. \tag{8.29}$$

In the particular case of linear plots like Figure 8.3 the intercept at $T = 0$ K can be thought of as the energetic component, f_E, of the total force, f. Table 8.1 lists f_E/f ratios determined via equation (8.29) for some common elastomers. It is of interest to observe that the negative value of f_E/f for polyethylene has a simple qualitative explanation in the preference of trans to gauche conformations. In order to effect an elongation, there must be a shift in population of conformations toward the more extended ones. The trans is more extended than the gauche and is also of lower energy; extension lowers the energy resulting in a negative value of f_E. In contrast, in polyoxyethylene for example, the preference of O—C—C—O bonds for the less extended gauche conformations requires an increase in energy for their extension and leads to a positive value of f_E.

In contrast to an energy-elastic material, there will be a change in temperature on stretching an ideal rubber adiabatically. It is seen from equation (8.17), since the enthalpy does not change with stretching, that $dH = C_P\,dT$ at constant pressure. From the relation in equation (8.16), that $dH = f\,dl$, it follows that a temperature change

$$dT = (f/C_P)\,dl \qquad\qquad (8.30)$$

accompanies a change in length dl. Atomistically, the energy-elastic vs entropy-elastic cases are contrasted by observing in the energy case that all of the work of stretching beyond P–V work is stored in the energies associated with local atom displacements. No temperature change to accommodate this work is necessary. In the ideal entropy-elastic case, there is no energy storage in local atomistic displacements. To accommodate the work done with the stretch with no heat flow, the temperature must rise.

8.2 Classical molecular theory of rubber elasticity

If a macroscopic collection of free, randomly coiled polymer molecules is deformed in shape, the resulting stress will quickly decay if the thermal motion is rapid enough that the molecules can relax by bond rotations back to their undeformed equilibrium states. Thus a permanent deformation will result by virtue of the segmental diffusion of the chains. If a steady shearing stress is applied, this segmental diffusion will result in permanent deformation proportional to the length of time of application of the stress. Since there will also be dissipation of energy, the latter can be described as viscous flow. However, the decay of stress in shape deformation or viscous flow under steady stress can be prevented by means of *crosslinking* to form a *network*. If crosslinks are introduced (either randomly or at end-groups, Chapter 2), each chain *segment* between two crosslinks can still undergo random coiling. There are still a large number of paths or conformational sequences that will allow a segment to reach from the crosslink point at one end to the crosslink at the other. The crosslinks, being of a permanent chemical nature, prevent the relaxation of stress and resultant permanent deformation. It is convenient to view the deformation of the network as due to the movement of the crosslinks in response to the macroscopic change in shape. The retractive force arises because there will be, on the average, fewer paths for each segment that span the crosslinks at its end in the deformed state than in the undeformed one. Qualitatively this is easy to see. If the deformation were so severe that a segment were required to be nearly at its maximum extension (planar zig-zag, for example), then it is obvious that there would be very few paths available compared to the large number available for spanning a distance which was small compared to the maximum extension of the segment. The fewer configurations result in lower entropy, higher free energy, and hence a retractive force.

The molecular theory is based on deducing the free energy of the network and how it depends on deformation. Thus the model has two aspects. One of these is a set of assumptions concerning the computation of the free energy of the network. The other is knowledge of the network itself. Networks are formed chemically by a number of routes. Formation can be the result of random attack of crosslinking reagents along preexisting chains, it can come by addition of multifunctional monomers in a largely difunctional monomer polymerization and so forth. In any case there will result a collection of chain segments of varying lengths connecting various crosslink points, distributed through space in the specimen, and perhaps dangling unconnected chain ends or even some entirely linear chains. The chain segments will be intertwined in various ways and some of these ways could present rather permanent entanglements. Regardless of the set of assumptions about how to compute the free energy a complete description of the network is, in principle, necessary.

The classical theory is based on treating the chain segments in the network as Gaussian phantom chains and computing the free energy from the statistics associated with such chains. After some early efforts in identifying rubber elasticity with such a statistical description (Kuhn, 1934, 1936; Meyer, von Susich and Valko, 1932; Guth and Mark, 1934), both James and Guth (Guth and James, 1941; James and Guth, 1942, 1943a) and Wall (1942, 1943; see also Treloar, 1943) derived the form of the stress–strain relation now associated with the Gaussian statistical model. By assuming that the deformation of individual chains follows that of the macroscopic sample, Wall deduced not only the form of the stress–strain relation but also a value for the multiplicative constant. James and Guth (1943a) emphasized the fact that the network is actually a coupled system of chains and derived equations, later elaborated on (James, 1947; James and Guth, 1947), for the deformation of the network junctions. They found that evaluation of the multiplying constant required careful consideration and deduced values for the latter that differed from that for the individual chain hypothesis. Thus there are two versions of the theory historically. The individual chain deformation version or *affine* model is somewhat simpler and will be considered first. If the reader is interested solely in how the *form* of the free energy and stress–strain relations arise, the affine version is sufficient for that.

8.2.1 The affine model

The affine phantom chain model (Wall, 1942, 1943; Flory and Rehner, 1943a; Wall and Flory, 1951; Flory, 1953), usually known more simply as the 'affine' model, treats the network chain segments as independent entities in the deformation process. An idealized network is illustrated in Figure 8.4. In Figure 8.5 the same network is shown but with each chain segment spanning network junctions replaced by a vector connecting these points. In both cases an undeformed state and a deformed one are depicted. The affine model makes the assumption that each segment vector is displaced in proportion to the macroscopic strain, i.e., it is displaced *affinely*. The vector marked **R** in Figure 8.5 in the deformed state is

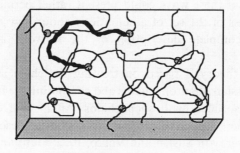

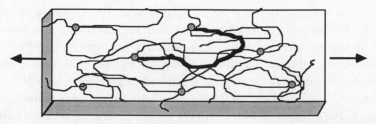

Fig. 8.4 An idealized network. Chain segments connect crosslinks which act as network junction points. The upper depiction is for an undeformed state and the lower one is for a uniaxial stretch in the direction indicated. The bold face chain spans the same junction points in both but its configuration changes continually over time in both states.

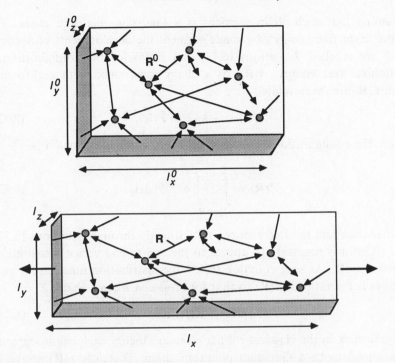

Fig. 8.5 The network of Figure 8.4 but with each chain segment replaced by a vector spanning the junctions. The bold face chain of Figure 8.4 is indicated by $\mathbf{R}^0$ in the undeformed state and by $\mathbf{R}$ in the deformed one.

assumed to be related to its counterpart in the undeformed state, $\mathbf{R}^0$, as

$$\mathbf{R} = \lambda_x R_x^0 \mathbf{i} + \lambda_y R_y^0 \mathbf{j} + \lambda_z R_z^0 \mathbf{k}, \tag{8.31}$$

where λ_x, λ_y, λ_z are the macroscopic *stretch ratios* in the x, y, z directions given by

$$\lambda_x = l_x/l_x^0, \qquad \lambda_y = l_y/l_y^0, \qquad \lambda_z = l_z/l_z^0$$

and where l_x, l_y, l_z are the macroscopic dimensions in the deformed state, the superscript 0 values are for the undeformed state and R_x^0, R_y^0, R_z^0 are the components of $\mathbf{R}^0$. Inherent in the affine assumption is the presumption that the network junction points are fixed in any given state of macroscopic deformation and they do not wander or fluctuate in their positions. The configurations of the chain segments joining them, of course, must be constantly changing, visiting the many conformations that leave their individual end-to-end vectors, $\mathbf{R}$, fixed.

The evaluation of the free energy of the network is accomplished by assuming that each chain segment is a Gaussian phantom chain. To evaluate the free energy of a single segment the developments of Section 6.3.1 are recalled. Equations (6.47) and (6.48) for the configurational Helmholtz free energy, $A(\mathbf{R})$, of a chain constrained with end-to-end vector, $\mathbf{R}$, are recapitulated here as

$$A(\mathbf{R}) = -k_B T \ln Z(\mathbf{R}),\qquad(8.32)$$

where the configurational partition function, $Z(\mathbf{R})$, of the chain is

$$Z(\mathbf{R}) = \int \cdots \int_{\sum \mathbf{l} = R} e^{-\beta E(\mathbf{l})}\, d\{\mathbf{l}\}.\qquad(8.33)$$

Rather than find the free energy, $A(\mathbf{R})$, directly through equations (8.32) and (8.33) it is necessary to appeal to the end-to-end vector distribution function, $p(\mathbf{R})$. As seen equation (6.45), the distribution function for end vectors is the ratio of $Z(\mathbf{R})$ to that for an *unconstrained* chain, Z, or

$$p(\mathbf{R}) = Z(\mathbf{R})/Z.\qquad(8.34)$$

As indicated, in the classical rubber elasticity theory each chain segment is assumed to be a Gaussian phantom chain. Therefore $p(\mathbf{R})$ above is given by the Gaussian chain function, $P(\mathbf{R})$, equation (6.33) or

$$p(\mathbf{R}) = \left(\frac{B}{\pi}\right)^{3/2} e^{-BR^2},\qquad(8.35)$$

where

$$B = \frac{3}{2\langle R^2\rangle_0} = \frac{3}{2C_\infty l^2 N}.$$

Therefore $Z(\mathbf{R})$ can be found from $p(\mathbf{R})$ as

$$Z(\mathbf{R}) = p(\mathbf{R})Z.\qquad(8.36)$$

It is not important to know the configurational partition function, Z, of a free chain in equation (8.36). It is independent of deformation and only the effect of deformation on the free energy need be evaluated.

At this point an aside is appropriate. The entropic nature of the elastic response has been emphasized in the physical interpretation of rubber elasticity. This is indeed correct. However, it is also to be emphasized that the statistical thermodynamic treatment above is in terms of free energy. Any energetic effects embodied in the end-to-end vector distribution function, $p(\mathbf{R})$, will be properly carried over into the free energy. There are, in fact, such effects. The characteristic ratio, C_∞, that appears in the Gaussian distribution, as has been seen in Chapters 6 and 7, responds to the energetics of the various conformational states. Thus the requirement indicated above for the occurrence of rubber elasticity does not actually require that the intramolecular energy be independent of the conformational changes induced by the state of stretch. Intramolecular energy changes are properly accounted for within the applicability of the Gaussian distribution.

Returning to the evaluation of the free energy, let equation (8.32) for a deformed state of a chain with end vector, **R**, be written, through the use of equation (8.36), as

$$A(\mathbf{R}) = -k_B T \ln Z(\mathbf{R})$$

$$= -k_B T \ln p(\mathbf{R}) Z. \tag{8.37}$$

From equation (8.31), the square of the chain segment vector, **R**, can be expressed in terms of components of the *undeformed* chain segment vector **R**^{0} as

$$R^2 = \lambda_x^2 (R_x^0)^2 + \lambda_y^2 (R_y^0)^2 + \lambda_z^2 (R_z^0)^2 \tag{8.38}$$

Substitution of the Gaussian distribution function, equation (8.35), into $A(\mathbf{R})$ gives

$$A(\mathbf{R}) = -k_B T \{ \ln Z + \ln(B/\pi)^{3/2} - B[\lambda_x^2 (R_x^0)^2 + \lambda_y^2 (R_y^0)^2 + \lambda_z^2 (R_z^0)^2] \}.$$

$$\tag{8.39}$$

This expression is for a single chain segment, with vector, **R**, spanning two junction points. It now remains to find the free energy of the network. Since the chains are assumed to be phantom and non-interfering, the network free energy is simply the sum of that for the individual chain segments. However, since the above relation contains the components R_x^0, R_y^0, R_z^0 of the undeformed spanning vector, **R**^{0}, performing this sum requires knowledge of all the individual vectors of the undeformed network.

Let **R**$_{ij}^0$ be a chain vector spanning the network junctions i and j in the undeformed network. The constant B appearing in the Gaussian function, equation (8.35), depends on N, the number of bonds in the chain segment. Since i and j designate this chain uniquely, B, through N, depends on i and j and thus must be denoted as B_{ij}. Then the free energy of the deformed network, A_{net}, is found by summing equation (8.39) over all chain segments. The summation can be over *all* values of pairs of junctions i, j or over all values of i and j such that $i < j$, if the convention is followed of setting $B_{ij} = 0$ for a pair of junctions not spanned by a chain.

$$A_{\text{net}} = -k_B T \sum_{i<j} [\ln Z + \ln(B_{ij}/\pi)^{3/2} - B_{ij}(R_x^0)^2_{ij}\lambda_x^2$$

$$- B_{ij}(R_y^0)^2_{ij}\lambda_y^2 - B_{ij}(R_z^0)^2_{ij}\lambda_z^2]. \qquad (8.40)$$

In performing the summation the first two terms in the square brackets are independent of the deformation, λ_x, λ_y, λ_z, and the result can be represented by a constant, C. Therefore equation (8.40) can be written as,

$$A_{\text{net}} = -k_B T v[C - \langle BR_x^2\rangle_{\text{net}}\lambda_x^2 - \langle BR_z^2\rangle_{\text{net}}\lambda_y^2 - \langle BR_z^2\rangle_{\text{net}}\lambda_z^2], \quad (8.41)$$

where

$$\langle BR_x^2\rangle_{\text{net}} = v^{-1}\sum_{i<j} B_{ij}(R_x^0)^2_{ij}, \qquad \langle BR_y^2\rangle_{\text{net}} = v^{-1}\sum_{i<j} B_{ij}(R_y^0)^2_{ij}$$

and

$$\langle BR_z^2\rangle_{\text{net}} = v^{-1}\sum_{i<j} B_{ij}(R_z^0)^2_{ij},$$

where $\langle\ \rangle_{\text{net}}$ denotes an average value of BR_x^2 etc. over the segments or strands in the network as defined by the sums above divided by the total number of chain segments, v, in the network. It is to be supposed that the undeformed crosslinked network was created under conditions such that it is isotropic and therefore $\langle BR_x^2\rangle_{\text{net}} = \langle BR_y^2\rangle_{\text{net}} = \langle BR_z^2\rangle_{\text{net}}$. Thus if $\langle BR^2\rangle_{\text{net}}$ is defined as $\langle B(R_x^2 + R_y^2 + R_z^2)\rangle_{\text{net}}$ then, leaving a factor of $\frac{1}{2}$ in front for later convenience,

$$A_{\text{net}} = \tfrac{1}{2}k_B T v[\tfrac{2}{3}\langle BR^2\rangle_{\text{net}}(\lambda_x^2 + \lambda_y^2 + \lambda_z^2 - 3)], \qquad (8.42)$$

where

$$\langle BR^2\rangle_{\text{net}} = v^{-1}\sum_{i<j} B_{ij}(R_{ij}^0)^2 \qquad (8.43)$$

and where the constant has been eliminated in favor of taking the free energy of the network as zero in the undeformed state, where λ_x, λ_y, $\lambda_z = 1$ and $\lambda_x^2 + \lambda_y^2 = \lambda_z^2 = 3$.

The above equations are the final result of later versions (Flory, Hoeve and Ciferri, 1959) of the affine model for the free energy. Thus the *form* of the free energy dependence on deformation, through $\lambda_x^2 + \lambda_y^2 + \lambda_z^2$, does not depend on the details of the network but the multiplicative factor, $\langle BR^2\rangle_{\text{net}}$ does. A simplifying assumption has often been made about the network (Flory, 1950). If it is supposed that the network vectors have the same distribution as the end-to-end vectors of *free chains* of the same length then R^2 and $\langle R^2 \rangle_0$ appearing in $B = 3/(2\langle R^2 \rangle_0)$ have the same N and spatial dependence and $\langle BR^2 \rangle_{\text{net}}$, equation (8.43), reduces to $\frac{3}{2}$. Under this circumstance equation (8.42) becomes

$$A_{\text{net}} = \tfrac{1}{2}vk_{\text{B}}T(\lambda_x^2 + \lambda_y^2 + \lambda_z^2 - 3). \tag{8.44}$$

It is in this form that the classical affine molecular theory has often been presented. It has also been proposed (Flory, 1950, 1953) that an additional term, $\frac{1}{2}vk_{\text{B}}T \ln V$, where V is the system volume, be included in equation (8.44).

8.2.2 *The phantom network model*

In the coupled phantom chain model (James and Guth, 1943a; James, 1947; James and Guth, 1947), usually called in the literature the 'phantom network' model, the free energy is also based on the Gaussian function. However, the chain segments are not assumed to deform independently but are taken as a coupled system. The connection to the macroscopic strain is established by requiring that *some* of the network junction points be constrained to follow the sample shape. These points could, for example, be those lying on the surface of the specimen. The rest of the network junction points are free to move under the potential implied by the Gaussian distribution. Their behavior is described statistically in terms of *average* position and *fluctuation* in average position. This contrasts with the affine model where the concept of enforcing the affine deformation of each network vector requires that there be no fluctuation at all about the average positions.

Since the chain segments of the network are considered as coupled together, a partition function for the entire network is formed. This is accomplished in terms of the network junction points rather than the chain segment vectors, **R**, directly. Let $\mathbf{r}_i$ be the position vector relative to an arbitrary origin of the ith network *junction*. A chain segment vector $\mathbf{R}_{ij}$ spanning junctions i and j is given by $\mathbf{r}_j - \mathbf{r}_i$. Since the assumption of phantom chains is made, the chains are independent of each other and the probability function, $p(\mathbf{r})$ for the instantaneous configuration, $\mathbf{r}$, of all of the network junctions, $\mathbf{r} = [\mathbf{r}_1, \mathbf{r}_2, \ldots, \mathbf{r}_i, \ldots, \mathbf{r}_N]$, is just the product of the probability functions, $p(\mathbf{R}_{ij})$, for the individual chains, or, using equation (8.35),

$$p(\mathbf{r}) = \prod_{\text{chains}} p(\mathbf{R}_{ij})$$

$$= \prod_{i<j} \left(\frac{B_{ij}}{\pi}\right)^{3/2} e^{-B_{ij}|\mathbf{r}_j - \mathbf{r}_i|^2}. \tag{8.45}$$

The partition function, $Z(\mathbf{r})$, for the instantaneous configuration, $\mathbf{r}$, is, through equation (8.36),

$$Z(\mathbf{r}) = C \, e^{-\sum_{i<j} B_{ij}|\mathbf{r}_j - \mathbf{r}_i|^2}, \qquad (8.46)$$

where C is a constant. As above, the product in equation (8.45) and the sum in (8.46) are regarded as over *every pair* of points in the network. Pairs of points that are not actually connected by chains are handled by taking $B_{ij} = 0$.

The instantaneous configuration, $\mathbf{r}$, can be expressed in terms of the *average positions* of the junctions, $\bar{\mathbf{r}}_i$, and the *fluctuations* about the average positions, $\Delta \mathbf{r}_i = \mathbf{r}_i - \bar{\mathbf{r}}_i$. The average positions turn out to be equal to the most probable positions, those that maximize $Z(\mathbf{r})$ in equation (8.46). The latter is maximized by minimizing the argument in the exponential, $\sum_{i<j} B_{ij}|\mathbf{r}_j - \mathbf{r}_i|^2$. This expression can also be written in quadratic form as a product of $\mathbf{r}_j \cdot \mathbf{r}_i$ vectors by expanding $|\mathbf{r}_j - \mathbf{r}_i|^2$ as $\mathbf{r}_j^2 + \mathbf{r}_i^2 - 2\mathbf{r}_j \cdot \mathbf{r}_i$. This allows writing it in matrix form as

$$\sum_{i<j} B_{ij}|\mathbf{r}_j - \mathbf{r}_i|^2 = \mathbf{r}^{\mathrm{T}} \mathbf{K} \mathbf{r}, \qquad (8.47)$$

where $\mathbf{r}$ is a column vector containing the $\mathbf{r}_i$, $\mathbf{r}_i^{\mathrm{T}}$ is its transpose and the matrix $\mathbf{K}$ (called the *Kirchoff* matrix) has elements

$$K_{ij} = -B_{ij} \text{ for } i \neq j,$$

$$K_{ii} = \sum_{j(\neq i)} K_{ij}.$$

To avoid the trivial (and catastrophic) result that each chain vector, $\mathbf{R}_{ij} = \mathbf{r}_j - \mathbf{r}_i = 0$ on minimization of equation (8.47) and to introduce the connection to the macroscopic strain it is assumed that *some* of the network junctions are *constrained* in the sense of having no fluctuation, $\Delta \mathbf{r}$, but *follow affinely the macroscopic strain*. The fixed junctions could be conveniently thought of as lying on the sample surface, for example. Denote the constrained points by subscripts $\alpha, \beta, \ldots$ and the other free junctions by $\ldots p, q \ldots$. Let $\mathbf{r}(\alpha)$ represent the vector containing the $\ldots \mathbf{r}_\alpha \ldots$ constrained point positions and $\mathbf{r}(p)$ the one containing the free junction positions $\ldots \mathbf{r}_p \ldots$. Then the quadratic form of equation (8.47) can be written as

$$\mathbf{r}^{\mathrm{T}} \mathbf{K} \mathbf{r} = \mathbf{r}_\alpha^{\mathrm{T}} \mathbf{K}(\alpha, \alpha) \mathbf{r}_\alpha + 2\mathbf{r}(p)^{\mathrm{T}} \mathbf{K}(p, \alpha) \mathbf{r}(\alpha) + \mathbf{r}(p)^{\mathrm{T}} \mathbf{K}(p, p) \mathbf{r}(p), \quad (8.48)$$

where $\mathbf{K}$ has been partitioned into submatrices $\mathbf{K}(\alpha, \alpha)$, $\mathbf{K}(\alpha, p)$, $\mathbf{K}(p, p)$ containing the elements appropriate to the vectors indicated. Differentiation with respect to the x_p, y_p, z_p components of each $\mathbf{r}_p$ position vector in the $\mathbf{r}(p)$ vector and setting the result to zero leads to a set of linear equations for the positions of the junction points $\bar{\mathbf{r}}(p)$ that minimize equation (8.48)

$$\mathbf{K}(p, p)\bar{\mathbf{r}}(p) = -\mathbf{K}(p, \alpha)\mathbf{r}(\alpha). \tag{8.49}$$

Solving the above set formally in terms of the inverse, $\mathbf{K}(p, p)^{-1}$, of $\mathbf{K}(p, p)$ and anticipating that the average positions are the most probable, gives the average junction positions, $\bar{\mathbf{r}}(p)$, in terms of the constrained point positions, $\mathbf{r}(\alpha)$

$$\bar{\mathbf{r}}(p) = -\mathbf{K}(p, p)^{-1}\mathbf{K}(p, \alpha)\mathbf{r}(\alpha). \tag{8.50}$$

Notice that the average junction positions, $\bar{\mathbf{r}}(p)$, are linear functions of the constrained points, $\mathbf{r}(\alpha)$. Thus if the latter move affinely under deformation, as assumed, then the average junction positions will also. Thus the important conclusion that both the *constrained points and the average positions of the free junctions move affinely* is reached. It follows that the average chain vectors $\bar{\mathbf{R}}_{ij}$ spanning junctions i and j and given by $\bar{\mathbf{r}}_j - \bar{\mathbf{r}}_i$ deform affinely also.

The complete quadratic form, $\mathbf{r}^T\mathbf{Kr}$, of equation (8.47), which includes all the vectors, constrained and free, can be written through equation (8.48) in terms of the average junction positions and their fluctuations, $\mathbf{r}(p) = \bar{\mathbf{r}}(p) + \Delta\mathbf{r}(p)$ as

$$\mathbf{r}^T\mathbf{Kr} = \mathbf{r}(\alpha)^T\mathbf{K}(\alpha, \alpha)\mathbf{r}(\alpha) + 2[\bar{\mathbf{r}}(p) + \Delta\mathbf{r}(p)]^T\mathbf{K}(p, \alpha)\mathbf{r}(\alpha)$$
$$+ [\bar{\mathbf{r}}(p) + \Delta\mathbf{r}(p)]^T\mathbf{K}(p, p)[\bar{\mathbf{r}}(p) + \Delta\mathbf{r}(p)]. \tag{8.51}$$

This equation, when rearranged and when equation (8.49) is utilized, gives

$$\mathbf{r}^T\mathbf{Kr} = \bar{\mathbf{r}}^T\mathbf{K}\bar{\mathbf{r}} + \Delta\mathbf{r}(p)^T\mathbf{K}(p, p)\Delta\mathbf{r}(p), \tag{8.52}$$

where $\mathbf{K}$ is the complete Kirchoff matrix for all vectors and $\bar{\mathbf{r}}$ includes *both* the average positions of the network junction, $\bar{\mathbf{r}}(p)$ and the constrained points, $\mathbf{r}(\alpha)$. The partition function for an instantaneous configuration of the network, equation (8.46) in terms of equation (8.52), is

$$Z(\mathbf{r}) = C \, e^{-(\bar{\mathbf{r}}^T\mathbf{K}\bar{\mathbf{r}} + \Delta\mathbf{r}(p)^T\mathbf{K}(p,p)\Delta\mathbf{r}(p))}$$
$$= C \, e^{-\bar{\mathbf{r}}^T\mathbf{K}\bar{\mathbf{r}}} \, e^{-\Delta\mathbf{r}(p)^T\mathbf{K}(p,p)\Delta\mathbf{r}(p)}. \tag{8.53}$$

The above equation can be used to find the average of $\mathbf{r}_i = \bar{\mathbf{r}}_i + \Delta\mathbf{r}_i$ over the fluctuations, $\Delta\mathbf{r}_i$, by integration of $\mathbf{r}_i Z(\mathbf{r})$ over $\Delta\mathbf{r}(p)$ at fixed $\bar{\mathbf{r}}_i$ and normalizing. Thus $\langle\mathbf{r}_i\rangle = \langle\bar{\mathbf{r}}_i + \Delta\mathbf{r}_i\rangle = \bar{\mathbf{r}}_i + \langle\Delta\mathbf{r}_i\rangle$. However, evaluation with equation (8.53) leads to $\langle\Delta\mathbf{r}_i\rangle = 0$. This completes the identification of $\bar{\mathbf{r}}_i$ with the average of $\mathbf{r}_i$. Equation (8.53) also shows that an average-square fluctuation $\langle\Delta r_i^2\rangle$ will not be zero but will be independent of the $\bar{\mathbf{r}}_i$.

Thus the conclusion is reached that the junction fluctuations are independent of the state of strain. It also follows that the *instantaneous* vectors, r_i, do not deform affinely. Since in $r_i = \bar{r}_i + \Delta r_i$, the average vector $\bar{r}_i$ deforms affinely but the average-square value of the fluctuation Δr_i is independent of deformation, the rms r_i cannot deform affinely. This is in contrast to the affine model where the fluctuations are assumed to be zero and the affine assumption applied to the instantaneous vectors.

Equation (8.53) can be integrated over the fluctuations $\Delta r(p)$ to give a partition function for the network in terms of the constrained points and the *average positions* of the junctions, $\bar{r}$,

$$Z(\bar{r}) = C' e^{-\bar{r}^T K \bar{r}}$$

$$= C' e^{-\sum_{i<j} B_{ij}|r_j - r_i|^2}. \tag{8.54}$$

Using the chain vectors rather than the junction position vectors through

$$\bar{R}_{ij}^2 = |\bar{r}_i - \bar{r}_i|^2 \tag{8.55}$$

the free energy of the network, equation (8.37), becomes

$$A_{net}(r) = -k_B T\left(\ln C' - \sum_{i<j} B_{ij} \bar{R}_{ij}^2 \right). \tag{8.56}$$

Using the fact that the average chain vector positions in equation (8.55) move affinely, $\sum_{i<j} B_{ij} \bar{R}_{ij}^2$ can be written, on invoking isotropy of the undeformed network and in analogy with the development of equation (8.42), in terms of the macroscopic strains as

$$\sum_{i<j} B_{ij} \bar{R}_{ij}^2 = (1/3) \sum_{i<j} B_{ij} (\bar{R}_{ij}^0)^2 (\lambda_x^2 + \lambda_y^2 + \lambda_z^2), \tag{8.57}$$

where $(\bar{R}_{ij}^0)^2$ is the value of $\bar{R}_{ij}^2$ in the *undeformed network*. Thus the free energy based on zero in the undeformed state is, introducing the total number of chain segments, v,

$$A_{net}(\bar{r}) = \tfrac{1}{2} k_B Tv[(2/3)\langle B\bar{R}^2 \rangle_{net}(\lambda_x^2 + \lambda_y^2 + \lambda_z^2 - 3)] \tag{8.58}$$

and where in analogy to equation (8.43)

$$\langle B\bar{R}^2 \rangle_{net} = v^{-1} \sum_{i<j} B_{ij} (\bar{R}_{ij}^0)^2.$$

8.2.3 Comparison between the two versions

To summarize, both versions assume that the chains are Gaussian phantom chains. The affine model *a priori* assumes that the junction positions are fixed in any state of deformation and that they deform affinely under deformation. The phantom network model considers the junctions to be coupled together in the network. They are allowed to fluctuate. When some of the network points, those associated with the sample boundaries perhaps, are constrained to deform affinely, it follows that the average junction positions deform affinely. The junctions fluctuate about their mean positions with an rms displacement that is independent of deformation.

The result for the free energy of coupled phantom chains, equation (8.58), bears a close formal resemblance to the affine model result, equations (8.42) and (8.43). The group of terms inside the square bracket multiplying the strain term $(\lambda_x^2 + \lambda_y^2 + \lambda_z^2)$ in equations (8.42) and (8.43) is often called the 'front factor'. The affine model front factor, F_{aff} is thus given by

$$F_{aff} = \tfrac{2}{3}\langle BR^2 \rangle_{net} = \tfrac{2}{3}v^{-1} \sum_{i<j} B_{ij}(R_{ij}^0)^2 \qquad (8.59)$$

and by the phantom network model as

$$F_{pn} = \tfrac{2}{3}\langle B\bar{R}^2 \rangle_{net} = \tfrac{2}{3}v^{-1} \sum_{i<j} B_{ij}(\bar{R}_{ij}^0)^2, \qquad (8.60)$$

where in both,

$$B_{ij} = 3/(2\langle R_{ij}^2 \rangle_0) = 3/(2C_\infty l^2 N) \quad \text{for connected junctions}$$

$$\neq 0 \qquad \text{for unconnected junctions} \qquad (8.61)$$

and $\langle R_{ij}^2 \rangle_0$ is the unperturbed dimension of a *free* phantom chain of the same number of bonds, N, as in the chain connecting i and j. In the affine model case, the sum is over the square of each chain vector, assumed to be fixed, in the undeformed network. As it stands, in equation (8.59), there is no prescription for what the values of the chain vectors should be. It is just implied that any real undeformed network consists of some distribution of network junctions and vectors and that these are fixed in space. In contrast, the coupled phantom chain model, equation (8.60), involves a sum over the square of the *average* chain vectors in the undeformed network. It provides a prescription, equation (8.50), for their determination if the topology of the network is known. That is, if it is known what the connected junctions are and how many bonds there are in each connecting chain then the B_{ij} are known. This, in turn, allows the determination of the average chain vectors.

James and Guth (1947) performed a calculation for a network based on a cubic lattice and tetrahedral functionality in which they found $F_{pn} = \frac{1}{2}$. Under the argument championed by Flory concerning the distribution of vectors in the undeformed network, $F_{aff} = 1$, see equation (8.44). Thus there was disagreement about what the front factor should be. Since it is not known in general what the distribution of chain vectors (or the connectivity leading to the B_{ij}) is a direct approach to evaluation of the sums in equations (8.59) and (8.60) has not been pursued. However, some very useful further elaboration is possible (Duiser and Staverman, 1965; Graessley, 1975a,b; Ronca and Allegra, 1975; Flory, 1976). The affine and phantom networks can be compared by establishing *the difference between the front factors in the two versions*. This proceeds by expressing the average over the square of the mean chain vectors in equation (8.60) in terms of the average over the square of the instantaneous chain vectors and the average over the fluctuations from the mean positions (Graessley, 1975a,b; Staverman, 1982). Thus for $\mathbf{R}_{ij} = \mathbf{r}_j - \mathbf{r}_i$ and in terms of the fluctuation $\Delta\mathbf{R}_{ij} = \Delta\mathbf{r}_j - \Delta\mathbf{r}_i$,

$$\mathbf{R}_{ij} = \bar{\mathbf{R}}_{ij} + \Delta\mathbf{R}_{ij}$$

and because the linear terms vanish on averaging it follows that

$$\langle B_{ij}\bar{\mathbf{R}}_{ij}^2\rangle_{\text{net}} = \langle B_{ij}R_{ij}^2\rangle_{\text{net}} - \langle\langle B_{ij}\Delta R_{ij}^2\rangle\rangle_{\text{net}}. \qquad (8.62)$$
$$\;\;\;\text{phantom}\qquad\qquad\text{affine}\qquad\qquad\text{fluctuation}$$

In the fluctuation term, $\langle\langle\Delta R_{ij}^2\rangle\rangle_{\text{net}}$, $\langle\ \rangle$ represents the statistical mechanical average of ΔR^2 for each segment and $\langle\ \rangle_{\text{net}}$ represents as usual the average of this over all segments in the network. It is especially apparent in equation (8.62) that the difference between phantom network and affine versions lies in the neglect of the fluctuation term in the latter and that the phantom network front factor will be less than that for the affine version. In terms of junction position vectors, $\mathbf{r}$, $\Delta\mathbf{R}_{ij} = \Delta\mathbf{r}_j - \Delta\mathbf{r}_i$

$$\langle B_{ij}\Delta R_{ij}^2\rangle_{\text{net}} = v^{-1}\sum_{i<j}B_{ij}\Delta R_{ij}^2 = v^{-1}\sum_{i<j}B_{ij}(\Delta\mathbf{r}_j - \Delta\mathbf{r}_i)^2.$$

In analogy with equation (8.47) the above may be written as a quadratic form in the Δrs as $\langle B_{ij}\Delta R_{ij}^2\rangle_{\text{net}} = v^{-1}\Delta\mathbf{r}^{\mathsf{T}}\mathbf{K}\Delta\mathbf{r}$ and thus

$$\langle\langle B_{ij}\Delta R_{ij}^2\rangle\rangle_{\text{net}} = v^{-1}\langle\Delta\mathbf{r}^{\mathsf{T}}\mathbf{K}\Delta\mathbf{r}\rangle. \qquad (8.63)$$

It is apparent from equation (8.53) that the free energy contribution of the fluctuation term is

$$A_\Delta = -k_{\text{B}}T\ln Z_\Delta = k_{\text{B}}T\langle\Delta\mathbf{r}^{\mathsf{T}}\mathbf{K}\Delta\mathbf{r}\rangle.$$

The quadratic form can be diagonalized to a sum of squares by a unitary transformation. There are μ terms in the sum of squares where μ is the *number of network junctions*. The sum of squares has the same properties as the kinetic energy in classical statistical mechanics. The average value will be given by equipartition, or

$$A_\Delta = \tfrac{3}{2}\mu k_B T. \tag{8.64}$$

Thus $\langle \Delta \mathbf{r}^T \mathbf{K} \Delta \mathbf{r} \rangle = \tfrac{3}{2}\mu$ and the *difference* between the front factors is, according to equations (8.59) and (8.60), given by

$$F_{\text{aff}} - F_{\text{pn}} = \mu/\nu. \tag{8.65}$$

Individual identification of F_{aff} and F_{pn} is often established by making an *additional* assumption, one that has already been introduced in conjunction with equation (8.44). That is, if the network vectors have the same distribution as a set of *free chains* of the same length then because each $B_{ij} = 3/(2\langle R_{ij}^2 \rangle_0)$, equation (8.61), then

$$F_{\text{aff}} = 1. \tag{8.66a}$$

Under this circumstance it follows from equation (8.65) that

$$F_{\text{pn}} = 1 - \mu/\nu, \tag{8.66b}$$

where to recapitulate, ν is the number of chain segments in the network and μ is the number of junctions.

It is implicit that ν and μ refer to chains and junctions that are able to participate in deformation. Dangling chains or completely free chains would not, nor more generally, would junctions and chains that are not part of cycles in the network. Thus ν, μ are to be interpreted as the number of *elastically active* chains and junctions, i.e., those for which closed paths can be traced. However, in the phantom network case further insight can be achieved (Flory, 1976). Consider a network as shown in Figure 8.6. It may be seen that the number of independent cycles is equal to the number of cuts required, ξ, required to reduce the network to an *acyclic tree*, a branched structure containing no cyclic paths and that $\xi = \nu - \mu + 1$. The number of cuts, ξ, is called the cycle rank in graph theory. It is to be noticed that the phantom network free energy, equations (8.58), (8.60) and (8.66b), is proportional to $\nu F_{\text{pn}} = \nu - \mu$. Thus since in practice $\xi \gg 1$ the phantom network free energy is proportional to the cycle rank. The latter can be calculated from the total chain and junction count regardless of whether these are individually elastically active or not.

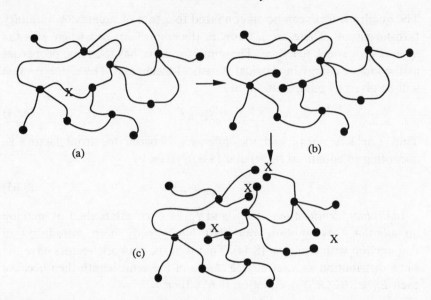

Fig. 8.6 Cycle rank and elastically active chains and network points. (a) A network containing a number of network points or junctions (chain ends are included as network points) as represented by filled circles. A variety of functionalities are shown including mono- (dangling chain ends), di-, tri- and tetra-. A cut is introduced at (X) in the network to produce (b). A cycle is broken and one new chain and two new network points are created. In (c) an *acyclic tree* is produced by introduction of three more cuts.

It is useful to relate the expression for the front factor to the functionality of the network junctions. For a perfect network, one in which all of the junctions and chains are elastically active, the number chains, v, is equal to half the junction crosslink functionality, f_{cr}, times the number of junctions, or

$$v = f_{cr}\mu/2.$$

Thus, in these terms

$$F_{aff} = 1$$
$$F_{pn} = 1 - 2/f_{cr}. \tag{8.67}$$

Corrections for imperfect networks under varying circumstances have been derived (Dossin and Graessley, 1979; Pearson and Graessley, 1978, 1980; Queslel and Mark, 1985).

Both the affine and phantom network models in their original contexts were intended to treat the same physical situation, an elastomer consisting of a network of Gaussian chains. In this context the phantom network treatment is clearly superior. The affine deformation of the average junction positions follows from the model rather than being imposed on individual non-fluctuating chain vectors as an additional assumption. It is clear that the affine model is not correct in this sense. It is also clear that the difference in result between the two versions is simply the free energy of junction fluctuation. Therefore, were it not for other factors, there would be little reason beyond historical interest and pedagogically somewhat simpler mathematics (elimination of the optimization of junction positions step) for presenting the affine version. However, it will be seen later that attempted improvements on the classical theory have involved concepts related to junction fluctuations. In this sense the relation between the nonfluctuating affine model and the phantom network is instructive.

8.2.4 Stress–strain relations

Since the unique properties of elastomers are embodied in their mechanical behavior, derivation of stress–strain curves from the molecular theory is called for. This is accomplished via the connections already alluded to concerning the free energy and the retractive force. In a tensile experiment for example, equation (8.4) gives such a relation. Generally, a stress, force per unit area, is given by the derivative of the free energy per unit volume with respect to a strain. For a tensile experiment the engineering tensile stress, σ_e, the force in the x direction divided by the original cross-sectional area, is given by equation (8.4) in terms of the strain $\lambda_x = l_x/l_x^0$ as

$$\sigma_e = \frac{f}{(l_y^0 l_z^0)} = \left(\frac{1}{l_x^0 l_y^0 l_z^0}\right)\left(\frac{\partial A}{\partial \lambda_x}\right)_{T,V} = V_0^{-1}\left(\frac{\partial A}{\partial \lambda_x}\right)_{T,V}. \tag{8.68}$$

In applying this relation to the free energy expression of the molecular theory, with the front factor written generically as F, or

$$A_{net} = \tfrac{1}{2}\nu F k_B T(\lambda_x^2 + \lambda_y^2 + \lambda_z^2 - 3), \tag{8.69}$$

it is necessary to express the strain ratios in the transverse directions, y, z, in terms of that in the stretch direction, x. This can be done by invoking the constant volume condition, $\lambda_x \lambda_y \lambda_z = 1$, along with the condition that the contraction in the transverse directions should be isotropic or $\lambda_y = \lambda_z$. This leads to $\lambda_y = \lambda_z = \lambda_x^{-1/2}$ and to

$$A_{net} = \tfrac{1}{2}\nu F k_B T(\lambda_x^2 + 2\lambda_x^{-1} - 3). \tag{8.70}$$

Application of equation (8.68) and designating λ_x more simply by λ gives

$$\sigma_e = \bar{v}Fk_B T(\lambda - \lambda^{-2}) \qquad (8.71)$$

for the stress–strain relation for tensile deformation and where $\bar{v}$ has been defined as the number of elastically active chains per unit volume, $\bar{v} = v/V_0$.

It is to be concluded from the classical molecular theory that there is but one material parameter required to characterize the elastomer. That parameter is the number of chains per unit volume, $\bar{v}$, or, on invoking the functionality, the crosslink density. Thus a rubber can be made more or less stiff, within limits, by manipulating the crosslink density. The upper limit for the stiffness achievable is set by the requirement that the chain segments must be long enough that the statistical behavior associated with long chains is observed. At the lower limit a coherent network must be present. There is nothing in the theory to indicate dependence on the chemical structure of a particular type of chain. However, it is to be kept in mind that in order for the statistical behavior associated with long chains to be operative, visitation of the many configurations of a chain segment that spans two network junctions must be present. This implies thermal motion must be rapid. Thus the glass temperature is an important parameter specific to various chemical structures in permitting rubbery behavior.

8.3 Experimental behavior

A brief description of mechanical behavior in bulk elastomers is given here. It is, of course, of considerable interest to examine the extent to which the molecular theory is in accord with these experiments. In this context it has already been pointed out that there are two important aspects of the theory that need addressing. These are the ability of the functional form of the stress–strain curves derived from the theory to represent the data and the extent to which the front factors given by either of the versions are quantitatively correct. Swelling equilibrium studies have also played an important role in rubber elasticity concepts and this subject is also taken up.

8.3.1 Stress–strain curves

An example of the application of the classical tensile stress–strain relation, equation (8.71), to elongational data is shown in Figure 8.7. In principle,

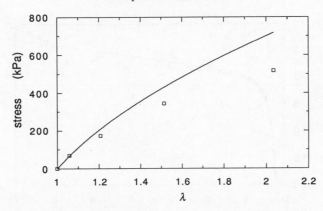

Fig. 8.7 Engineering stress for a rubber sample plotted vs elongation ratio (open squares) compared with the calculated stress from classical molecular theory, equation (8.71) (curve). The calculated curve was adjusted to fit the data near the origin at low elongation.

the concentration of chains, $\bar{v}$, could be known from the density of crosslinks and their functionality. However, this number is difficult to determine *a posteriori* and requires carefully controlled crosslinking reactions of known chemistry to determine it through the synthesis reactions leading to the network. Much of the time $\bar{v}F$ must be regarded as an adjustable parameter. Quantitative determinations of $\bar{v}$ and the implications for the actual values thus inferred for the front factor, F, from comparisons of stress–strain data will be taken up below. In Figure 8.7, the value of $\bar{v}Fk_BT$ has been adjusted to fit the data at low elongation. It may be seen that the theoretical curve has the general shape of the experimental data, the slope decreases with increasing elongation. However, the data fall systematically below the theory as elongation increases. This phenomenon is found to be quite general in bulk elastomers and represents a shortcoming of the classical molecular theory.

It is of interest to note that uniaxial stress–strain curves are not confined to tension or values of the elongation ratios greater than one but may include as well compression and elongation ratios less than one. Figure 8.8 shows strain–strain data over a range of elongation ratios that include both tension and compression. Also shown is the fit of the classical theory, equation (8.71). It can be seen that the fit in the compression region is good. From this more comprehensive viewpoint, the deviations in the classical theory seem less severe.

Much effort has made to explain the deviations from the classical

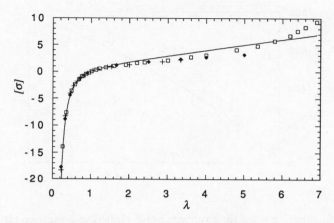

Fig. 8.8 Stress–strain curves for several elastomers in tension–compression. The curve is for the relation, $\sigma_e = (\lambda - \lambda^{-2})$. The stress data for the different materials, $[\sigma]$, have been scaled to fit this relation near $\lambda = 1$: crosses natural rubber (Rivlin and Saunders, 1951); open squares, natural rubber (Treloar, 1944); filled diamonds, PDMS (Flory and Pak, 1979). After Higgs and Gaylord (1990; by permission Butterworth Heinemann Ltd ©).

molecular theory and to improve it. Before discussing some of these efforts in the succeeding section (Section 8.4) it is useful to describe here a phenomenological approach that has found wide use in representing the deviations. First some results of macroscopic elasticity theory are recapitulated. Description of deformation requires, in general, six strains, three tensile and three shear strains. However, by choice of axes, i.e., the principal axes, the strain tensor can be diagonalized. Consonant with this there are three descriptors that are invariant to the choice of axes. The three strain invariants, I_1, I_2, I_3, can be written in terms of principal axis strains λ_1, λ_2, λ_3 as $I_1 = \lambda_1^2 = \lambda_2^2 + \lambda_3^2$, $I_2 = \lambda_1^{-2} + \lambda_2^{-2} + \lambda_3^{-2}$ and $I_3 = \lambda_1\lambda_2\lambda_3$. Thus it is seen after making the identification of x, y, z with 1, 2, 3 that the free energy from the molecular theory is a function of the first strain invariant, I_1. It is an empirical observation that addition of a term in the second invariant I_2 to give a free energy function, known as the Mooney–Rivlin equation (Mooney, 1940, 1948; Rivlin, 1948)

$$A = C_1 I_1 + C_2 I_2, \tag{8.72}$$

where C_1 and C_2 are adjustable constants, tends to represent fairly well the deviations from the $C_1 I_1$ term alone of the molecular theory. For the example of tensile deformation, it follows from equation (8.68) that

$$\sigma_e = 2C_1(\lambda_x - \lambda_x^{-2}) + 2C_2(1 - \lambda_x^{-3})$$

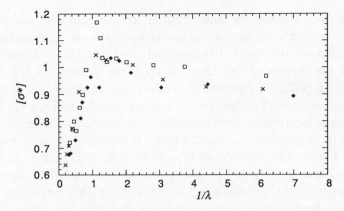

Fig. 8.9 Mooney plots of reduced stress, $\sigma^* = \sigma_e/(\lambda - \lambda^{-2})$ vs $1/\lambda$, for several elastomers in tension–compression, same data as for Figure 8.8. Stress data for different materials are scaled, $[\sigma^*]$, to fit near $\lambda = 1$ as in Figure 8.8. After Higgs and Gaylord (1990; by permission Butterworth Heinemann Ltd ©).

and on dropping the subscript x,

$$= 2(\lambda - \lambda^{-2})(C_1 + C_2/\lambda). \tag{8.73}$$

This suggests plotting elongational data as the quantity, $\sigma_e/(\lambda - \lambda^{-2})$, called the *reduced stress*, vs $1/\lambda$, a procedure known as a Mooney plot. Examples of such plots are shown in Figure 8.9. It may be seen that in the tension region the data are, in fact, rather linear and in compression the slope is nearly zero as is consistent with the good fit apparent in Figure 8.8. In general Mooney plots are useful for emphasizing the deviations from the classical relation equation (8.71).

8.3.2 *Stiffness and crosslink density*

The classical theory predicts that the stress is directly proportional to the number of elastically active chains and hence, in perfect networks, to the crosslink density. The proportionality constant is given by the theory, but is different for the two versions. As indicated, the question of comparing front factors from the classical theory with those from stress–strain data is non-trivial because experimental knowledge of the number of elastically active chains is difficult to achieve. However, considerable interest has been taken in preparing model networks where the crosslinking reactions are understood and their extent can be monitored. The deduction of front factors is also obviously complicated by the fact that the theoretical relations that establish the connection of stress–strain with the density of chain segments do not do a quantitative job in representing the stress–strain relation itself. Experimental work in this connection has centered on expressing the effect of crosslink density on the initial modulus. According to equation (8.71) the initial tensile modulus, $E_{11} = (\partial \sigma_e / \partial \lambda)_{\lambda = 1} = 3\nu F k_B T$. Since for an incompressible material the small strain tensile modulus is three times the shear modulus, the initial shear modulus, G is given by $G = \bar{\nu} F R T$ where $\bar{\nu}$ is expressed in moles per unit volume. Figure 8.10 shows data (Gottlieb *et al.*, 1981) taken on poly-(dimethylsiloxane) (PDMS) networks that are end-linked by agents of several functionalities, specifically 3, 4 and several high functionality ones. The initial shear moduli are plotted against $\bar{\nu} R T$. Thus the slope of these plots is the front factor, F. Recall that according to equation (8.67) for the affine version $F = 1$ and for the phantom network $F = 1 - 2/f_{cr}$. These are quite different dependencies. For the phantom network, F ranges from $\frac{1}{3}$ for the trifunctional crosslinked networks to nearly 1 for the high functionality ones, in contrast to the affine case where there is no dependency on functionality at all. According to Figure 8.10, there is indeed dependence on functionality and it approximates that predicted for the phantom network case. The occurrence of an intercept of finite G from the extrapolation to zero crosslink concentration is of great importance also as will be seen below.

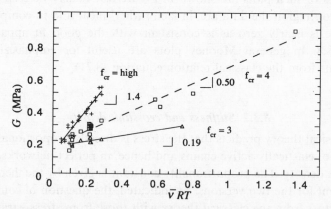

Fig. 8.10 Initial shear modulus vs strand density times RT for end-linked PDMS networks with different functionalities, f_{cr}, of the crosslinks. The strand density was computed directly from the molecular weight of the starting oligomers and the density and thus assumes a perfect network with no corrections for dangling ends, only data where these corrections would be relatively modest were used (after Gottlieb *et al.*, 1981 with permission from *Macromolecules*, © 1981, American Chemical Society).

8.3.3 Swelling

In a favorable solvent, a collection of linear chains can be dispersed on a molecular scale into the random coil state. Dilute solutions, for example, are useful in the characterization techniques of Chapter 3. If, however, the same molecules have been crosslinked to form a network then molecular dispersion to a dilute solution is no longer possible. Since the same basic affinity for the solvent exists the process of dissolution is replaced by that of swelling. The free energy associated with the network will be present in conjunction with the free energy of solution similar to that obtaining in the linear molecule case. As the network expands, the elastic free energy associated with it will increase for the same reasons that it does under mechanical deformation. Thus the network will swell when immersed in solvent until the decrease of solution free energy with dilution is balanced by the increase of the elastic free energy of the network and equilibrium is reached. If (1) the solution and the elastic free energy are indeed separable into additive terms and (2) adequate models exist for both free energy contributions, then equations relating the degree of swelling to solution and network parameters can be developed.

The condition of additivity of elastic free energy of the network and of mixing of solvent and polymer may be written

$$\Delta A = \Delta A_{\text{net}} + \Delta A_{\text{mix}}, \tag{8.74}$$

where it is implied that ΔA_{mix} is that for a solution of uncrosslinked molecules of the same composition as the swollen network. The chemical potential of the solvent designated as component 1, is therefore

$$\mu_1 = \left(\frac{\partial A}{\partial n_1}\right)_{T,P} = \left(\frac{\partial A_{\text{net}}}{\partial n_1}\right)_{T,P,n_2} + \left(\frac{\partial A_{\text{mix}}}{\partial n_1}\right)_{T,P,n_2}. \tag{8.75}$$

The classical theory of swelling uses the Flory–Huggins expression for the contribution from the mixing term, utilizing equation (9.29), or,

$$\mu_1 = \mu_1^0 + RT(\ln \phi_1 + \phi_2 + \chi\phi_2^2) + (\partial A_{\text{net}}/\partial n_1)_{T,P,n_2}, \tag{8.76}$$

where μ_1^0 is the chemical potential of the pure solvent, ϕ_1, ϕ_2, are the volume fractions of the solvent and polymer respectively, χ, is the interaction parameter and the effective DP, x, of the polymer is taken to be very high. To find the contribution from the rubber elasticity term it is necessary to express the deformation in terms of the degree of swelling. For isotropic swelling, let $\lambda_s = \lambda_x = \lambda_y = \lambda_z$ and thus $\lambda_s^3 = V/V^0$ where V is the swollen network volume and V^0 is the unswollen polymer volume.

V is approximated by $n_1 V_1^0 + V^0$ where V_1^0 is the molar volume of the pure solvent. Since $\phi_2 = V^0/(n_1 V_1^0 + V^0)$, it follows also that $\lambda_s^3 = \phi_2^{-1}$. Applying $\lambda_x = \lambda_y = \lambda_z = \lambda_s$ to the network free energy results in

$$A_{net} = \tfrac{3}{2}\nu F k_B T \lambda_s^2$$

$$= \tfrac{3}{2}\nu F[(n_1 V_1^0 + V^0)/V^0]^{2/3}. \tag{8.77}$$

Differentiation of A_{net} with respect to n_1 gives

$$(\partial A_{net}/\partial n_1)_{T,P,n_2} = \bar{\nu} F R T V_1^0 \phi_2^{1/3}, \tag{8.78}$$

where $\bar{\nu} = \nu/V^0$, the number of elastically active chains per unit volume in the unswollen polymer. At equilibrium the chemical potential of the solvent in the swollen crosslinked sample, μ_1, will be equal to the chemical potential of the pure solvent μ_1^0. Applying this to equation (8.76) and utilizing equation (8.78) gives

$$\ln \phi_1' + \phi_2' + \chi \phi_2'^2 = -\bar{\nu} F V_1^0 \phi_2'^{1/3}, \tag{8.79}$$

where the values of the volume fractions of polymer in a sample in equilibrium *with pure solvent* are denoted as ϕ_1', ϕ_2'. The number of chain segments per unit volume in the network can be expressed in terms of an average molecular weight per segment or between crosslinks, M_c, through $\bar{\nu} = \rho^0/M_c$ where ρ^0 is the density of the unswollen polymer. Thus this equation provides a relation between the maximum degree of swelling, i.e., equilibrated in pure solvent, $\lambda_s'^3 = \phi_2'^{-1}$ and the degree of crosslinking as expressed by the average molecular weight of the segments. The latter, being related to the number of elastically active chains, is subject to estimation from the original linear polymer molecular weight as was commented on in conjunction with equation (8.67). If the degree of swelling $\phi_2'^{-1}$ is high, as is often observed, expansion of the $\ln \phi_1 = \ln(1 - \phi_2)$ term gives, for an affine network where $F = 1$,

$$(\phi_2'^{-1})^{5/3} = (\tfrac{1}{2} - \chi)M_c/(\rho^0 V_1^0), \tag{8.80}$$

a result known as the Flory–Rehner (1943b) equation, see also Flory (1950, 1953).

The above developments invoke a specific model for both the free energy of mixing and for the elastic free energy of the network. The model for the mixing can be displaced in favor of experimental measurements of the solvent activity in the swollen network. According to equation (3.1) the solvent chemical potential can be expressed in terms of measured vapor pressures over the solution (the swollen rubber) as

$\mu_1 = \mu_1^0 + RT \ln(P_{1c}/P_1^0)$ where P_{1c} and P_1^0 are the vapor pressures of the solvent over the *crosslinked* swollen rubber and over the pure solvent respectively. Use of this relation in equation (8.75) results in

$$\mu_1^0 + RT \ln(P_{1c}/P_1^0) = (\partial A_{\text{mix}}/\partial n_1)_{T,P,n_2} + (\partial A_{\text{net}}/\partial n_1)_{T,P,n_2} \quad (8.81)$$

and its application to a solution of *uncrosslinked* linear polymer of the *same composition* gives

$$\mu_1^0 + RT \ln(P_{1\text{un}}/P_1^0) = (\partial A_{\text{mix}}/\partial n_1)_{T,P,n_2}, \quad (8.82)$$

where $P_{1\text{un}}$ is the vapor pressure of the solvent over the uncrosslinked polymer solution. The original additivity hypothesis dictates that the A_{mix} derivatives in the two equations be equal at the same composition and thus

$$RT \ln(P_{1c}/P_{1\text{un}}) = (\partial A_{\text{net}}/\partial n_1)_{T,P,n_2}. \quad (8.83)$$

Therefore under the additivity hypothesis the elastic term can be evaluated experimentally as a function of solvent composition by a series of pairs of measurements of vapor pressures over the swollen networks and their uncrosslinked counterparts.

From equation (8.78) evaluating the elastic free energy of the network according to the classical rubber elasticity theory gives the following,

$$RT \ln(P_{1c}/P_{1\text{un}}) = \bar{v}FRTV_1^0 \phi_2^{1/3} = \bar{v}FRTV_1^0 \lambda_s^{-1}. \quad (8.84)$$

Thus according to the classical theory $\lambda RT \ln(P_{1c}/P_{1\text{un}})$ should be independent of the solvent composition of the swollen rubber. Figure 8.11 shows such a plot (Brotzman and Eichinger, 1982). It may be seen that this is far from the case. Thus an inadequacy of the classical rubber elasticity theory is indicated.

In the above, it was seen how vapor pressure measurements can be used to evaluate experimentally the mixing free energy in order to evaluate the elastic free energy contribution. In analogy with this, mechanical experiments can be used to measure the elastic contribution and thus evaluate the mixing term contribution. This requires an assumption about the mechanical behavior but it is a more general one than adopting a specific model. This more general assumption is that the mechanical strain energy function is additively separable as functions in the three stretch directions λ_x, λ_y, λ_z, a condition proposed by Valanis and Landel (1967). Under the Valanis–Landel assumption the elastic contribution can be directly measured. In the case of the swelling equilibrium in pure solvent,

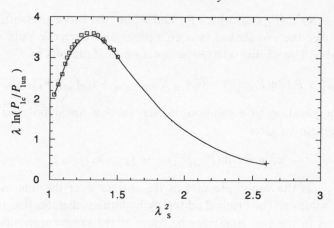

Fig. 8.11 Test of swelling equilibrium theory from solvent activity measurements. According to classical molecular theory and the additivity hypothesis, equation (8.84), the quantity $\lambda_s RT \ln(P_{1c}/P_1)$ plotted as ordinate should be independent of the degree of swelling, λ_s. The point, $+$, is the swelling equilibrium point with pure solvent (from Brotzman and Eichinger, 1982 with permission from *Macromolecules*, © 1982, American Chemical Society).

equation (8.79) becomes

$$RT(\ln \phi'_1 + \phi'_2 + \chi\phi'^2_2) = -V^0_1 w'(\lambda')/\lambda'^2, \qquad (8.85)$$

where $w'(\lambda)$ is the derivative of the elastic free energy with respect to λ and is determined experimentally via mechanical measurements. Results show the inadequacy of the Flory–Huggins left-hand side. The inferred χ parameters are found to be a function of the degree of crosslinking in a series of experiments as a function of the latter (McKenna, Flynn and Chen, 1988).

Another test of the additivity hypothesis expressed in equation (8.74) can be made from mechanical measurements, in this case utilizing stress–strain curves. In reducing the elastic free energy of the network to a function of the degree of swelling the free energy was scaled using $\lambda^3_s = \phi^{-1}_2$. The additivity hypothesis requires that the network elastic free energy depend only on this geometric factor and not specifically on the presence of solvent. This requires, for example, that the elastic free energy be determinable for the swollen state characterized by λ_s from knowledge of the elastic free energy as a function of deformation in the unswollen state. Thus stress–strain behavior in the swollen state should be determined by that in the unswollen one. This has been tested for deformation in

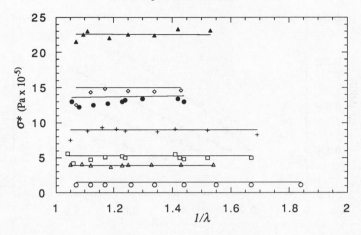

Fig. 8.12 Model-independent test of the additivity hypothesis. Reduced stress vs $1/\lambda$ for elongation of natural rubber swollen to equilibrium in methyl ethyl ketone. Each set of points is for a different degree of crosslinking with dicumyl peroxide. The solid curves are calculated from a strain energy function determined from mechanical measurements on each of the unswollen rubbers (from McKenna *et al.*, 1989 with permission *Macromolecules*, © 1989, American Chemical Society).

compression (McKenna, Flynn and Chen, 1989). Torsional measurements were used to determine the strain energy function for dry rubber. Then this function was used to calculate the uniaxial compression stress–strain behavior for swollen specimens. Here it was found that a good representation of the swollen state stress–strain relations can be obtained from the dry state measurements, Figure 8.12. This would indicate that, even though the molecular theories used to represent ΔA_{net} and ΔA_{mix} are inadequate, the additivity hypothesis embodied in $\Delta A = \Delta A_{net} + \Delta A_{mix}$ may still be valid.

Finally, an interesting empirical observation with respect to swelling and stress–strain behavior can be made. It is found that the deviations of the stress from the classical theory as uniaxial elongation increases (Figure 8.9) become less severe as swelling increases. Figure 8.13 shows Mooney plots for elongation for samples at varying degrees of swelling. It is to be seen that the slope decreases markedly with increasing swelling indicating an approach to the classical theory behavior in highly swollen systems. It may also be noticed in Figure 8.12 where the samples are highly swollen (to equilibrium with the solvent) that the Mooney plot slopes, *albeit* in compression, are effectively zero.

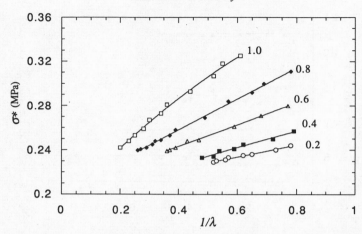

Fig. 8.13 Mooney plot (reduced stress vs $1/\lambda$) for sulfur crosslinked polybutadiene samples swollen with 1,2,4-trichlorobenzene to the various volume fractions of polymer indicated (from Brotzman and Mark, 1986 with permission from *Macromolecules*, © 1986, American Chemical Society). The curves shown smooth the data.

8.4 Further development of molecular theory

From the discussions in Section 8.3 it may be fairly concluded that the classical theory captures the physical essence of the statistical behavior of long chain molecules that leads to rubber-like elasticity. However, there are significant quantitative deficiencies with respect to the stress–strain relations and other properties. Accordingly there has been great effort made in probing the reasons for inadequacy and in developing better theory. In this context it is useful to review the basis of the classical theory, the assumption of Gaussian chains. The Gaussian statistics require long chains, extensions that are short compared to the contour length and phantom behavior. The chain lengths in a typical rubber network, a hundred or more bonds on average, are of sufficient length that the chain statistics are reasonably well approximated by the Gaussian function, see Section 6.2.3. Similarly the deviations in the classical theory appear even at moderate extensions where the Gaussian approximation should be adequate, see Section 6.3.4. With respect to phantom behavior, in the case of bulk systems of *uncrosslinked* chains there is compelling evidence that surrounding chains act as a theta solvent for a given chain and that unperturbed behavior and chain dimensions are observed. Thus, *per se*, phantom behavior is not an unrealistic assumption. However, consideration of network formation on crosslinking leads to serious reservations.

The network model as enunciated requires phantom behavior of the connecting chains. Visitations of the conformations required to validate the model of phantom behavior may well be impossible due to the topological constraints of the network. Two chains may intertwine in such a way that they are effectively permanently constrained from passing through each other whereas in an uncrosslinked melt with sufficient time this is not the case. Thus attention is focused on the entangling effects of network formation on chain behavior. There have been two types of approach to addressing inclusion of entanglement effects in the theory. One of these seeks to include the effects of topological constraints in the behavior of the existing network junctions, those created chemically on network formation. This is the *constrained junction-fluctuation* approach. The other group of theories seeks to model more explicitly the constraining effects of the entanglements along the chains themselves. These will be called generically, *entanglement models*. The constrained junction-fluctuation approach is directly related to the classical theory and the issues of affine vs phantom network behavior already discussed and is taken up first.

8.4.1 Constrained junction-fluctuation models

It is recalled that the phantom network treatment, Section 8.2.2, is the direct transcription of the basic idea of the classical theory, the response of a network of phantom Gaussian chains. The affine version, Section 8.2.1, imposes the additional condition that the network junctions do not fluctuate about their mean positions. The resulting stress–strain curves are of the same form but when evaluated under the same assumptions about network topology the front factors are different. In the phantom network model the junction fluctuations are Gaussian and thus spherical in symmetry and are independent of deformation. The constrained junction-fluctuation approach is based on the presumption that the effect of topological constraints will be to alter in some way the junction fluctuations. Ronca and Allegra (1975) who first proposed a model based on the concept assumed that the fluctuations are distorted by deformation. An attractive interaction between the network junctions, compared to the undistorted fluctuation case, is introduced by assuming that the junction fluctuations do not remain spherical on elongation but are distorted affinely into ellipsoids at small deformations. This results in affine behavior of the instantaneous vectors as well as the average ones and thus the affine model is recovered at low elongations. The distortion of the fluctuations is given a form that does not allow the distortion to continue indefinitely with elongation but approaches an asymptotic limit. This limitation gives the phantom network result at high elongations. Since the magnitudes of the fluctuations are not disturbed, only their isotropies, and the transition between the low strain and the high strain behavior is determined by minimization of the free energy, no new parameters are introduced. Their result for the free energy is

$$A_{net} = \tfrac{1}{2}(v - \mu)k_B T I_1 - \tfrac{1}{2}\mu k_B T \ln(I_3^2/I_1^3), \qquad (8.86)$$

where $I_1 = \lambda_1^2 + \lambda_2^2 + \lambda_3^2$ and $I_3 = \lambda_1\lambda_2\lambda_3$. At constant volume where $I_3 = 1$, the above gives for the engineering stress for a tensile test,

$$\sigma_e = V_0^{-1}k_B T[3\mu I_1^{-1} + (v - \mu)](\lambda - \lambda^{-2}). \qquad (8.87)$$

At $\lambda \to 1, I_1 \to 3$ and, as stated, the affine result, equation (8.65a), is found and at large λ the phantom network relation, equation (8.65b), is approached. It was observed above (Section 8.3.1) that experimentally the deviations from the classical theory were in the direction of the measured stress falling further below the calculated values as the strain increases. The transition between the affine result (front factor = 1 for tetrafunctional crosslinks) and the phantom network result (front factor = $\tfrac{1}{2}$ tetrafunctional) is thus in the right direction for conforming with experiment. In tension the theory is in reasonable agreement. In compression experimentally the variation of reduced stress with strain is much less than in tension. The theory finds this also but not to the degree found experimentally.

Another version of constrained junction-fluctuation theory has been developed by Flory (1977) and coworkers (Erman and Flory, 1978). The effect of network topological constraints is assumed to be such that the average phantom network junction positions are displaced by the constraints. An adjustable parameter, κ, that measures the strength of this displacement effect as the ratio of the mean-square network junction fluctuation to the mean-square size of the constraint domain fluctuation is introduced. In a later version (Flory and Erman, 1982) another parameter, ζ, that allows the constraint domain size to depend on deformation was introduced. The free energy is given by,

$$A_{net} = \tfrac{1}{2}(v - \mu)k_B T I_1$$

$$+ \tfrac{1}{2}\mu k_B T \sum_{i=1,3} \{(1 + g_i)b_i - \ln[(b_i + 1)(g_i b_i + 1)]\}, \qquad (8.88)$$

where

$$b_i = (\lambda_i - 1)(1 + \lambda_i - \zeta\lambda_i^2)/(1 + g_i)^2,$$

$$g_i = \lambda_i^2[\kappa^{-1} + \zeta(\lambda_i - 1)],$$

$$I_1 = \lambda_1^2 + \lambda_2^2 + \lambda_3^2.$$

This expression reduces to the phantom network result for $\kappa = 0$ and to the affine result for $\kappa \to \infty$. At high elongations, like the Ronca–Allegra theory, the phantom network result is approached but as $\lambda \to 1$ the degree of approach to the affine result depends on κ. With the two adjustable parameters, κ, ζ, available in addition to $v - \mu = v(1 - 2/f_{cr})$, the Flory–Erman theory has been found to be able to fit a variety of data (Erman and Flory, 1982; Gottlieb and Gaylord, 1983, 1984, 1987; Brotzman and Mark, 1986), see Figure 8.14. For the parameters chosen, the degree of approach to the affine result at $\lambda \to 1$ is rather complete (Brotzman and Mark, 1986). The question of network junction functionality dependence is an important one and will be addressed further below. It is to be noticed here that the junction fluctuation suppression theory makes two predictions with respect to the stress–strain behavior. The first is that since affine behavior is found at low elongation there is no dependence of initial modulus on junction functionality. Second, since phantom behavior is found at high elongations, the difference between the low strain modulus

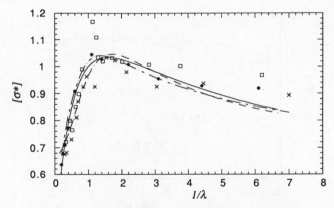

Fig. 8.14 Fit of theories to elongation data. Mooney plots of scaled reduced stress. Points are the same data as in Figure 8.9. The chain curve is the constrained junction model (Flory and Erman, 1982) with $\kappa = 10$, $\zeta = 0$; the dashed curve is the slip-link replica model with $\eta = 0.82$, $N_s/\mu = 3/2$ (Ball, Doi, Edwards and Warner, 1981); and the solid curve is the localization model with $G_e/3 = 3G_{class}$ (after Gaylord and Douglas, 1987, 1990).

and the high strain value, and thus the Mooney–Rivlin C_2, will be greater for lower functionality networks than for high ones. The question of junction functionality and initial modulus will be discussed below.

8.4.2 Entanglement theories

Entanglement theories are based on the premise that the topological constraints trapped in the network have a direct effect on elasticity, one not modelled solely by altering the behavior of existing junction points. That is, the constraints contribute to the stiffness in addition to the chemically introduced crosslinks. These theories are motivated by the experimental observation that when the stiffness, as represented by the initial modulus, is plotted vs crosslink density a non-negligible intercept is present on extrapolation to zero concentration of the latter. This was alluded to in Section 8.3.2, see Figure 8.10. Uncrosslinked melts, as noted, display viscous flow rather than rubber elasticity. However, if the molecular weight is high, the viscosity is extremely high and the viscous flow regime is delayed to long times. It is well known (Ferry, 1981) that in such high molecular weight polymers a well-developed rubbery plateau region is displayed in plots of shear modulus vs log time, see Figure 8.15. In long chains the density of entanglements should be independent of chain length and indeed the plateau modulus is independent of molecular

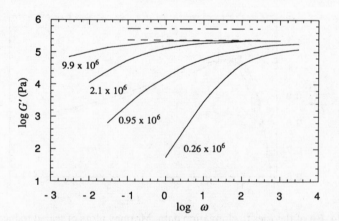

Fig. 8.15 Plateau modulus. Dynamic small strain shear modulus vs angular frequency for four high molecular PMDS samples (full curves, molecular weight values shown). Also shown are two end-linked networks (oligomer mol wts: 18 000 (dashed line) and 4000 g/mole (chain line)) (from Gottlieb *et al.*, 1981 with permission from *Macromolecules*, © 1981, American Chemical Society).

weight. This strongly suggests that, in fact, entanglements contribute directly to the modulus.

If the contributions to the initial shear modulus, G, of the network junctions and the entanglements are regarded as additive (Langley, 1968; Dossin and Graessley, 1979), then one way of expressing the initial modulus that accommodates, in principle, both the entanglements and the possibility of junction fluctuation suppression is (Pearson and Graessley, 1980)

$$G = (\bar{v} - h\bar{\mu})k_{\mathrm{B}}T + G_0 T_{\mathrm{e}}, \tag{8.89}$$

where the first term is the phantom network contribution, G_0 is the contribution from entanglements in the high molecular weight bulk polymer limit and T_{e} is the fraction of such trapped entanglements active in a network that may not be complete or perfect. An empirical factor, h, with possible values between 0 and 1 multiplies the fluctuation contribution, $-\mu k_{\mathrm{B}}T$, and is introduced to allow for behavior spanning complete junction fluctuation suppression and complete participation. The T_{e} factor can be estimated under favorable circumstances from the premise that the entanglement probability between two chains is the product of probabilities of each being an elastically active chain. Figure 8.16 is a plot similar to Figure 8.10. However, it represents an attempt to account in non-perfect networks for varying degrees of entanglement trapping through the factor T_{e} and also to correct $\bar{v}$ for elastically inactive chains in incomplete networks. These corrections allow the inclusion of a wider set of data than in Figure 8.10. It appears that the dependence of initial modulus on crosslink functionality of the type represented by equation (8.89) is discernible here as well as in Figure 8.10. The trend of the slopes follows that expected for non-zero h. That is, for the phantom network relation, $h = 1$, and $v - \mu = v(1 - 2/f_{\mathrm{cr}})$, perfect networks with the values $f_{\mathrm{cr}} = 3$, 4, and higher, give slopes of $\frac{1}{3}$, $\frac{1}{2}$ and ~ 1. The tetrafunctional slope is found to be higher than 0.5 so that some suppression or $h < 1$ would be indicated. Figure 8.17 shows similar plots for randomly cross-linked, and therefore tetrafunctional, networks of PMDS and and PBD (polybutadiene). For PBD the slope is found to be very near 1 so that $h \sim 0$ is indicated. In all cases, however, there is non-negligible intercept. Thus two conclusions seem justified. (1) It appears to be well established that trapped entanglements, described by a G_0 value characteristic of the polymer chain type, contribute directly to the modulus. (2) There is some evidence that the junction-fluctuation contribution may be present and not completely suppressed in the initial modulus.

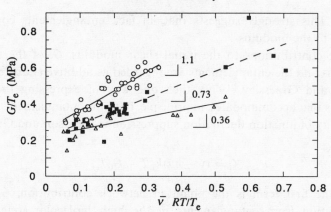

Fig. 8.16 Dependence of initial shear modulus, G, on crosslink density and functionality for end-linked PMDS networks. G is divided by T_e, the fraction of melt entanglements trapped in the network and plotted against the concentration of elastically active chains, $\bar{v}$, times RT/T_e. Unlike Figure 8.10, the concentration of elastically active chains has been determined using corrections for incompleteness of the network and includes more data than the former figure. Triangles are for trifunctional crosslinks, squares for tetrafunctional and circles for high functionality. Slopes for the linear fit to each set are shown (after Gottlieb *et al.*, 1981 with permission *Macromolecules*, © 1981, American Chemical Society).

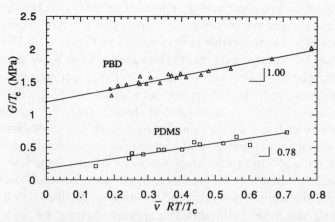

Fig. 8.17 Dependence of initial shear modulus, G, on crosslink density for tetrafunctional randomly crosslinked PBD and PMDS networks. G/T_e is plotted vs $v \, RT/T_e$ as in Figure 8.15 (after Dossin and Graessley, 1979; Gottlieb *et al.*, 1981 with permission *Macromolecules*, © 1979, 1981 American Chemical Society).

If it is accepted that these experiments indicate that entanglements contribute directly to stiffness and that phantom network behavior is at least approximately observed in the initial modulus, then the basis for the constrained junction-fluctuation theories is to some degree compromised. In the latter case, the initial modulus is predicted to possess affine behavior with no network functionality dependence and no additive effect on the modulus of permanently trapped entanglements is incorporated. Thus the motivation for models with more explicit inclusion of the topological constraints exists.

If the entanglements are also to be responsible for the deviations in the stress–strain relations from classical behavior then the theory should include more than simply ascribing to the entanglements crosslink-like behavior. Such models presumably would associate more mobility or degrees of freedom with topological constraints than implied by fixed crosslinks. One such model is based on the notion of the *slip-link* (Ball *et al.*, 1981). An entangled, yet to some degree mobile, pair of chains is represented as in Figure 8.18. The introduction of the slip-link constraint is handled statistically mechanically by averaging the free energy through a general *replica formalism*. The result is the following expression for the free energy,

$$A = \tfrac{1}{2}\mu k_{\mathrm{B}} T(\lambda_1^2 + \lambda_2^2 + \lambda_3^2) + \tfrac{1}{2} N_{\mathrm{s}} k_{\mathrm{B}} T \sum_{i=1}^{3} \left[\frac{\lambda_i^2(1 + \eta)}{1 + \eta\lambda_i^2} + \ln(1 + \eta\lambda_i^2) \right],$$

$$(8.90)$$

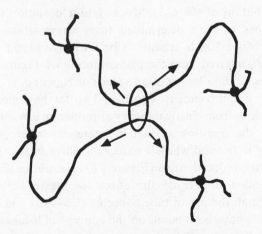

Fig. 8.18 Topological constraint modelled by a slip-link. The chains are forced to be in contact but are free to slide through the link.

where the λ_i are the extension ratios, μ is the number of crosslinks, N_s is the number of slip-links and η is a parameter characterizing the extent of slippage along the chain contour lengths. The free energy is seen to be additive in the phantom network result (for a tetrafunctional network that is implied by the treatment and for which $v - \mu = \mu$) and the effect of the N_s slip-links. The slip parameter, η, can be estimated in theory by free energy minimization and by assuming that each slip-link can slide on average half the distance between a crosslink or another slip-link. This result in $\eta = 0.234$. The above relation leads to $3d_0 k_B T$ for the initial tensile modulus, where

$$d_0 = \mu + N_s(1 + 6\eta + 3\eta^2)/(1 + \eta)^2. \qquad (8.91)$$

For the above value of η this gives $(\mu + 1.69N_s)k_B T$ for the initial shear modulus. Therefore, in principle, it is possible to determine N_s from data analysis similar to that connected with Figure 8.16 and equation (8.89) through

$$G_0 = N_s(1 + 6\eta + 3\eta^2)/(1 + \eta)^2 k_B T$$

$$= 1.69N_s k_B T \quad \text{(for } \eta = 0.234), \qquad (8.92)$$

where G_0 is either the plateau shear modulus or determined through plots associated with equation (8.89). However, such G_0 values are not always available.

It has been found that values of η determined from stress–strain curve fitting (Thirion and Weil, 1984; Higgs and Gaylord, 1990) are reasonably consistent with the theoretical estimate. The ratio N_s/μ determines the relative contribution of the C_2/λ Mooney plot deviation from classical behavior. Values of N_s/μ determined from stress–strain curve fitting (Mooney plot slopes) also appear to be fairly consistent with what is known about N_s inferred from the plateau modulus. Figure 8.14 shows a fit to experimental data for adjusted values of N_s/μ and η.

All the theory improvements considered so far have been concerned with the deviations from classical theory at moderate extensions and have not addressed the question of higher extensions where the Gaussian picture is likely to fail and where a hardening effect rather than softening is observed in the reduced stress (Figure 8.8). The above slip-link theory has been extended to include the effects of inextensibility at higher extensions through the use of tube concepts (Edwards and Vilgis, 1986).

Other theories have been based on the concept of localization in tubes as the primary effect in modeling of topological constraints and have been successful in fitting data (Gaylord, 1982; Gaylord and Douglas, 1987;

Higgs and Gaylord, 1990). In the localization model the source of confinement or constraint is associated with finite chain volume. The confinement free energy is proportional to the square of the segment density and thus to the plateau modulus. Local tube sections are randomly distributed in the undeformed state. In deformation the tube diameters change so as to preserve tube volume. The free energy expression that results is given by the sum of the classical free energy of the network and the entanglement or localization contribution, or,

$$A_{net} = \tfrac{1}{2}G_{class}(\lambda_x^2 + \lambda_y^2 + \lambda_z^2 - 3) + (G_e/3)(\lambda_x + \lambda_y + \lambda_z - 3), \quad (8.93)$$

where G_{class} is vFk_BT and G_e is an adjustable constant representing the entanglement effect. This equation is identical to one derived previously by Edwards and Stockmayer (1973) from a somewhat different viewpoint. The latter authors modeled restricted access to the configurational space of a chain by regarding the end-to-end vector as the sum of two vectors. One of these is free to access configurational space during deformation but the other is frozen into a fixed value. Over the network, the latter fixed values form a Gaussian distribution. The resulting bivariate Gaussian distribution, including both types of vectors, leads to equation (8.93).

The fit to data from this equation is illustrated in Figure 8.14. In a later version (Gaylord and Douglas, 1990) the entanglement contribution, G_e, is regarded as proportional to both the plateau modulus and the crosslink density, or

$$G_e = \gamma(vk_BT) + G_N^*, \quad (8.94)$$

where γ is a constant and G_N^* is the plateau modulus of the uncrosslinked melt. The linear dependence of G_e on crosslink density with a finite G_N^* intercept has been confirmed experimentally in randomly crosslinked natural rubber specimens (McKenna *et al.*, 1991). The functional form of equation (8.92) was also found to fit the stress–strain data very well through the tension and compression regions. It has also been applied to swollen rubbers (Douglas and McKenna, 1992).

A modification of the constrained junction-fluctuation theory has been introduced by Erman and Monnerie (1989). In this model the fluctuations of the radii of gyration of the network chains are assumed to be hindered rather than the junctions. This has the effect of spreading out the effects of constraints along the chain contours. In this respect the theory is somewhat similar to the localization models above. It has been applied to stress–strain data for solution cross-linked PDMS by Erman and Mark (1992).

8.5 The chemistry of crosslinking

Since the physical basis of rubber elasticity lies in establishment of a network, it is appropriate to consider how networks can be achieved synthetically. There are several basic strategies. One starts with a pre-existing collection of high molecular weight polymer chains and introduces chemical linkages, i.e., crosslinks, between them at random positions in the chains. The stiffness of the network, which is determined by the average strand length, is thus determined by the number of crosslinks introduced. Therefore the amount of crosslinking agent and/or time and efficiency of reaction will be key variables in determining the stiffness. This method requires that the chains be susceptible to reaction in the desired manner. Chains lacking this chemical propensity can often be modified by copolymerization with relatively small amounts of comonomer units that do have the requisite capacity for reaction. In this case the number of reactive comonomer units may well be the parameter determining the modulus since reaction with the latter can be carried out until near completion. Another strategy is to use a preexisting polymer of the segment length desired in the final network and to end-link these chains with a multifunctional linking agent. Thus the stiffness is determined by the starting polymer or, as usually described, 'oligomer', since the molecular weight will be rather low. It is also possible to construct an elastomeric network in a single step via condensation polymerization with a mixture of difunctional and monofunctional monomers as described in Chapter 2.

Several routes are available for the introduction of random crosslinks into a polymer, resulting in different physical characteristics. The crosslinks can range in size from a single carbon–carbon bond to a long chain and often differ in chemical microstructure from the polymer being crosslinked. There are also particulate crosslinks in which the polymer chains are chemically attached to a foreign particle. Additionally phase separation in the form of crystals or hard amorphous domains produces crosslinked systems.

Rather than classify crosslinks according to the chemical route followed, classification will be based on reference to the physicochemical character of the crosslink produced. With few exceptions it will be found that the methods of random crosslinking for elastomers fall neatly into one or other of the sections of the classification scheme.

The subgroups are as follows:

(i) single bond crosslinks;

(ii) more complex crosslinks of similar chemical character to the polymer;
(iii) more complex crosslinks of different chemical character from the polymer;
(iv) particulate crosslinks;
(v) domain links;
(vii) crystallization.

The last three have sometimes been referred to as 'virtual' crosslinks.

8.5.1 *Single bond crosslinks – peroxides*

Single covalent links between chains can only be produced through the use of free radical attack on the original polymer to produce randomly distributed crosslinks. Two techniques commonly used for free radical production are degradation of peroxides and irradiation. The peroxide method is taken up first.

The general scheme of peroxide degradation is illustrated in **8.1**,

$$R\text{-}O\text{-}O\text{-}R \rightarrow 2\,R\text{-}O\,\cdot$$

8.1

oxygen–oxygen bond scission usually being achieved by raising the temperature (De Boer and Pennings, 1976). The free radicals produced usually terminate through abstraction of a hydrogen from the polymer chain. A crosslink results when two so-produced polymeric radicals terminate by combination thereby creating a single bond crosslink, **8.2**.

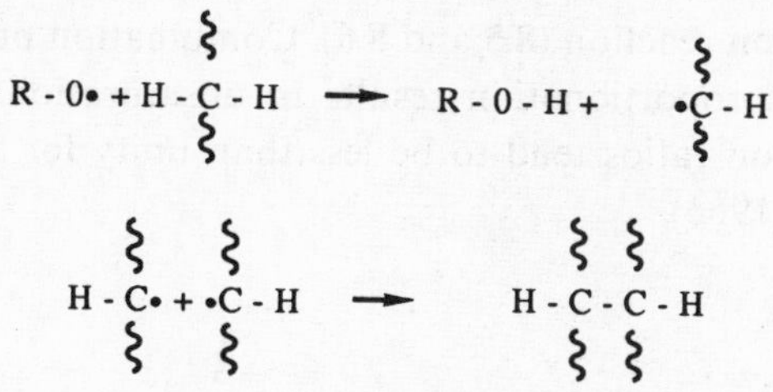

$$R\text{-}O\cdot + H\text{-}C\text{-}H \longrightarrow R\text{-}O\text{-}H + \quad \cdot C\text{-}H$$

$$H\text{-}C\cdot + \cdot C\text{-}H \longrightarrow H\text{-}C\text{-}C\text{-}H$$

8.2

The most commonly encountered peroxide used for crosslinking purposes is dicumyl peroxide, **8.3**. Dissociation kinetics are well known for this system (Ainberg, 1964; Harpell and Walrod, 1973) the reaction normally being carried out somewhere in the range 150–200 °C. A complication

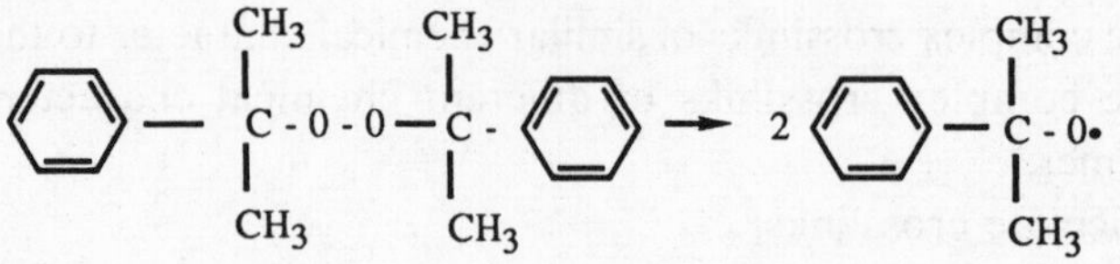

$$\underset{CH_3}{\overset{CH_3}{\phi - C - O - O - C - \phi}} \longrightarrow 2 \; \phi - \underset{CH_3}{\overset{CH_3}{C - O\cdot}}$$

8.3

results from the ability of the cumyloxy free radical to rearrange ejecting an acetophenone molecule thereby creating a methyl radical, see **8.4**.

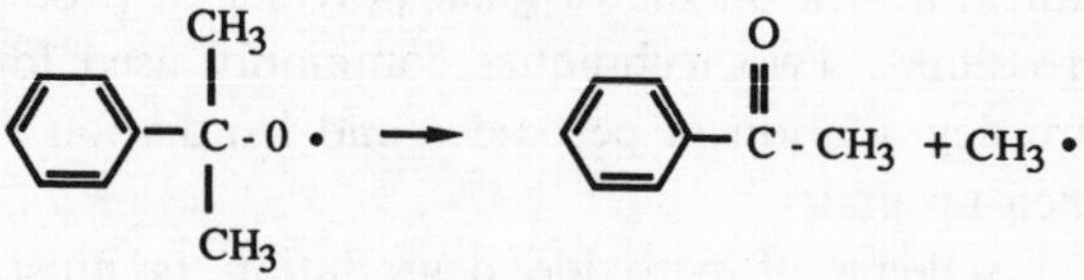

$$\phi - \underset{CH_3}{\overset{CH_3}{C - O\cdot}} \longrightarrow \phi - \overset{O}{\underset{}{C}} - CH_3 + CH_3\cdot$$

8.4

Both the cumyloxy and methyl radicals can abstract hydrogen atoms from hydrocarbon chains (e.g., cis(polyisoprene) or polyethylene) producing 2-phenyl-2-propanol and methane, respectively. The latter can result in unwanted bubbles and even foam formation unless the process is carried out under elevated pressures. A less important but significant side reaction produces α-methyl styrene.

After the hydrogen has been abstracted from the polymer chain the polymeric radicals can terminate through either a combination or disproportionation reaction (**8.5** and **8.6**). Combination produces a cross-link whereas disproportionation results in unsaturation. Combination/disproportionation ratios tend to be less than unity for secondary alkyl radicals (Pryor, 1966).

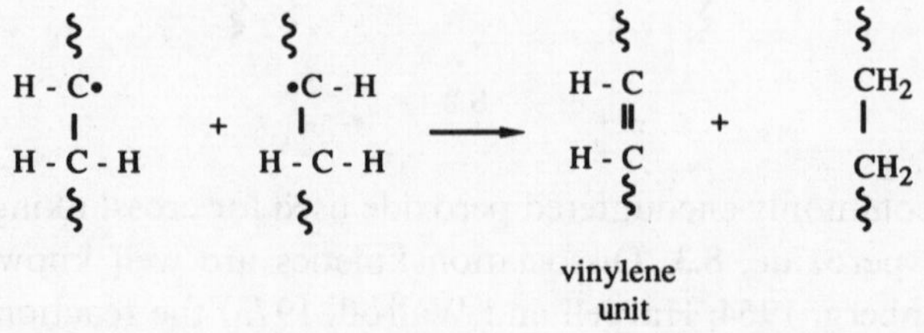

$$\begin{array}{ccc}
\overset{\wr}{\underset{\wr}{H-C\cdot}} + \overset{\wr}{\underset{\wr}{\cdot C-H}} \longrightarrow & \overset{\wr}{\underset{\wr}{\underset{H-C}{\overset{H-C}{\parallel}}}} + & \overset{\wr}{\underset{\wr}{\underset{CH_2}{\overset{CH_2}{\mid}}}} \\
H-C-H \quad\quad H-C-H & & \\
\end{array}$$

vinylene
unit

8.5

$$
\begin{array}{ccccccc}
\text{\{} & & \text{\{} & & \text{\{} & & | \\
H\text{-}C\bullet & & \bullet C\text{-}H & & H\text{-}C & & CH_2 \\
| & + & | & \longrightarrow & \| & + & | \\
R\text{-}C\text{-}H & & H\text{-}C\text{-}H & & R\text{-}C & & CH_2 \\
\text{\{} & & \text{\{} & & \text{\{} & & | \\
\text{branchpoint} & & & & \text{vinylidene} & & \\
& & & & \text{unit} & &
\end{array}
$$

8.6

An alternative internal rearrangement, similar to that of the cumyloxy radical, produces unsaturation and is known as β-scission since the chain breaks at the bond two atoms distant from a branch point. It tends to occur almost exclusively in branched chain radicals (Patel and Keller, 1975), see **8.7**.

$$
\begin{array}{ccc}
\overset{\bullet}{\text{\char`\~\char`\~} \, C\text{-}CH_2\text{-}CH_2 \text{\char`\~\char`\~}} & & \text{\char`\~\char`\~} \, C=CH_2 + \bullet CH_2 \, \text{\char`\~\char`\~} \\
| & \longrightarrow & | \\
R & & R \\
\text{branchpoint} & & \\
& \text{pendant vinyl} &
\end{array}
$$

8.7

Other complications can result from the presence of oxygen dissolved in the polymer since peroxy radicals are easily produced through addition. The crosslink junctions produced can be regarded as being branch points having four arms for the purposes of rubber elasticity studies.

The final product of a crosslinking reaction contains both the gel network and extractable (sol) molecules. This is determined in part by the initial molecular weight distribution since the probability of a molecule reacting with a free radical is directly proportional to its length. Small molecules tend therefore to be incorporated into the network to a lesser degree. The presence of a chain scission reaction results in sections of chain being broken off from the network also contributing to the sol fraction. For crosslinking agent concentrations sufficient to incorporate all initial molecules into the network the sol fraction is due to chain scission alone. The molecular weight distribution of the sol fraction is therefore dependent on the concentration of crosslinking agent used, as is the value of the sol fraction itself. The amount of scission reaction occurring can be estimated using the Barton (1968) modification of the Charlesby–Pinner equation (1959), originally derived for crosslinking/scission by irradiation,

$$
S + S^{1/2} = p_0/q_0 + 1/q_0 X_N r, \tag{8.95}
$$

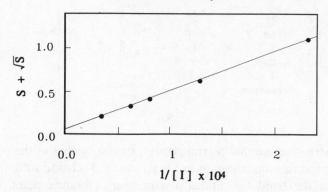

Fig. 8.19 Charlesby–Pinner–Barton plot of $S + S^{1/2}$ vs reciprocal initiator concentration for 200 °C cure of low density polyethylene with dicumyl peroxide (Karakelle, 1986).

where S is the soluble fraction, p_0 and q_0 are the proportion of monomer units involved in a crosslink or main-chain scission, respectively, per unit radiation dose (r) and X_N is the number average degree of polymerization. For chemical crosslinking r becomes the number of free radicals and the p_0 and q_0 are on a per radical basis. The equation implicitly assumes that the presence of crosslinks does not influence scission and that the molecular weight distribution is exponential or 'most probable', see equation (2.24), in character. A plot of $S + S^{1/2}$ vs $1/r$ or $1/[\text{initiator}]$ will give as an intercept the ratio, p_0/q_0, and hence the amount of chain scission occurring. Such a plot is shown in Figure 8.19 for low density polyethylene crosslinked using dicumyl peroxide; the level of scission was 5%. Extrapolation to $S = 1$ gives the gel point.

Estimation of the number of crosslinks produced is a difficult matter. Several techniques are available, none of which is perfect. For systems such as the one currently under discussion the best technique is equilibrium swelling. The basic principle behind this technique is an extension of the Flory–Huggins theory of solution to an insoluble network (see Section 8.3.3). Originally derived by Flory and Rehner (1943b), the approach balances the osmotic force of swelling with the rubber elastic retractive force of the swollen network as described in Section 8.3, hence the expression, equilibrium swelling. In order to extract a value for the crosslink density, expressed as the molecular weight between crosslinks, M_c, either the Flory–Rehner equation (8.80),

$$(\phi_2'^{-1})^{5/3} = (1/2 - \chi)M_c/\rho^0 V_1^0,$$

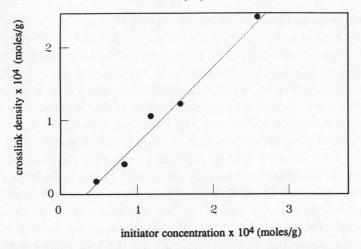

Fig. 8.20 Crosslink density vs initiator concentration for the cure of low density polyethylene with dicumyl peroxide (Karakelle, 1986). Note the negative intercept, indicating ~6% chain scission.

or the full equation (8.79)

$$\ln \phi_1' + \phi_2' + \chi_1 \phi_2'^2 = -\nu F V_1^0 \phi_2'^{1/3}$$

is used. The major problem connected with their use is the choice of the polymer solvent interaction parameter, χ. If ϕ_2' is determined as a function of initiator concentration then the efficiency of the reaction can be estimated. This is carried out by plotting crosslink density vs initiator concentration and taking a half of the slope since each initiator molecule generates two radicals. Such a plot is shown in Figure 8.20 for low density polyethylene crosslinked by dicumyl peroxide.

8.5.2 *Irradiation*

Irradiation is often carried out on polymers in the semi-crystalline state as a means of ensuring dimensional stability during the accidental melt down of a finished product. It is not generally an efficient means of crosslinking large objects because of absorption effects. Studies have rarely been carried out as a function of temperature or in the melt because of the difficulty of specimen temperature control in a nuclear reactor chamber. Common sources used are α-radiation and electron beams.

There are several complications arising from the use of radiation. First, the level of chain scission increases. Secondly, additional reactions are possible between the α quanta and the bonds. Hydrogen radicals (atoms) are liberated creating free radicals on the chain which themselves may attack double bonds and other radicals. An additional problem is that of the relative efficiency of crosslinking in the amorphous vs crystalline phases. Although it would be reasonable to suggest that crosslinking would be more efficient in the crystals because of the closeness of packing, this turns out not to be the case. The interchain spacing in the crystal is much greater than the carbon–carbon bond length. Considerable lateral motion of the chains would be required to complete the bond and this freedom of adjustment is denied in the crystal. In the amorphous phase the long-chain mobility facilitates reaction. There seems to be agreement that crosslinks are not present in the crystalline phase. It has been suggested (Patel and Keller, 1975) that radicals are formed in the crystalline phase but that they migrate to the fold surface and that crosslinks are concentrated in the fold surface.

It is likely that in solid state irradiation under low chain mobility conditions, radical migration occurs prior to the production of a crosslink. Migration is believed to occur through a series of hydrogen abstraction reactions and may be intra (**8.8**) or intermolecular (**8.9**) in nature. It is

$$\sim CH_2-\overset{\bullet}{C}H-CH_2\sim \quad\longrightarrow\quad \sim\overset{\bullet}{C}H-CH_2\text{-}CH_2\sim$$

8.8

$$\sim CH_2-\overset{\bullet}{C}H-CH_2\sim$$
$$+ \qquad\qquad\longrightarrow\qquad\qquad$$
$$\sim CH_2-CH_2\text{-}CH_2\sim$$

$$\sim CH_2-CH_2\text{-}CH_2\sim$$
$$+$$
$$\sim CH_2-\overset{\bullet}{C}H-CH_2\sim$$

8.9

believed that the details of the mechanism do not involve free hydrogen atoms since other radicals would react with them in preference to another polymeric radical and such a side reaction has not been observed.

The overall crosslinking reaction can be written as in **8.10**. There should therefore be a correspondence between crosslink formation and hydrogen

$$\sim CH_2-CH_2\text{-}CH_2\sim$$
$$+$$
$$\sim CH_2-CH_2\text{-}CH_2\sim$$

$$\longrightarrow$$

$$\sim CH_2-CH-CH_2\sim$$
$$\mid \qquad\qquad\qquad + H_2$$
$$\sim CH_2-CH-CH_2\sim$$

8.10

evolution. However, side reactions have prevented a valid materials balance equation from being confirmed experimentally.

When polymers other than polyethylene are irradiated more complications arise. For instance in the case of polypropylene the pendant methyl groups contain a high concentration of hydrogen and have a significant influence on the process. In the case of polydienes the mechanism remains basically as described for polyethylene, however, isomerization of the double bond can occur. The latter reaction is not of significance in cis(polyisoprene) (natural rubber) but is important for polybutadiene and its copolymers (e.g., SBR). α-Radiation is believed to produce an excited state of the orbitals of the double bond (**8.11**). The trans form predominates

8.11

resulting in a 2:1 trans:cis ratio which is determined by the relative populations of the cis and trans forms in the excited state. This is in contrast to the influence of ultraviolet radiation in the presence of a photosensitizer where the photosensitizer molecule becomes physically attached creating a free radical (**8.12**). Here the trans:cis ratio is determined by the energetics of bond rotation and hence the potential energy for rotation in the free radical state. Polybutadiene also undergoes isomerization in the presence of peroxides, unlike cispolyisoprene.

In all cases, however, the crosslink remains a single bond joining the two chains together. However, if the carbon–hydrogen bond scission responsible for initiation of the process occurs in a pendant methyl group the chain-to-chain crosslink can consist of two or three carbon–carbon bonds dependent on whether one or two methyl groups are involved.

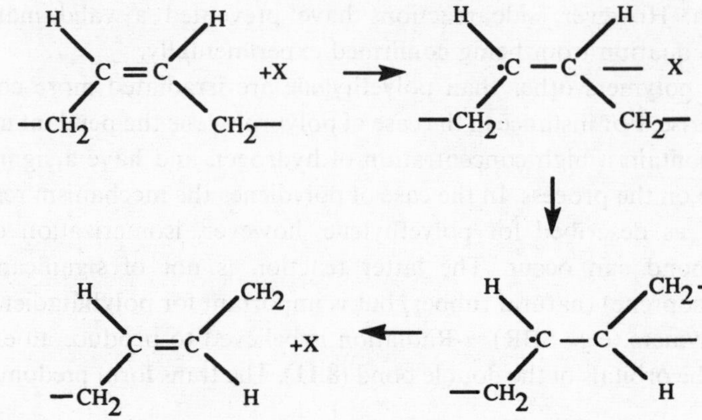

8.12

8.5.3 More complex crosslinks of similar chemical character to the polymer

Systems such as these are not commonly encountered and, when they are, they tend to be produced by copolymerization. An example would be styrene–divinyl benzene where the comonomer has the structure shown in **8.13**. In this case the relative placement of the two monomer units in

$$CH_2 = CH - \langle \rangle - CH = CH_2$$

8.13

a polymer chain will be determined by the reactivity ratios of the system (see Section 4.3). For a system such as this one, where the chemical nature of the two monomers is similar, the placement would be close to random.

8.5.4 More complex crosslinks of different chemical character

The best-known, but possibly least-understood mechanistically, example of this class is sulfur crosslinking of rubbers (Saville and Watson, 1976). The case of natural rubber, cis(polyisoprene), has been investigated intensively because of its commercial importance. Crosslinks are produced at the methylene unit adjacent to the main-chain carbon atom to which the methyl group is attached (see **8.14**). The value of 'x' may vary from one to eight and is determined by the activator system used. A major complication is the formation of heterocyclic rings in the main chain through an important side reaction (**8.15**). Usually the value of 'y' is

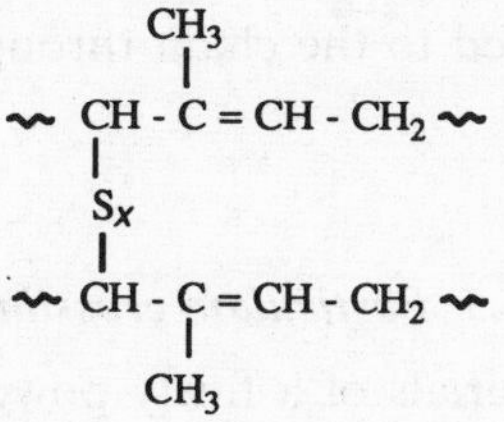

$$
\begin{array}{c}
CH_3 \\
| \\
\sim CH - C = CH - CH_2 \sim \\
| \\
S_x \\
| \\
\sim CH - C = CH - CH_2 \sim \\
| \\
CH_3
\end{array}
$$

8.14

$$
\begin{array}{c}
CH_3 \\
| \\
C - CH_2 \\
\sim CH_2 \diagdown \qquad \diagup CH_2 \\
S_y - CH \\
| \\
C = CH_2 \sim \\
| \\
CH_3
\end{array}
$$

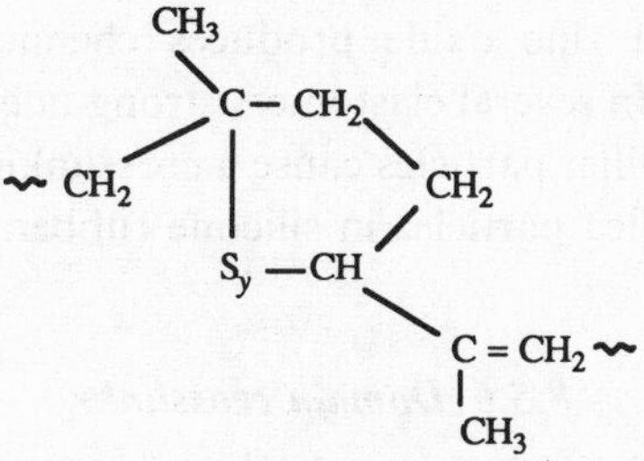

8.15

unity and there can be as many heterocycles formed as there are crosslinks. It is the formation of these chain stiffening rings that causes the glass transition temperature to increase drastically, not the insertion of the crosslinks.

The original Goodyear process consists of 10 parts of sulfur to 100 parts of natural rubber, the cure being carried out at 140 °C for 4–10 hours. A measure of the effectiveness of the reaction is the efficiency factor, E, which is defined as the number of sulfur atoms needed to produce one crosslink. For the Goodyear process E is between 40 and 100! There are 6–10 sulfur atoms per crosslink, and some unreacted sulfur, but the bulk of the remaining atoms is present in heterocycles. A more effective process uses 100 parts natural rubber, 2.5 parts sulfur and 0.5 parts of an accelerator together with small amounts of zinc oxide and either lauric or stearic acid. A cure time of 30–60 min at 140 °C gives $E \sim 15$–20 with crosslink populations of S_6 (80%), S_2 (10%) and S_1 (10%). The most effective process, which is expensive because of the amount of accelerator involved, uses 100 parts natural rubber, 0.4 parts sulfur and 6 parts accelerator together with zinc oxide and lauric acid. A cure time of 30–60 min at 140 °C in this case produces $E = 2$–3 and 85% monosulfur and 15% disulfur crosslinks. Obviously it also creates close to one heterocycle for each crosslink. A small number of branches having an accelerator residue attached to the chain through a sulfur atom are also produced.

8.5.5 *Particulate crosslinks*

Here solid inorganic materials of a finely powdered form serve as both crosslinks and filler particles. In polychloroprene conventional cross-linking does not function because of the chlorine groups and crosslinking is achieved through the addition of zinc oxide. Reaction between the polychloroprene and zinc oxide produces chemical crosslinks at the particulate interface. In several elastomers strong interfacial forces between the polymer and the filler particles cause a crosslinking effect. An example of this behavior is silica particles in silicone rubber.

8.5.6 *Domain crosslinks*

Polymers such as polyurethanes and ionomers (e.g., neutralized copoly-mers of ethylene with methacrylic acid) are phase separated because of polarity differences. Ionomers are believed to have small ionic domains of, as yet, undetermined size and structure. The polyethylene chain sections between methacrylate groupings link the domains as an essentially semi-crystalline polyethylene phase. In polyurethanes based on polyether or polyester soft segments the hard short aromatic urethane sections are believed to form small needle-like domains. In both systems these 'internal composite' particles serve as the crosslinks thus providing elastomeric behavior. The 'A–B' block copolymer of styrene and butadiene is also an interesting example. Phase separation of the two segments occurs. When the monomer ratio is such that the polystyrene segments are the minor component the domains of the latter are hard high T_g spheres dispersed in a rubbery matrix of low T_g polybutadiene. Such materials, e.g., 'Kraton©' rubber, have the interesting property that they can be formed by conventional thermoforming techniques such as injection moulding as the hard styrene domains reversibly soften above the T_g of the latter. This is in obvious contrast to the irreversibly chemically linked networks. This phenomenon is found in many domain crosslinked structures.

8.5.7 *Crystallization*

The existence of elasticity in a semi-crystalline polymer such as poly-ethylene is a direct result of individual polymer chains being present in more than one crystal. Otherwise the polymer would behave as a slurry consisting of crystallites floating in a viscous liquid. The crystals therefore serve as a means of 'crosslinking' the molecules of a viscous liquid. If the degree of crystallinity is very low rubber-like elasticity may be manifested. Chlorinated polyethylene or vinyl acetate/ethylene copolymers would be examples.

8.6 Structures of elastomers

8.6.1 Glass formation and rubbery behavior

In order to exhibit elastomeric behavior a polymeric material must be a dynamically mobile network. In order to possess the requisite segmental mobility the polymer obviously must not be vitrified and hence must have a glass temperature considerably below the temperature of use. This means therefore that the list of elastomers is essentially a list of polymers that have glass temperatures well below room temperature. Usually this means a T_g value of the order of $-50\,°C$ or lower. There is no theory or model of how chemical structures of polymers are related to T_g or that explains why some polymers have exceptionally low T_g values or others have high ones. There are some comments that could be useful. It is intuitively attractive to suppose that low barriers to internal rotation about skeletal bonds would be conducive to segmental mobility and that large complex rigid chemical groups either in the main chain or as side group substituents would lead to large swept-out volumes for segmental reorientation and thus result in high T_g values. To some degree these notions seem to be useful. Most elastomers do have relatively chemically simple flexible chain structures. The carbon–carbon double bonds found in a number of hydrocarbon elastomer backbones are, of course, not flexible but the single bonds adjacent to them have rather low barriers to rotation. The question of the role of bond rotational barrier *per se* is not so simple. For example, polyisobutylene is very sterically crowded and probably has a higher rotational barrier than about other carbon–carbon single bonds. However, it has a low T_g and is an important rubber. This disubstituted 'vinylidene' effect in giving low glass temperatures is general and obtains in polyvinylidene chloride and fluoride as well. This points to simplicity of the chain cross-section or surface as an important parameter.

8.6.2 Crystallization and elastomers

A not so immediately obvious but essential requirement that an elastomer must possess is that it does not crystallize under normal use conditions.

Any chemically and sterically regular structure can, in principle, crystallize. The kinetics of crystallization are aided by backbone structures that are simple and flexible. Thus the factors that mitigate for low glass temperatures also increase the likelihood that crystallization will take place. Thus prevention of crystallization plays an important role in rubber technology. There are a number of examples where copolymerization is employed to introduce necessary irregularity. Ethylene is copolymerized with propylene or reacted with chlorine for this purpose. Vinylidene fluoride homopolymer is highly crystalline but when copolymerized with perfluoropropylene is amorphous with a fairly low T_g. Phosphazene rubbers have mixed side group substituents to suppress crystallinity. Polybutadiene has a great deal of structural irregularity due to the presence of both cis and trans double bond configurations and to 1,2 addition side groups. However, there are several cases where crystallization is possible but fortuitously is not a serious issue. Both natural rubber and polyisobutylene have regular structures and can crystallize. However, in both cases the melting points are rather low, in the vicinity of room temperature. The undercoolings required for crystallization are substantial and in ordinary use they remain amorphous. Both crystallize readily when subjected to cooling in a highly stretched state. The stress hardening at very high elongations in natural rubber is sometimes attributed to the onset of crystallization and is thought to be a highly desirable property since it results in a major increase in tear strength.

8.6.3 Some common elastomers

The chemical structures of some common elastomers are listed below along with a few comments about their synthesis and uses.

(1) Cis(polyisoprene) (natural rubber, NR, Hevea rubber), **8.16**. A widely found natural product in plants but the principal source is from the *Hevea brasiliensis* tree.

8.16

(2) Polybutadiene (PBD), **8.17**. 1,3-butadiene polymerizes in free radical

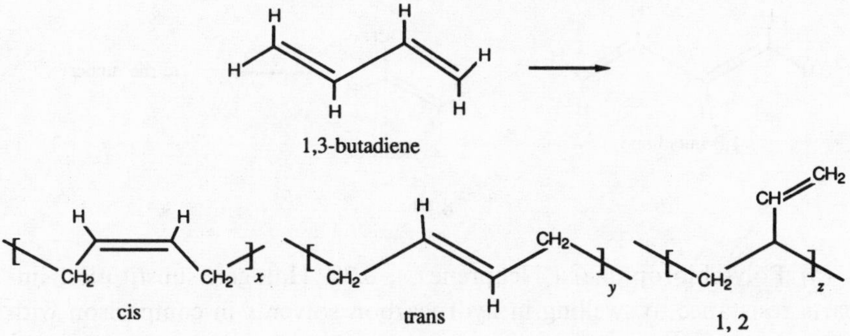

1,3-butadiene

cis   trans   1, 2

8.17

emulsion polymerization by predominantly 1,4 addition to give a polymer containing both cis, x, and trans, y, configurations of the double bond. However $\sim 20\%$ of the units, z, are from 1,2 addition. The cis/trans composition depends on polymerization temperature, ranging from $x = \sim 55\%$, $y = \sim 25\%$ at higher temperature, $100\,°C$, to higher values of trans, $x = \sim 78\%$, $y = 6\%$ at lower temperature, $-10\,°C$. Coordination catalysts can give nearly pure cis 1,4 or trans 1,4 addition depending on the choice of catalyst.

(3) 'SBR', 'Synthetic' or 'GRS' (Government Reserve Stock) rubber, **8.18**. This was developed during World War II as a substitute for natural rubber. It is a random copolymer of 1,3-butadiene and styrene ($\sim 75, 25\%$ by weight) and has similar microstructural propensities with respect to cis/trans content and 1,4 vs 1,2 addition as polybutadiene.

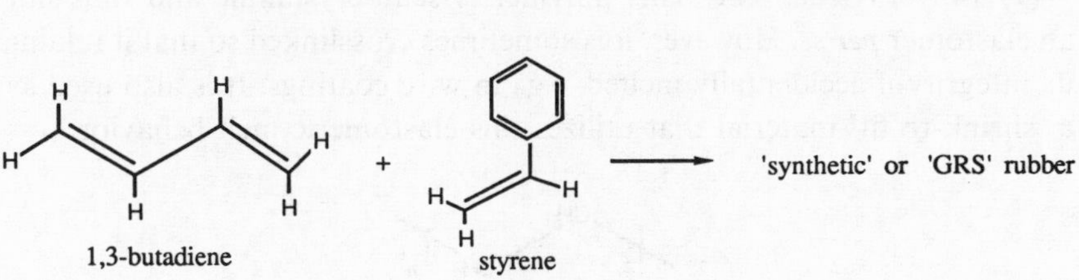

1,3-butadiene   styrene   'synthetic' or 'GRS' rubber

8.18

(4) 'Nitrile' rubber, **8.19**. This is a copolymer of 1,3-butadiene and acrylonitrile. The polar cyano or nitrile groups impart resistance to solvent swelling.

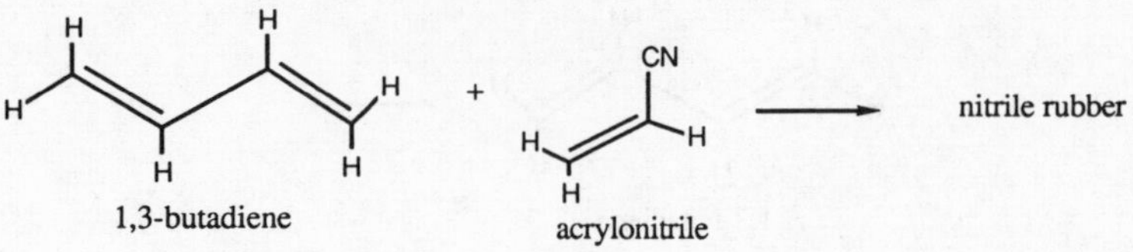

1,3-butadiene + acrylonitrile ⟶ nitrile rubber

8.19

(5) Poly(chloroprene) ('Neoprene®'), **8.20**. Halogen substitution imparts resistance to swelling in hydrocarbon solvents in comparison with natural rubber.

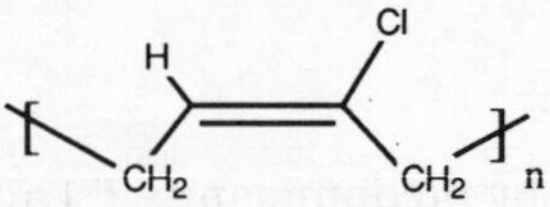

8.20

(6) Polyisobutylene (PIB), 'butyl rubber' (with a few per cent coisoprene for crosslinking), **8.21**. It is made from cationic polymerization of isobutene. It possesses the property of low gas permeability and is used for inner tubes and tire liners.

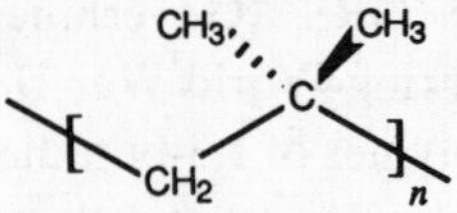

8.21

(7) Polyethylene, **8.22**. This polymer is semi-crystalline and thus not an elastomer *per se*. However, it is sometimes crosslinked so that it retains its integrity if accidentally melted, e.g., in wire coatings. It is also used as a 'shrink-to-fit' material that utilizes this elastomeric melt behavior.

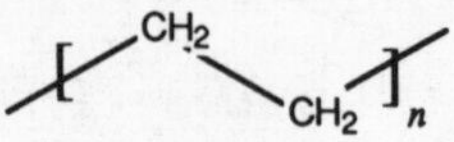

8.22

(8) Ethylene propylene rubber, **8.23**. A copolymer of ethylene and propylene, often with a small amount of termonomer to provide crosslinking sites. Hence, EPDM rubber, 'ethylene-propylene-diene monomer', where the diene is often 1,4-hexadiene or ethylidene norbornene.

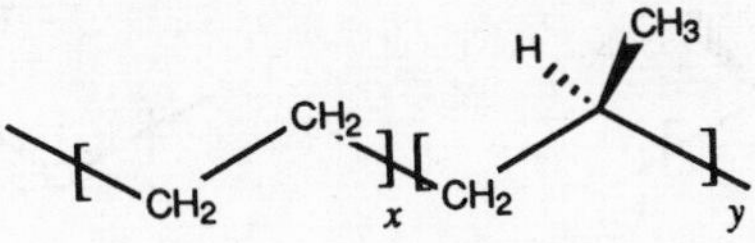

8.23

(9) Chlorinated polyethylene, **8.24**. Chlorine rapidly reacts with polyethylene in solution to displace hydrogens. The crystallinity of the polyethylene is largely destroyed at intermediate chlorine contents, $\sim 40\%$ by weight, and the material is elastomeric, at higher chlorine contents the glass temperature increases.

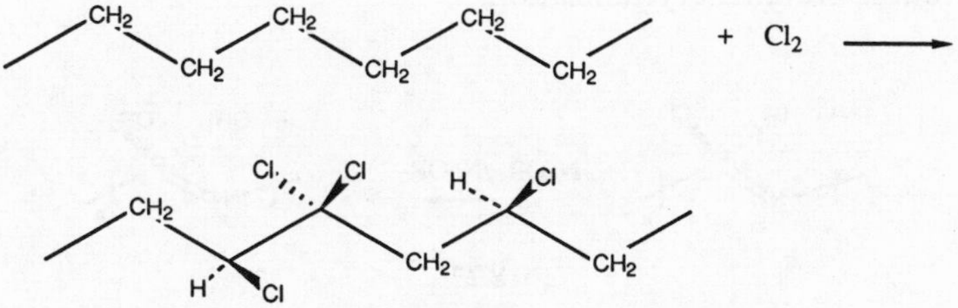

8.24

(10) Chlorosulfonated polyethylene ('Hypalon®'), **8.25**. If both sulfur dioxide and chlorine are reacted with polyethylene, some sulfonyl groups are introduced in addition to the chlorine substitution. A rubber with outstanding environmental resistance is produced.

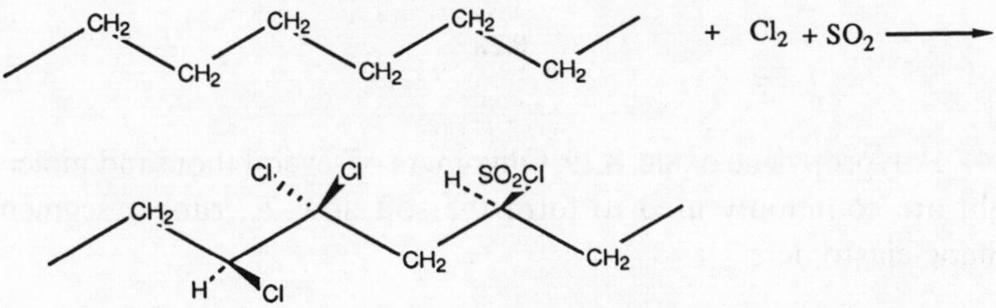

8.25

(11) The copolymer of vinylidene fluoride and hexafluoropropylene ('Viton®'), **8.26**, possesses outstanding thermal stability and resistance to solvent swelling and chemical attack.

8.26

(12) Phosphazene rubber, **8.27**. Poly(dichlorophosphazene) is a highly crystalline polymer but is hydrolytically unstable due to the labile chlorines. However, displacement of the chlorines with a mixture of perfluoroalkoxy groups where R_1, R_2 might be C_2F_5, C_3F_7, for example, results in a very thermally stable and chemically resistant polymer with a very low glass temperature. The randomness introduced by the R_1, R_2 substitution prevents crystallization.

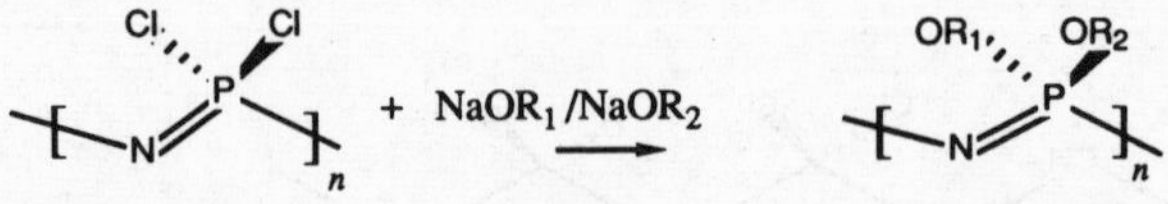

8.27

(13) Poly(dimethylsiloxane) (PDMS), 'silicone rubber', **8.28**. This finds wide use as a sealant and has good high temperature stability.

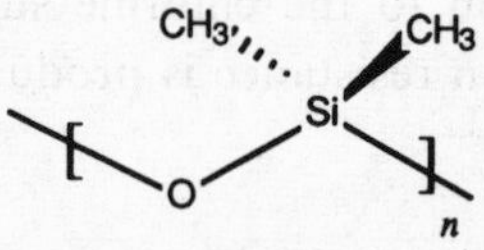

8.28

(14) Polypropylene oxide, **8.29**. Oligomers of several thousand molecular weight are commonly used to form the soft, low T_g, rubber segment in urethane elastomers.

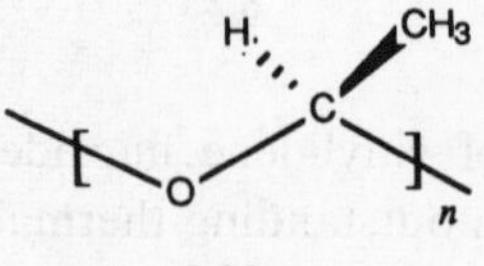

8.29

8.7 Further reading

The book by Treloar (1975) is a classic and a standard reference on rubber elasticity. There are a number of reviews concerning subjects taken up here. There are general reviews by Mark (1985), Erman and Mark (1989), a review about swelling by Queslel and Mark (1985), reviews of theory by Eichinger (1983), Heinrich, Straube and Helmis (1988) and Edwards and Vilgis (1988) and reviews concerning the correspondence of various theories with experiment by Gottlieb and Gaylord (1983, 1984, 1987), Brotzman and Mark (1986) and Higgs and Gaylord (1990). An edited volume of contributions commemorating the work of Eugene Guth gives valuable discussions of a wide variety of issues (Mark and Erman, 1992). For issues relating to the effect of entanglements a review by Graessley (1974) is very instructive. There is an introductory book by Mark and Erman (1988). Since rubbery materials have been widely used for well over a century many early books exist on their chemistry. It is interesting to compare a book from the 1920s (e.g., Weber, 1926) with one from the 1930s (e.g., Memmler, 1934). The earlier one was written with explanations in terms of the 'colloidal state' whereas in the latter, polymers had become accepted by chemists as a reality.

Nomenclature

A = Helmholtz free energy

A_{net} = elastic free energy of a network

A_{mix} = free energy of mixing

B = constant in Gaussian function = $3/(2\langle R_0^2 \rangle)$, B_{ij} = value for chains with ends i, j

C_P = constant volume heat capacity

E = energy

E_{11} = tensile modulus

f = retractive force, f_E, energetic contribution, f_S, entropic contribution

f_{cr} = functionality of a network crosslink

F = 'front factor' in network elastic free energy function, F_{aff} = value for affine model, F_{pn} = value for phantom network model

G = Gibbs free energy

H = enthalpy

I_1, I_2, I_3 invariants of the elastic strain tensor

$\mathbf{K}$ = Kirchoff matrix for network connectivity

l = specimen length; also bond length

N = number of bonds in a polymer chain

n_1, n_2 = moles of solvent, polymer

$p(\mathbf{R})$ = probability distribution function for a chain with end-to-end vector $\mathbf{R}$

P = pressure

P_1 = vapor pressure of solvent, superscript 0 = pure solvent, subscript cr = in crosslinked, un = in uncrosslinked materials

$\mathbf{r}$ = position coordinate of a network junction point, $\bar{\mathbf{r}}$ = average value, $\Delta\mathbf{r}$ = fluctuation from average value

$\mathbf{R}_{ij}$ = vector spanning network junctions i, j that are connected by a chain, $\bar{\mathbf{R}}_{ij}$ = average value, superscript zero = value in undeformed network

S = entropy

T = temperature

V = volume, V^0 = volume of unswollen polymer, V_1^0 = molar volume of solvent

Z = partition function of a chain; $Z(\mathbf{R})$ partition function of a chain restricted to end-to-end vector value = $\mathbf{R}$

ϕ_1, ϕ_2 = volume fraction of solvent, polymer; ′ indicates value in sample equilibrated in pure solvent

λ = elongational ratio, subscripts x, y, z ratios for these directions, subscripts 1, 2, 3 values in principal axes, λ_s = ratio under isotropic swelling

μ_1, μ_2 = chemical potential of solvent, polymer, superscript 0 = value in reference state of pure component

μ = number of (elastically active) junctions in a network

v = number of (elastically active) chain segments in a network, $\bar{v}$ = their concentration (moles/volume)

σ = (tensile) stress, σ_e engineering stress (force/original cross sectional area)

ξ = cycle rank of network

χ = Flory–Huggins interaction parameter

Problems

8.1 Using the data in Figure 8.3 resolve the total retractive stress, f, for natural rubber at 300 K into energetic and entropic components. At $\lambda = 1.42$, 1.87 and 2.58 compute f, f_E and f_S.

8.2 According to the classical molecular theory, equation (8.70), what are the values of C_1, C_2 in the Mooney–Rivlin relation, equation (8.72)?

8.3 The following data were obtained with a rubber band, a set of weights and a ruler (cross-section of rubber band 0.11 in. by 0.04 in.).

mass (g)	length (in)
0	5.15
20	5.45
50	6.22
100	7.80
150	10.48

(a) Compute the engineering stress, σ_e, in Pa at each stretch ratio, λ.
  (i) Plot σ_e vs λ.
  (ii) Plot reduced stress $\sigma_e/(\lambda - 1/\lambda^2)$ vs λ.
  (iii) Plot reduced stress $\sigma_e/(\lambda - 1/\lambda^2)$ vs $1/\lambda$, a 'Mooney' plot.
(b) From the value of C_1 in (a)(iii) above and the result for C_1 in problem 8.1, determine the number of crosslinks per cubic meter (assuming tetrafunctional crosslinks) and from this compute the number-average DP of the segments. Assume the material is natural rubber and the density is 0.91 g/cm^3.
(c) Gough (1805) describes the following experiment with a Caoutchouc slip (rubber band). 'Hold one end of the slip, thus prepared, between the thumb and fore-finger of each hand; bring the middle of the piece into slight contact with the edges of the lips; taking care to keep it straight at the time, but not to stretch it much beyond its natural length: after taking these preparatory steps, extend the slip suddenly; and you will immediately perceive a sensation of warmth in that part of the mouth which touches it, arising from an augmentation of temperature in the Caoutchouc: for this resin evidently grows warmer the further it is extended ... The increase in temperature, which is perceived upon extending a piece of Caoutchouc may be destroyed in an instant, by permitting the slip to contract again'.
For the rubber band in (a) and (b), calculate the temperature change for Gough's experiment. Assume a 2 × stretch and $C_P = 1.9$ J/(g K).

9

Solutions

The ability of polymers to dissolve in various media has great practical importance and can be of either positive or negative benefit. Processing of polymers is often aided by forming solutions but formed polymers in use would most often benefit from being impervious to the environmental effects of potential solvents. Solutions also form an important arena for the characterization of polymers. For example, the various means for molecular weight determination rely on solution measurements. Thus there is good reason to understand the factors governing solubility and to understand the molecular organization of solutions.

Solutions in general, not just polymer solutions, are obviously of high importance and a great deal of attention has been fixed on understanding them. In any such endeavour it is very useful to have a simple theory that conceptually encompasses many of the phenomena observed even if it is not necessarily quantitatively accurate. That role for solutions of simple organic molecules has been filled by the 'regular solution' model. In the case of polymer molecules their long chain connectivity requires significant modification of the regular solution model. Thus an appropriate first task here is briefly to review the theory of regular solutions and then to introduce the Flory–Huggins modification for polymer solutions.

9.1 Regular solutions of simple non-electrolytes

The regular solution model is based on assuming the spatial disposition of two kinds of molecules about each other in a two-component mixture is random and separately evaluating the energy and entropy of mixing on this basis. The energy is determined by counting the contacts between molecules after having assigned values to the various types of pair contact

330

energies. The entropy is determined from the number of ways of randomly mixing objects over fixed locations.

The entropy of mixing per one total mole of a two-component solution is thus expressed as the entropy of 'random mixing' of two classes of objects (see Appendix A9.1),

$$\Delta S^m/N = -R(x_1 \ln x_1 + x_2 \ln x_2) \tag{9.1}$$

where N is the total number of moles of the two kinds, $N = N_1 + N_2$, R is the gas constant, and x_1, x_2 are the respective mole fractions.

The enthalpy of mixing can be often approximated in the absence of specific effects such as hydrogen bonding, complex formation and so forth, as

$$\Delta H^m/N = (x_1 V_1^0 + x_2 V_2^0)\phi_1\phi_2(\delta_1 - \delta_2)^2, \tag{9.2}$$

where V_1^0, V_2^0 are the molar volumes of pure 1 and 2 and ϕ_1 and ϕ_2 are volume fractions, e.g.,

$$\phi_1 = \frac{N_1 V_1^0}{N_1 V_1^0 + N_2 V_2^0} \qquad \phi_2 = \frac{N_2 V_2^0}{N_1 V_1^0 + N_2 V_2^0}. \tag{9.3}$$

δ_1, δ_2 are 'solubility parameters'

$$\delta_1 = CED_1^{1/2} \qquad \delta_2 = CED_2^{1/2} \tag{9.4}$$

where the *cohesive energy density, CED,* of component 1 is

$$CED_1 = \frac{\Delta E_1^{\text{vap}}}{V_1^0} = \frac{\Delta H_1^{\text{vap}} - RT}{V_1^0} \tag{9.5}$$

with an analogous expression for CED_2. The quantities ΔE^{vap}, ΔH^{vap} are the energy and enthalpy of vaporization of pure component 1 or 2. Equation (9.2) reduces in the case of a '*strictly regular solution*' of *equal-sized* molecules ($V_1^0 = V_2^0$) to,

$$\Delta H^m/N = Bx_1 x_2 \tag{9.6}$$

where

$$B = [(\Delta E_1^{\text{vap}})^{1/2} - (\Delta E_2^{\text{vap}})^{1/2}]^2. \tag{9.7}$$

Equation (9.6) can be derived from a simple lattice model for the solution (see Appendix A9.1) and equation (9.7) from the *geometric mean rule* for finding the unlike pair contact energy, ε_{12}, from the like ones, $\varepsilon_{11}, \varepsilon_{22}$, as

$$-\varepsilon_{12} = (\varepsilon_{11}\varepsilon_{22})^{1/2}, \tag{9.8}$$

where the minus sign is occasioned by the fact that the energies are negative. If the geometric mean rule does not hold, then B is given more generally (Appendix A9.1) in terms of the contact exchange energy, $\omega = \varepsilon_{12} - \varepsilon_{11}/2 - \varepsilon_{22}/2$, as

$$B = zN_A\omega, \tag{9.9}$$

where z is the contact coordination number and N_A is Avogadro's number.

The mole fractions occur directly as the composition variables in equation (9.6) for equal sized molecules because, as the lattice derivation shows, both types of molecule have the same number of contacts. The volume fractions in equation (9.2) appear in an approximate attempt to correct for the greater number of contacts in the larger molecule (Hildebrand and Scott, 1950).

The Gibbs free energy of mixing per one total mole is

$$\Delta G^m/N = \Delta H^m/N - T\Delta S^m/N. \tag{9.10}$$

The resulting free energy for a strictly regular solution is obtained from equations (9.1) and (9.6) as

$$\Delta G^m/N = Bx_1x_2 + RT(x_1 \ln x_1 + x_2 \ln x_2). \tag{9.11}$$

9.2 Phase behavior

The question of solubility is to be framed in terms of phase behavior. That is, it needs to be established whether or not phase separation takes place and, if so, the compositions of the resulting phases need to be determined. In this section the general thermodynamic criteria, independent of particular models, are considered. In addition, they are illustrated for the regular solution.

From the general thermodynamic relations

$$\Delta G^m/N = x_1(\mu_1 - \mu_1^0) + x_2(\mu_2 - \mu_2^0) \tag{9.12}$$

and

$$x_1\,d\mu_1 + x_2\,d\mu_2 = 0 \tag{9.13}$$

the following two general relationships may be derived

$$\mu_1 - \mu_1^0 = \frac{\Delta G^m}{N} + x_2\frac{d\Delta G^m/N}{dx_1} \tag{9.14}$$

and

$$\mu_2 - \mu_2^0 = \frac{\Delta G^m}{N} + x_1\frac{d\Delta G^m/N}{dx_2}. \tag{9.15}$$

These equations have the graphical interpretation shown in Figure 9.1.

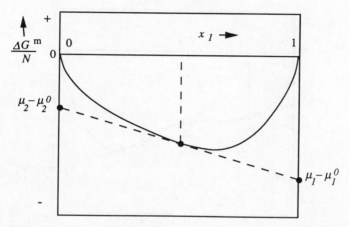

Fig. 9.1 Schematic rendering of the free energy of mixing (solid curve) vs composition (mole fraction, x_1). For a selected composition (vertical dashed line) the tangent to the free energy curve has intercepts at $x_1 = 0, 1$ that define the chemical potentials of the two components relative to their pure states.

The enthalpy of mixing of simple liquids in the absence of specific interaction effects is usually positive. For example, the geometric mean rule gives a positive value for the enthalpy in the case of the regular solution, equation (9.2) or (9.6). The entropy of mixing is expected to be positive and therefore its contribution, $-T\Delta S^m$, to the free energy is expected to be negative. As temperature is lowered the relative importance of the positive enthalpy term increases and the opposite effects of these two terms can result in an upwards bulge in the free energy curve, as in Figure 9.2. A common line can be drawn to the free energy curve that is tangent at two points, called *binodals*, of composition x_1' and x_1''. In the interval between the tangent points *any* point on the tangent line also represents the additive free energy of a physical mixture of two phases, one of composition x_1' and the other of composition x_1''. The relative proportions of the two phases are determined by the lever rule. In this region the free energy along the tangent line and hence that of two separate phases is below the free energy curve for the single phase solution. Further, these two phases are in equilibrium since the chemical potentials of a component in each phase are the same (same intercepts from the common tangent line). Thus the separated phases of composition, x_1', x_1'' are more stable than the single phase in this interval of composition. In order for this condition to occur there must be points of inflection in the free energy curve. These inflection points are called the *spinodal points*.

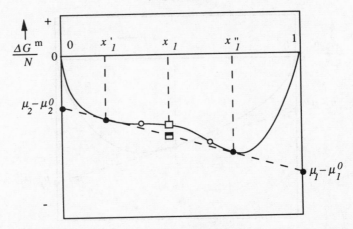

Fig. 9.2 Phase separation. Free energy of a homogeneous solution vs composition (solid curve). If inflection points (*spinodals*, open circles) exist on the curve, then a common tangent line to two points (*binodals*), at compositions, x'_1, x''_1, may be drawn. For a composition, x_1, between x'_1 and x''_1, the free energy of the homogeneous single phase solution (open square) is above that of a phase-separated physical mixture of two solutions, one of composition x'_1 and the other, x''_1 (black and white square). The composition, x_1, as marked (the open square), lies between the spinodal points. In this region the curvature of the free energy is negative and the solution is unstable with respect to small composition fluctuations. It decomposes spontaneously, a process called spinodal decomposition. If x_1 were to lie between a spinodal and a binodal point, the free energy curvature would be positive and the solution would be metastable toward composition fluctuations. Nucleation and growth would be associated with the phase separation.

Thus the condition for phase separation is the existence of spinodals and, given their occurrence, the common tangent points establish the phase compositions.

There can exist a critical temperature below which pairs of spinodal points exist (phase separation), but above which they do not (no phase separation). The disappearance of the spinodals as temperature increases to the critical temperature is a process of coalescence of the spinodal pair. The condition for existence of the spinodal pair, i.e., the existence of points where

$$\frac{d^2 \Delta G^m / N}{dx_1^2} = 0, \qquad (9.16)$$

is also seen from equation (9.14) to be capable of expression as

$$d\mu_1 / dx_1 = 0. \qquad (9.17)$$

The condition for the existence of the critical temperature, where these points coalesce, is therefore given by

$$d^2\mu_1/dx_1^2 = 0. \tag{9.18}$$

The above considerations may be illustrated for the regular solution. Application of equations (9.14) and (9.15) to equation (9.11) gives the following relations for the chemical potentials for the strictly regular solution,

$$\mu_1 = \mu_1^0 + RT \ln x_1 + Bx_2^2, \tag{9.19}$$

$$\mu_2 = \mu_0^2 + RT \ln x_2 + Bx_1^2. \tag{9.20}$$

For this model, equation (9.11), the solution is symmetrical, see Figure 9.3. This simplifies the process of finding the double tangent points as the composition of the two phases can be found from the condition that the slope is zero,

$$\frac{d\Delta G^m/N}{dx_1} = 0 \quad \text{(symmetrical solution)}. \tag{9.21}$$

An expression for the value of the critical temperature, T_c, may be found. Application of the conditions expressed in equations (9.17) and (9.18) to equation (9.19) leads to the relation $T_c = B/2R$.

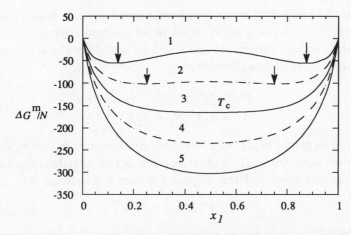

Fig. 9.3 Free energy of mixing of a symmetrical, 'strictly' regular solution. Five temperatures are shown 300, 350, 389, 450, 500 K, labelled 1–5 respectively. Phase separation occurs at the first two temperatures, phase compositions (binodals) are indicated by arrows. Temperature 3 is the critical solution temperature.

9.3 Mean-field lattice theory for polymers (Flory–Huggins model)

Consideration of the derivation in Appendix A9.1 of the equations for a strictly regular solution indicates that the energy of mixing term can be interpreted appropriately for polymer solutions if contacts are taken to involve chain beads or monomeric units and solvent molecules. However, the entropic term is clearly not appropriate. The chain connectivity precludes the random mixing of chain beads and solvent molecules. Yet the chains are flexible and can adopt multiple configurations which lead to an entropic contribution. The mean-field lattice approximation to this contribution constitutes the Flory–Huggins model (Flory, 1942, Huggins, 1942a, b, c).

9.3.1 Free energy of mixing

As indicated, the lattice theory for simple liquids is modified by reconsidering the entropy of mixing. It is now required that the number of configurations on a lattice (comprising $N = N_1 + N_2 X$ sites) of N_1 moles of solvent molecules and N_2 moles of polymer molecules each with X connected units be evaluated. When this is done (Appendix A9.2) the entropy of mixing is found to be approximated in mean-field by,

$$\Delta S^m = -R[N_1 \ln(N_1/N) + N_2 \ln(N_2 X/N)]. \tag{9.22}$$

Because the number of moles of polymer molecules in a high molecular weight polymer is not a good indicator of solution concentration, it is convenient to use, rather, volume fractions based on chain monomeric units rather than chains themselves, or

$$\left. \begin{aligned} \phi_1 &= N_1/(N_1 + N_2 X), \\ \phi_2 &= N_2 X/(N_1 + N_2 X). \end{aligned} \right\} \tag{9.23}$$

These can be placed in accord with the volume fractions ϕ_1 and ϕ_2 defined along with equation (9.2) by regarding the degree of polymerization, X, as the ratio of molar volumes of pure polymer and solvent, or

$$X \cong V_2^0/V_1^0 = v_2^0 M/V_1^0 \tag{9.24}$$

and where M is the polymer molecular weight and v_2^0 the specific volume of pure polymer. In this notation

$$\Delta S^m = -R(N_1 + N_2 X)[\phi_1 \ln \phi_1 + (\phi_2/X) \ln \phi_2]. \tag{9.25}$$

With the volume fractions and DP defined as in equations (9.23) and (9.24) the heat of mixing is assumed to be the same as in the lattice theory of equal sized molecules, equation (9.3), or

$$\Delta H^m = NB\phi_1\phi_2 \tag{9.26}$$

and B is given by either equation (A9.7) or (A9.11) and $N = N_1 + N_2 X$. The free energy of mixing, divided by RT, thus is

$$\Delta G^m/RT = (N_1 + N_2 X)[\chi\phi_1\phi_2 + \phi_1 \ln \phi_1 + (\phi_2/X) \ln \phi_2], \tag{9.27}$$

where the *Flory–Huggins parameter*, χ, is equal to B/RT and is related to the coordination number, z, and contact exchange energy, ω, equation (A9.7),

$$\chi = B/RT = z\omega N_A/RT. \tag{9.28}$$

The chemical potential and activity of the solvent may be found through application of equation (9.14) as

$$(\mu_1 - \mu_1^0)/RT = \ln a_1 = \ln \phi_1 + (1 - 1/X)\phi_2 + \chi\phi_2^2 \tag{9.29}$$

and those of the polymer from equation (9.15) as,

$$(\mu_2 - \mu_2^0)/RT = \ln a_2 = \ln \phi_2 + (1 - X)\phi_1 + X\chi\phi_1^2. \tag{9.30}$$

Sources of experimental information for applying the model consist of dilute solution measurements of osmotic pressure (second virial coefficient), measurements of solvent activity in more concentrated solutions (via solvent vapor pressure or osmotic pressure) and solubility and precipitation phenomena including critical solution temperatures.

9.3.2 Phase behavior

It is interesting and of importance to examine the predictions of the Flory–Huggins theory with respect to solubility and to contrast them with the behaviour of simple non-electrolyte solutions of similar-sized molecules. The free energy of mixing relation of the polymer lattice theory, equation (9.27), bears some resemblance to the corresponding one for regular solutions, equation (9.11). Since the volume fractions in the polymer theory are equivalent to the mole fractions of regular solutions of equal-sized small molecules, the two expressions differ only in the occurrence of the degree of polymerization factor, X, dividing ϕ_2 in the entropy term. This is a significant difference however. Physically, it reflects the reduction in entropy of mixing of polymers with solvents due to the

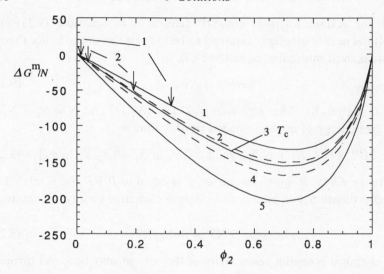

Fig. 9.4 Free energy of mixing from Flory–Huggins theory for a polymer, of DP = 100, with a solvent vs volume fraction polymer. Five temperatures are shown 300, 325, 332, 350, 400 K, labelled 1–5 respectively. Solubility limits (binodals) are shown for the first two temperatures, which are below $T_c = 332$ K.

constraint of chain connectivity. This has an important effect on the free energy function. The latter is rendered highly asymmetric with respect to solution composition, Figure 9.4.

Phase separation occurs below a critical temperature, T_c, that is, is derived from equation (9.29) by application of the conditions in equation (9.17)

$$-\frac{1}{\phi_1} + \left(1 - \frac{1}{X}\right) + 2\chi\phi_2 = 0 \tag{9.31}$$

and equation (9.18)

$$-\frac{1}{\phi_1^2} + 2\chi = 0. \tag{9.32}$$

These, in turn, give for the critical concentration ϕ_2^c and interaction parameter χ^c,

$$\phi_2^c = 1/(1 + X^{1/2}) \tag{9.33}$$

and

$$\chi^c = \frac{1}{2} + \frac{1}{2X} + \frac{1}{X^{1/2}}, \tag{9.34}$$

where χ^c is related to T_c through $\chi^c = B/RT_c$.

The composition of the phases reflects the asymmetry in the free energy function; see the binodals marked in Figure 9.4. The solvent-rich phase contains little polymer. The polymer-rich phase is highly swollen with solvent and, in fact, contains more solvent than polymer. These results are qualitatively in accord with experience. Amorphous polymers do tend to precipitate on cooling into a highly swollen gel and a relatively pure solvent phase.

The critical solution temperature is predicted to be molecular-weight-dependent, increasing toward a limiting value as the latter increases. This is in accord with experimental observations. The critical temperature in the limit of very high molecular weight is, from equation (9.28) and $\chi^c = \frac{1}{2}$, $T_c = 2B/R$. This contrasts with the value of $T_c = B/2R$ for a regular solution of small molecules (Section 9.2). The reduction in entropy of mixing in polymers is thus predicted to result in much higher precipitation temperatures for the same energy parameter B compared to small molecules. This is also qualitatively in accord with experience. Polymer–solvent pairs tend to have lower solubility and higher precipitation temperatures than for small molecule pairs with chemical structures similar to the polymer–solvent pair.

9.4 Experimental behavior of polymer solutions and the Flory–Huggins model

In the previous section it was seen that the Flory–Huggins model results in a phase behavior description for polymers in low molecular weight solvents that contains a considerable degree of realism when contrasted on a qualitative basis with regular solutions of small molecules. In this section, a more quantitative description of polymer behavior and its comparison with the Flory–Huggins model is undertaken. In this context, the simple theory will be found to have serious shortcomings.

9.4.1 Solvent activity in concentrated solutions

As developed in Chapter 3, solvent activity may be directly obtained via osmotic pressure, where from equation (3.15), $\Pi V_1^0 = -RT \ln a_1$, or via vapor pressure measurement where from equation (3.1), $RT \ln P_1/P_1^* = RT \ln a_1$. According to equation (9.29) a plot of

$$\ln(a_1/\phi_1) - (1 - 1/X)\phi_2 \text{ vs } \phi_2^2$$

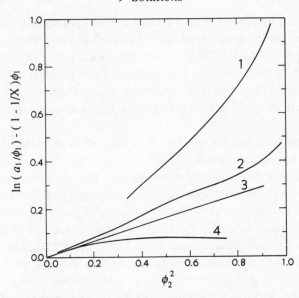

Fig. 9.5 Test of the Flory–Huggins theory. Plot of $\ln(a_1/\phi_1) - (1 - 1/X)\phi_2$ vs ϕ_2^2. The theory predicts this should be a straight line of slope $= \chi$. Systems are (1) polyisobutylene–benzene; (2) natural rubber–benzene; (3) polybutadiene–benzene; (4) polystyrene–chloroform. From Morawetz (1965, © 1965, by John Wiley & Sons, Inc., with permission).

should be a straight line of slope χ. Tests of equation (9.29) carried out in this manner show the equation to be useful. In many cases approximate straight line fits are obtained. On the other hand deviations are usually noticeable and in many cases are quite serious (Figure 9.5). A common way of plotting the deviations is to use equation (9.29) to define a value of χ at each experimental point and to plot these versus ϕ_2. Figure 9.6 is an example.

As formulated, on the strictly lattice basis, the interaction parameter χ times RT is a (contact exchange) *energy*. However, it is apparent that the real system will possess degrees of freedom (vibrational, orientational, etc.) that are not accounted for by a simple contact exchange energy. Even realizing that the lattice is only an approximation for liquids leads to the expectation that the average lattice dimension will be temperature-dependent through thermal expansion. One way to accommodate these effects partially and yet retain the simple lattice model is to let the interaction parameter be temperature-dependent. In that case χRT is to be regarded as an interaction exchange *free energy*. If this is allowed, then from its temperature dependence, the free energy, $RT\chi$, can be resolved

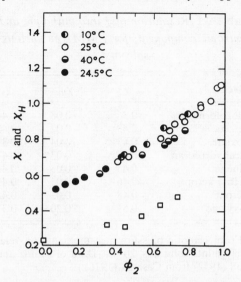

Fig. 9.6 Comparison of Flory–Huggins theory with experimental data. Effective χ parameter calculated from equation (9.29) is plotted against concentration. The system is polyisobutylene in benzene. The χ_H enthalpic component of χ was also derived and is plotted. From Flory (1970), data of Eichinger and Flory (1968).

into enthalpic and entropic terms

$$H_\chi = -RT^2\left(\frac{\partial \chi}{\partial T}\right)_P, \tag{9.35}$$

$$S_\chi = -R\left(\frac{\partial (T\chi)}{\partial T}\right)_P. \tag{9.36}$$

Or, from

$$\chi = (H_\chi - TS_\chi)/RT \tag{9.37}$$

enthalpic and entropic contributions to χ itself

$$\chi = \chi_H + \chi_S \tag{9.38}$$

can be defined where

$$\chi_H = H_\chi/RT, \tag{9.39}$$

$$\chi_S = -S_\chi/R. \tag{9.40}$$

When data are taken at a number of temperatures on the same polymer–solvent system it is generally found that χRT is indeed markedly temperature-dependent. In fact when χ is resolved into its enthalpic and entropic components it is found (Table 9.1) that the *entropic* term often

Table 9.1 *Resolution of χ into enthalpic and entropic components for PMMA in various solvents*[a]

solvent	χ	χ_H	χ_S
chloroform	0.36	−0.08	0.44
benzene	0.43	−0.02	0.45
dioxane	0.43	0.04	0.39
tetrahydrofuran	0.45	0.03	0.42
toluene	0.45	0.03	0.42
diethyl ketone	0.46	0.05	0.41
acetone	0.48	0.03	0.45
m-xylene	0.51	0.20	0.31

[a] From osmotic pressure measurements on relatively dilute solutions, 20 °C. (Data of Schulz and Doll (1952) from Casassa (1976).)

dominates. When it is considered that what originally was taken in the model to be a contact energy actually turns out to be entropic in character the model obviously has very serious shortcomings.

9.4.2 Solubility and critical solution temperatures

It has already been commented, in Section 9.3.2, that phase separation can occur and the relations defining the critical temperature and composition in terms of the χ parameter of the Flory–Huggins theory, equations (9.33) and (9.34), were presented. It is of interest to examine the quantitative validity of these relations. As it stands, equation (9.34) is a one-parameter relation between critical solution temperature and molecular weight, which can be written as,

$$1/T_c = (B/R)^{-1}(1/2 + 1/2X + X^{-1/2}). \qquad (9.41)$$

The slope and intercept of a plot of T_c^{-1} vs $(1/2 + 1/2X + X^{-1/2})$ are both determined by the single parameter, B. This is certainly not the case experimentally. However, it has already been seen above that the χ parameter has to be regarded as a free energy function (free energy divided by temperature). Thus a more appropriate formulation is obtained by replacing χ^c in equation (9.34) with $H_\chi/RT_c - S_\chi/R$, see equation (9.37), to obtain

$$H_\chi/RT_c - S_\chi/R = 1/2 + 1/2X + X^{-1/2}. \qquad (9.42)$$

If the value of T_c at infinite molecular weight is designated as θ, or

$$H_\chi/R\theta - S_\chi/R = 1/2, \qquad (9.43)$$

then the relation in equation (9.42) may be expressed as

$$1/T_c = 1/\theta + (H_\chi/R)^{-1}(1/2X + X^{-1/2}). \qquad (9.44)$$

In the above it has been implicitly assumed that the H_χ, S_χ parameters are temperature-independent. In general, it is found that this expression is quite successful in representing the experimental molecular weight dependence of the critical solution temperature, see Problem 9.3.

A plot of $1/T_c$ vs $(1/2X + X^{-1/2})$ establishes H_χ/R and θ from the reciprocals of the slope and intercept respectively. Through equation (9.43) they also establish S_χ/R and thus χ as a function of temperature via equation (9.37). Since χ at any given temperature is all that is required to define the free energy function at that temperature then all of the other properties, the spinodals and binodals, the critical concentration, ϕ_2^c, can be calculated from an experimental determination of T_c vs molecular weight. Figure 9.7 shows the result of such an exercise. It is noticed that the calculated value of T_c, i.e., the maximum temperature where phase separation exists, follow molecular weight well (as was implied in the above discussion). The calculated values of ϕ_2^c are realistic in decreasing

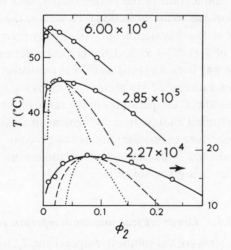

Fig. 9.7 Experimental and calculated phase diagrams for three polyisobutylene fractions of the molecular weights indicated. Dashed curves are calculated Flory–Huggins binodals, dotted curves are the spinodals. From Casassa (1976, © 1976 by John Wiley & Sons, Inc., with permission), data of Schultz and Flory (1952).

with increasing molecular weight. However, the calculated values are noticeably too low at high molecular weight. The calculated solubility gap is considerably too narrow and this is a serious shortcoming of the theory.

In the above, the parameter 'θ' is defined as the limiting temperature for T_c as $X \to \infty$. Thus the same notation is used as for the 'theta' temperature for solutions where unperturbed phantom conditions exist, as was developed in Section 3.3.3. This is intentional as a connection between the above θ temperature and the theta solvent condition can be made. The second virial coefficient A_2 of equation (3.6), for any solution, is a measure of the onset of intersolute interactions as concentration increases from the very dilute limit. Positive values of A_2 indicate net repulsive interactions between solute molecules relative to solute–solvent interactions and negative ones indicate attractive solute–solute interactions. Thus a temperature where A_2 goes through zero is indicative of the vanishing of distinction between solute–solute and solute–solvent interactions. In the polymer solution case this indicates the unperturbed, vanishing excluded volume condition. Thus the most fundamental definition of the theta temperature is that where the second virial coefficient disappears. It will be seen below that the Flory–Huggins lattice theory value for the second virial coefficient is $A_2 = (\frac{1}{2} - \chi)v_2^2/V_1^0$. In general, as will be seen, the application of the lattice theory, and thus this relation, in a dilute solution has many objections. However, the specific property that A_2 vanishes at $\chi = \frac{1}{2}$ *is* significant. As already seen, equation (9.34), this is the value of χ at $X \to \infty$ and the critical solution temperature at this circumstance has been assigned as θ. It is *consistent* and *appropriate* to regard θ values, established via plots of T_c invoking equation (9.44) as a function of X, as 'theta' temperatures in the sense of the already invoked concept of unperturbed chains. It is convenient in this context to think of the theta temperature experimentally as the precipitation temperature of a dilute solution ($\phi_2^c \to 0$) of a very high molecular weight specimen of a given polymer in a given solvent.

9.4.3 Lower critical solution temperatures

To this point, the term critical solution temperature, T_c, or CST, has meant the highest temperature at which phase separation occurs and above which complete miscibility obtains. The companion effect of precipitation occurring again as the temperature further increases is also commonplace in polymer solutions, see Figure 9.8. There is thus another critical solution

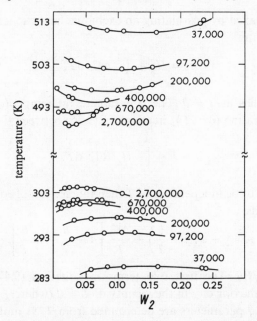

Fig. 9.8 Phase diagrams for polystyrene fractions of indicated molecular weights in cyclohexane. Precipitation temperatures vs weight fraction polymer. Notice the presence of both UCST and LCST phenomena (from Saeki, Kuwahara, Konno and Kaneko, 1975 with permission *Macromolecules*, © 1975, American Chemical Society).

temperature, a lowest temperature which bounds this region of immiscibility and below which the completely miscible region occurs. To distinguish these CSTs the latter one is called a *lower* critical solution temperature, *LCST*, and the former, to now, more familiar one, the *upper* critical solution temperature, *UCST*. Note that, somewhat displeasingly from a mnemonic viewpoint, the LCST has a larger numeric value than the UCST in the same system.

The LCST phenomenon is not accounted for by the simple lattice theory as it stands. However, the shortcomings and necessary modifications that have already been noted provide a clue to why the phenomenon occurs. It has been seen that it is necessary to make χRT temperature-dependent (or equivalently to introduce an entropic component into the exchange free energy) to fit activity data in concentrated solutions and also to represent critical solution temperature dependence on molecular weight. When this is done the enthalpy and entropy terms H_χ, S_χ in $RT\chi = H_\chi - TS_\chi$ are themselves regarded as temperature-independent.

This can be relaxed by postulating an exchange heat capacity (Eichinger, 1970) and writing

$$H_\chi = H_\chi^\theta + \int_\theta^T \Delta C_P \, dT, \tag{9.45}$$

where H_χ^θ applies at $T = \theta$. The χ parameter at temperature T can be found by integrating $(\partial\chi/\partial T)_P$ in equation (9.35) between θ and T to find,

$$\chi = \tfrac{1}{2} - \int_\theta^T H_\chi/RT^2 \, dT. \tag{9.46}$$

If ΔC_P is taken to be independent of T then integration of equations (9.45) and (9.46) yields

$$\chi = \tfrac{1}{2} - \psi\left(1 - \frac{\theta}{T}\right) + \frac{\Delta C_P}{R}\left[1 - \frac{\theta}{T} + \ln\frac{\theta}{T}\right], \tag{9.47}$$

where $\psi = H_\chi/R\theta$. The first two terms in equation (9.47) specify the temperature behavior of χ in the vicinity of $T = \theta$ (where $\chi = \tfrac{1}{2}$ at $T = \theta$) and the ψ and θ parameters are determined from T_c vs molecular weight behavior, equation (9.44), in this region. The introduction of the ΔC_P term extends the applicability over a wider temperature range. It is instructive to note that, for negative values of ΔC_P, χ as a function of temperature goes through a minimum and increases through $\chi = \tfrac{1}{2}$, the limiting critical condition again. Thus for ΔC_P negative there exist two critical solution temperatures, the conventional $T = \theta$, limiting UCST and another at higher temperature. The latter one is characterized by the onset of immiscibility with increasing temperature and is a LCST. For physically reasonable values of $\Delta C_P/R$ realistic values of LCST can be generated from ψ, θ parameters appropriate to $T = \theta$, Figure 9.9.

The physical significance of the negative ΔC_P values can be attributed to the exchange interaction energy becoming weaker as the system expands in volume as temperature increases. The positive exchange term decreases in magnitude as temperature goes up. The resultant negative ΔC_P has a larger effect on the free energy function, χ, though the entropy component, $-S_\chi/R$, increasing than from the enthalpic component, H_χ/RT, decreasing.

9.4.4 Dilute solutions, the second virial coefficient

As was seen in equation (3.6), for solutions where one component dominates the composition and is therefore designated as the solvent, the

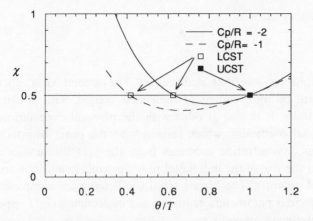

Fig. 9.9 Occurrence of the LCST in addition to the UCST as the result of negative ΔC_P contribution to temperature dependence of the χ parameter. Plotted from equation (9.47) for $\psi = 0.5$ and the indicated values of $\Delta C_P/R$.

activity of the latter can be developed as a power series,

$$-\ln a_1 = V_1^0\left[\frac{1}{M}C + A_2 C^2 + A_3 C^3 + \cdots\right]. \tag{9.48}$$

The Flory–Huggins activity of the solvent, from equation (9.29), can be expanded in powers of the polymer volume fraction, ϕ_2, as

$$-\ln a_1 = -\ln(1 - \phi_2) - (1 - 1/X)\phi_2 - \chi\phi_2^2$$
$$= \phi_2 + (1/2)\phi_2^2 + (1/3)\phi_2^3 + \cdots$$
$$+ (1/n)\phi_2^n + \cdots (1 - 1/X)\phi_2 - \chi\phi_2^2. \tag{9.49}$$

The weight concentration of polymer is $C = N_2 M/V$, where as usual N_2 and M are the number of moles of polymer and its molecular weight respectively and V is the solution volume. The volume of polymer, under the condition of no volume change on mixing, is $v_2^0 N_2 M$ where v_2^0 is the polymer specific volume, equation (9.24). Therefore, the concentration is related to volume fraction as

$$\phi_2 = v_2^0 C. \tag{9.50}$$

Introducing this relation into equation (9.49) and comparing with equation (9.48) gives the Flory–Huggins second virial coefficient as

$$A_2 = (\tfrac{1}{2} - \chi)\frac{(v_2^0)^2}{V_1^0} \tag{9.51}$$

and the higher ones as

$$A_n = \frac{(v_0^2)^n}{nV_1^0} \qquad n = 3, 4, \dots \qquad (9.52)$$

These results have serious deficiencies. The second virial coefficient is predicted to be independent of molecular weight. This is not the case experimentally. It is also at odds with the physical expectation that the second virial coefficient, which responds to the onset of intermolecular contacts as concentration increases from the very dilute region, should depend on the size of the individual polymer coils in addition to the total amount of polymer. The higher virial coefficients are completely devoid of the molecular parameters, neither χ nor molecular weight appears. This again is physically unrealistic.

The source of these problems is not hard to trace. There is a fundamental flaw in the theory that renders it inaccurate in dilute systems. In the counting process that was used in deriving the entropy (Appendix A9.2), it was assumed that the probability of finding a lattice site vacant for laying down the next segment was just equal to that expected on the basis of *random* distribution of previously placed segments. This may be a fair approximation in concentrated solutions where polymer molecules interpenetrate and overlap resulting in a fairly uniform segment distribution. However, it cannot be correct in dilute solution where the segment distribution is very non-uniform. The density is very high in the interior of polymer coils, lower at the edges and lowest between molecules.

9.5 The semi-dilute regime

It is apparent that the mean-field picture has serious shortcomings, the most profound being that it fails to represent properly the graininess of structure or spatial unevenness of polymer segment distribution that appears as dilution increases. It is very useful to consider criteria for when a solution may be considered to be dilute or concentrated. In so doing it becomes apparent that for polymer solutions a subtle phenomenon occurs that must be recognized. It is expected that the term dilute designates a region where interactions between solute molecules become negligible. In small molecule solutions, a single measure for monitoring approach to such condition suffices, whether it be volume fraction or weight concentration of solute. In polymers, the molecular coils of individual chains can overlap and interpenetrate even though the mean bead density

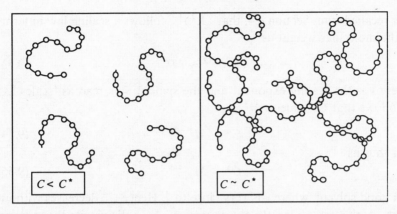

Fig. 9.10 Overlap threshold. In the dilute region, left-hand panel, both the bead density and the molecule density are low. As concentration increases, right-hand panel, the molecules start to overlap. If the molecular weight is high enough, the molecules can overlap when the bead density is still low; this is the semi-dilute region.

or weight concentration is low and in a region that in small molecule solutions would be considered as dilute, Figure 9.10. It is clear that the occurrence of overlap will depend on molecular weight. Many properties turn out to be sensitive to the presence or absence of molecular overlap and it is essential to recognize the latter.

9.5.1 *Overlap concentration in good solvents*

Molecules will start to overlap when the volume associated with the spatial extent of one molecule approaches the volume per polymer molecule of the solution. The spatial extent of one molecule can be expressed in terms of the end-to-end distance or the radius of gyration. If the solution were divided into N_2 cubic parcels, each containing a polymer molecule, then overlap would occur when the cube edges equal twice the radius of the molecule. If the rms end-to-end vector is taken as the measure of the radius, then, the solution volume per molecule at overlap, $V^*/N_2 = (2\langle R^2 \rangle^{1/2})^3$ and the *overlap concentration* C^*, are found as $N_2 M/(N_A V^*)$, or $C^* = M/[2^3 \langle R^2 \rangle^{3/2}]$. Since this concentration is not precisely defined, it is often expresssed without multiplying constants as

$$C^* \cong M/(N_A \langle R^2 \rangle^{3/2}) \qquad (9.52a)$$

or in terms of radius of gyration, s, as

$$C^* \cong M/(N_A \langle s^2 \rangle^{3/2}). \qquad (9.52b)$$

It is recalled from Section 6.4, that $\langle R^2 \rangle^{1/2}$ follows a scaling law in terms of the molecular weight as

$$\langle R^2 \rangle^{1/2} \sim M^v, \tag{9.53}$$

where v is the scaling exponent and the symbol $\sim$ is read as 'scales as'. It follows that C^* scales as

$$C^* \sim M^{1-3v} \tag{9.54}$$

and thus as

$$C^* \sim M^{-4/5} \tag{9.55}$$

in a good solvent, where v is very close to $\frac{3}{5}$. Hence C^* decreases without limit as the molecular weight increases. As anticipated, the overlap condition can be exceeded even in quite dilute solutions if the molecular weight is high enough. This gives rise to the concept and possibility of the *semi-dilute region*, where the concentration is above the overlap threshold,

$$C > C^*, \tag{9.56}$$

but is nevertheless very low; or in terms of the volume fraction, equation (9.50), $\phi_2 = v_2^0 C$,

$$\phi_2 \ll 1. \tag{9.57}$$

These two conditions define the semi-dilute region.

9.5.2 Osmotic pressure in the semi-dilute region

The semi-dilute region is an interesting one because the theoretical treatment of it is in many ways simpler than for the purely dilute region, where overlap does not occur. This is because molecular weight effects are suppressed and yet the solution is still dilute. The osmotic pressure is an important example of this situation.

In the truly dilute region, below the overlap threshold, it is apparent that the osmotic pressure depends on molecular weight. In fact, this forms an important method for molecular weight determination, Section 3.2.3. At infinite dilution, the osmotic pressure, Π, approaches equation (3.16), $\Pi = CRT/M$. Above C^*, due to the entangled nature of the solution, it would be expected that the solvent activity and hence osmotic pressure would no longer depend on the free molecule size but rather on some length that is set by the chain length between entanglement points. Both these features can be accommodated in a relation of the form

(des Cloizeaux, 1975)

$$\Pi/CRT = M^{-1}f(C/C^*), \qquad (9.58)$$

where $f(C/C^*)$ is a function with the properties that

$$f \to 1 \text{ for } C \to 0 \text{ in the dilute range,}$$

but

$$f \to (C/C^*)^p \text{ for increasing } C \text{ within the semi-dilute range.}$$

The scaling exponent, p, is to be a universal one, independent of the nature of the polymer or solvent. If Π is to be independent of M in the semi-dilute region, then according to equation (9.58) and using equation (9.54) for C^*, the relation

$$-1 - (1 - 3v)p = 0$$

or

$$p = -1/(1 - 3v) \qquad (9.59)$$

must be obeyed. Therefore the osmotic pressure scaling exponent is given by $p = \frac{5}{4}$ for $v = \frac{3}{5}$.

The relation in equation (9.58) can be arranged as

$$\log(\Pi M/CRT) = \log f(C/C^*).$$

The right-hand side should approach 0 as $C \to 0$ and $\frac{5}{4}\log(C/C^*)$ at $C > C^*$. A test of this is shown in Figure 9.11. The slope found in the semi-dilute region is 1.33 and is to be compared with the scaling prediction of 1.25. The agreement is good and would be even better if the slightly smaller value of $v \cong 0.58$ indicated in Section 6.4 were adopted.

9.5.3 The correlation length

The concept of the semi-dilute region is based on the presence of enmeshed molecules under the condition of the polymer bead density being low. This implies the existence of a typical mesh size, i.e., a typical chain length between mesh points, Figure 9.12. The length is called the *correlation length*, ξ. It is a measure of the length scale over which nearby beads may be expected to be connected in the same chain. A scaling form for the correlation length can be constructed as follows. Below the overlap threshold, the correlation length will be the molecular length, $\xi \cong \langle R^2 \rangle^{1/2}$. Thus according to equation (6.59) in a good solvent,

$$\xi \cong \langle R^2 \rangle^{1/2} \cong R_0 N^v \cong c'M^v; \quad \text{dilute region } C < C^*.$$

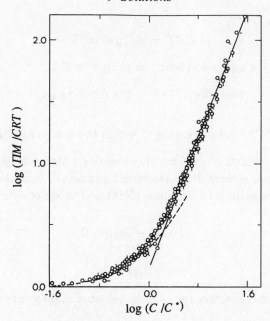

Fig. 9.11 The quantity $\log(\Pi M/CRT)$ is plotted against $\log(C/C^*)$ for p(α-methyl styrene) solutions in toluene. The various ticks indicate differing molecular weights. The slope in the semi-dilute region, $C > C^*$ is 1.33. From Noda, Kato, Kitano and Nagasawa (1981, with permission *Macromolecules*, © 1981, American Chemical Society).

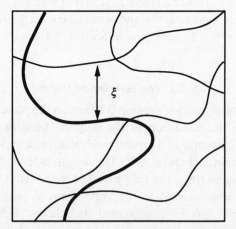

Fig. 9.12 The correlation length, ξ, in the semi-dilute region is a measure of the effective mesh size. Beyond this length a single chain, marked in bold for example, becomes screened by the other chains and its overall dimensions, $\langle R^2 \rangle$, are not perturbed by self-interaction.

For $C < C^*, \xi$ is to be independent of M. These requirements can be matched at $C = C^*$ by writing,

$$\xi \cong \langle R^2 \rangle^{1/2}; \qquad\qquad C < C^* \text{ dilute region}, \qquad (9.60a)$$

$$\xi \cong [\langle R^2 \rangle^{1/2}](\phi_2/\phi_2^*)^m; \quad C > C^* \text{ semi-dilute region}, \qquad (9.60b)$$

where $\phi_2^* = v_2^0 C^*$. In order for ξ to be independent of M above C^*, the M exponents of $\langle R^2 \rangle^{1/2}$ and $(\phi_2/\phi_2^*)^m$ must sum to 0. Therefore from equations (9.53) and (9.54), $v - m(1 - 3v) = 0$ or

$$m = v/(1 - 3v). \qquad (9.61)$$

For $v = \frac{3}{5}$, m takes the value $-\frac{3}{4}$.

It is also of interest to express how the molecular dimensions, $\langle R^2 \rangle$, depend on concentration in the semi-dilute region. At this point a digression is made. It is necessary to consider how the chain dimensions behave in a very concentrated solution, one approaching pure polymer, or $\phi_2 = 1$. Here the chains are solvent for themselves. The argument of Section 3.3.3 concerning the concept of theta conditions is recalled. In order for phantom or unperturbed conditions to prevail, the affinity for polymer–solvent interaction has to be exactly balanced by polymer–polymer and solvent–solvent interactions. In a melt, made up of only polymer molecules, provided that the distances involved are considerable, there is no way to distinguish between an interaction between a bead and another bead belonging to the same molecule and a bead in another molecule. Since the excluded volume effect is a relatively long-range one, this means chains in a melt will behave as if in theta condition. The dimensions will be those for the unperturbed state, $\langle R^2 \rangle^{1/2} = c' M^{1/2}$, where c' is a constant.

The effect of concentration on molecular dimensions is now taken up again. It is noted that this can be done in analogy with the correlation length by writing (Daoud *et al.*, 1975)

$$R(\phi_2)^2 \cong [\langle R^2 \rangle^{1/2}]^2 (\phi_2/\phi_2^*)^{m'}; \quad C > C^* \text{ semi-dilute region}, \qquad (9.62)$$

where $R(\phi_2)^2$ is the average-square end-to-end distance at the concentration ϕ_2 and $\langle R^2 \rangle^{1/2}$ continues to mean the concentration-independent value in a good solvent in dilute conditions. The argument is then made that the molecular weight dependence of $R(\phi_2)^2$ must be that of ideal or unperturbed chains. That is, at high molecular weight, a chain will traverse a number of correlation lengths. Thus effectively it will become screened by other chains and lose memory of its correlation with itself. Therefore

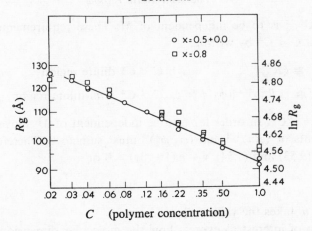

Fig. 9.13 Logarithmic plot of radius of gyration vs polymer concentration, in mole fraction of polymer, for polystyrene in toluene-d8. From neutron diffraction experiments. The various points designated by x are mole fractions of monomer labeled with deuterium (from King, Boyer, Wignall and Ullman, 1985, with permission *Macromolecules*, © 1985, American Chemical Society).

the condition, $v = \frac{1}{2}$ or $R(\phi_2)^2 \sim M$ can be imposed. Thus the molecular weight exponents in the right-hand side of equation (9.62) must sum to 1, or, $2v - (1 - 3v)m' = 1$, from which it follows that

$$m' = -(1 - 2v)/(1 - 3v)$$
$$= -\tfrac{1}{4} \quad \text{for} \quad v = \tfrac{3}{5}. \tag{9.63}$$

Thus scaling predicts that R^2 for a polymer of fixed molecular weight should contract with increasing concentration as $\phi_2^{-1/4}$. This is subject to experimental verification. Neutron diffraction studies on samples made up of a few completely deuterium-substituted chains in otherwise unlabeled chains allow determination of the radius of gyration. Results from such an experiment are shown in Figure 9.13. The experimental exponent is -0.16 and is to be compared to -0.25 from scaling.

9.6 The dilute regime

The dilute region is concerned with the behavior where molecular overlap does not dominate the state of the solution. There will be, however, interactions between the polymer molecules that will affect the thermodynamic behavior and be measurable, for example, through the second virial coefficient. In addition, within individual molecules the solvent will affect how the spatial dimensions depend on molecular weight. That is, the excluded volume phenomenon is associated with the 'goodness' of the solvent in the dilute regime. These are the features to be taken up.

9.6.1 The two-parameter model

Most of the theoretical efforts to model dilute solutions have been based on a simplification that treats only the polymer chains explicitly and regards the solvent as a continuum. Further, the polymer molecule is idealized in a fashion that includes only the essential features of the chain. The model is based on taking advantage of the fact that, in the absence of excluded volume, Gaussian statistical behavior describes the end-to-end distance of even relatively short chains (see Sections 6.2.3 and 6.3.4). The chain is divided up into segments that are long enough for each to be Gaussian in behavior. Each of these segments is then regarded as a bond in an equivalent freely jointed chain (Section 6.3.5), Figure 9.14.

In analogy with equations (8.33), (8.34) and (8.35), the probability distribution function for one of the chain segments, $1, 2, 3, \ldots$ in Figure 9.14 can be written as,

$$p(\mathbf{R}_i) = Z(\mathbf{R}_i)/Z, \tag{9.64}$$

where

$$Z(\mathbf{R}_i) = \int \cdots \int_{\sum l = R_i} e^{-\beta E(l)} \, d\{l\}$$

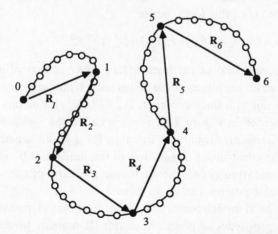

Fig. 9.14 The equivalent chain used in two-parameter theory. The real chain, open circle elements, is replaced by an equivalent freely jointed chain, filled circle beads, whose interactions are governed by the Gaussian distribution functions associated with each of the vectors, $\mathbf{R}_1, \mathbf{R}_2, \mathbf{R}_3, \ldots$.

represents integration over all of the bond vectors, l, in the segment R_i, subject to the constraint $(\sum l = R_i)$ that their sum be equal to a fixed value of R_i and where Z represents the complete integration over all values of R_i. Each segment is assumed to obey Gaussian statistics so that equation (9.64) becomes, cf. equations (6.33) and (8.35),

$$p(R_i) = \left(\frac{3}{2\pi l_K^2}\right)^{3/2} e^{-3R_i^2/2l_K^2}, \tag{9.65}$$

where l_K is the Kuhn length of the segment (Section 6.3.5) used to represent $\langle R_i^2 \rangle^{1/2}$. The complete probability function for the chain, $p(R)$, where R represents all of the R_i, is the product of functions $p(R_i)$ since the segments are independent of each other via the Gaussian assumption

$$p(R) = \prod_i p(R_i)$$

or, using equation (9.65),

$$p(R) = \left(\frac{3}{2\pi l_K^2}\right)^{N_K} e^{-(3/2l_K^2)\sum R_i^2}, \tag{9.66}$$

where N_K is the number of Kuhn segments in the real chain (Section 6.3.5). The above relation can be given the following interpretation. The configurational probability distribution of a system of particles that have an energy $U(R)$ is $p(R) = e^{-U(R)/k_B T}/Z$ where Z is the integration over R. Equation (9.66) is of this form, where

$$U(R) = k_B T [3/(2l_K^2)] \sum R_i^2. \tag{9.67}$$

Notice that a potential of the form $u(R) = \frac{1}{2}kR^2$ is that of a harmonic oscillator, or more simply, a spring. Thus each Kuhn segment behaves as if it were a spring with force constant, $2k_B T [3/(2l_K^2)]$. For this reason, the model, represented in Figure 9.15, is often called the *spring-bead* model. Because the Gaussian probability function for a Kuhn segment actually arises from the constrained integration of the individual bond vectors in the segment, equation (9.64), the harmonic potential of each segment is properly called a potential of mean force.

The spring-bead model, constructed from Gaussian elements, obviously then has the properties of phantom chains. It remains to introduce the interactions responsible for the excluded volume effect. This is done by inserting additional terms that arise from steric interactions between more distant pairs of the Kuhn beads into the potential of mean force. Thus

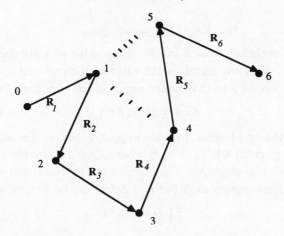

Fig. 9.15 The spring-bead model. Each of the vectors in Figure 9.15 defines a Kuhn segment. The latter have the same potential function as a spring (energy dependence on R^2). To model excluded volume, additional steric interactions between distant beads (||||) are invoked.

the complete potential is

$$U = k_B T[3/(2l_K^2)] \sum_i R_i^2 + \sum_{j>i} \sum_i u(\mathbf{r}_j - \mathbf{r}_i), \qquad (9.68)$$

where $u(\mathbf{r}_j - \mathbf{r}_i)$ is the potential of mean force arising from bead(i)–bead(j) contacts in the binary cluster approximation. This potential is usually approximated by a delta function

$$u(\mathbf{r}_j - \mathbf{r}_i) = k_B T\beta\delta(\mathbf{r}_j - \mathbf{r}_i) \qquad (9.69)$$

so that contact between i and j contributes an interaction of strength β but otherwise is zero. The excluded volume strength β is given by a binary cluster integral $\beta = \int 4\pi r^2(1 - e^{-v(r)})\,dr$. Since the effective interaction potential $v(r)$ will depend on the solvent as well as the polymer–polymer interaction, the excluded volume strength, β, will depend on temperature and the nature of the polymer and solvent. It has to be regarded here as a parameter appearing in the theory. It is evident through equations (9.68) and (9.69) that there are two parameters in the model, one of them is the effective segment length or Kuhn length, l_K, and the other is the excluded volume strength, β. When used to compute average properties the distribution functions based on (9.68) will lead to combinations of these parameters in the results.

9.6.2 Expansion factor

The effect of excluded volume on the dimensions of a polymer molecule in solution is often expressed as the ratio of the rms end-to-end vector for a given value of β to that under unperturbed conditions, or

$$\alpha_R^2 = \langle R^2 \rangle / \langle R^2 \rangle_0, \tag{9.70}$$

where α_R is the end-to-end distance *expansion factor*. The denominator is, of course, given by $\langle R^2 \rangle_0 = N_K l_K^2$ where N_k and l_K are the number and length of the Kuhn segments. The distribution function, $p(\mathbf{r}) = e^{-U/k_B T}/Z$ based on U from equations (9.68) and (9.69) can be written as

$$p(\mathbf{r}) = Z^{-1} \prod_i f(R_i) \, e^{-\beta \sum_i \sum_j \delta(\mathbf{r}_j - \mathbf{r}_i)}, \tag{9.71}$$

where

$$f(R_i) = Z^{-1} [3/(2\pi l_K^2)]^{3/2} \, e^{-(3/2 l_K^2)R_i^2}$$

and Z is the normalization. Therefore $\langle R^2 \rangle$ is given by,

$$\langle R^2 \rangle = Z^{-1} \int \cdots \int R^2 \prod_i f(R_i) \, e^{-\beta \sum_i \sum_j \delta(\mathbf{r}_j - \mathbf{r}_i)} \, d\mathbf{r}. \tag{9.72}$$

Historically there has been a great deal of effort put into using the formulation of equation (9.72) to calculate the expansion factor. This is, in general, an intractable problem. However, the region of weak expansion has been explored by means of perturbation expansions in powers of the excluded volume strength. That is, the exponential term in equation (9.72) is developed in a Mayer–Ursell cluster expansion and the resulting terms are averaged with the Gaussian functions, $f(R_i)$. The integration over the Gaussian functions gives rise to terms that are functions of the indices i, j in equation (9.69). The summations over these indices are replaced by integrations. Since the summations have upper limits of N_K, the number of Kuhn segments, this parameter will appear in the expansion. The result for the expansion factor is a series (Fixman, 1955; Yamakawa, 1971),

$$\alpha_R^2 = 1 + C_1 z + C_2 z^2 + C_3 z^3 + \cdots +, \tag{9.73}$$

in the dimensionless *excluded volume parameter*, z. The latter is given by,

$$z = (3/2\pi l_K^2)^{3/2} \beta N_K^{1/2}. \tag{9.74}$$

It has proven difficult to evaluate the terms in the series by the direct cluster expansion and they were limited to C_3 by this method. By applying a different expansion method it is now possible to extend the known coefficients to C_6 (Muthukumar and Nickel, 1984): $C_1 = 4/3$,

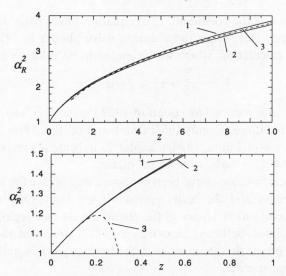

Fig. 9.16 End-to-end distance expansion factor versus the excluded volume parameter. Upper panel, high z region: curve 1 is the interpolation formula of equation (9.78) (Stockmayer, 1977); curve 2 is the formula of equation (9.79) (Fujita, 1990); dashed curve 3 is a renormalization group theory calculation, equation (9.81) (Douglas and Freed, 1984). Lower panel is the low z region. Curves 1 and 2 are the same as in the upper panel; curve 3 is the six-term series expansion of equation (9.73) (Muthukumar and Nickel, 1984).

$C_2 = -2.0753$, $C_3 = 6.2968$, $C_4 = -25.057$, $C_5 = 116.134$, $C_6 = -594.71$. It may be surmised from the coefficients that the series converges very slowly except for quite small values of z, see also Figure 9.16.

The asymptotic behavior at high z is now considered. This is the domain of good solvents. It has already been discussed in Section 6.4 how both renormalization group theory and computer simulations have been used to study the molecular weight dependence of $\langle R^2 \rangle$ in the good solvent limit. The exponent in $\langle R^2 \rangle^{1/2} = R_0 N^v$ is known to a fair degree of accuracy, being very slightly less than $\frac{3}{5}$. This leads to $N^{2v}/N = N^{2v-1}$ as the dependence of α_R^2 on N. Since z depends on N (i.e., N_K) as $N^{1/2}$, equation (9.74), α_R^2 depends on z as $z^{2(2v-1)}$. For $v = \frac{3}{5}$, this reduces to $z^{2/5}$ dependence. Thus the asymptotic form for $v = \frac{3}{5}$ is

$$\alpha_R^2 = cz^{2/5}; \quad z \to \text{large} \tag{9.75}$$

or as alternatively expressed

$$\alpha_R^5 = c'z; \quad z \to \text{large}, \tag{9.76}$$

where $c' = c^{5/2}$.

With respect to the coefficient, c, in equation (9.75) (or c' in equation (9.76)), it is not clear what the accurate value should be. The original Flory (1953) calculation, which was intended to be valid for all z, gave

$$\alpha_R^5 - \alpha_R^3 = 2.60z. \tag{9.77}$$

For large z, this reduces to equation (9.76) with $c' = 2.60$ ($c = 1.46$). Computer simulations (Domb, 1963; Domb and Barrett, 1976; Alexandrowicz, 1983; see also Fujita, 1990, Chapter 2) indicate a somewhat larger value, $c = 1.64$ ($c' = 3.44$), to be appropriate.

Interpolation functions have been proposed that bridge the low z series expansion region and the high z good solvent limit. Obviously these functions depend on the choice of the results for the two regions that are being interpolated between. Stockmayer (1977) proposed the function below, based on the first three terms in the low z expansion and expressed as a Padé approximate,

$$\alpha_R^5 = 1 + \left(\frac{10z}{3}\right)\left(\frac{1 + 2.953z}{1 + 3.509z}\right). \tag{9.78}$$

This relation gives the high z coefficient c' as 2.81 (or $c = 1.51$). Thus it is intermediate between the Flory and computer simulation values. Another Padé approximate function has been formulated by Fujita (1990, Chapter 2) to take advantage of the more extensive six term low z expansion,

$$\alpha_R^5 = \frac{1 + 11.367z + 36.992z^2 + 31.411z^3}{1 + 8.034z + 12.069z^2}. \tag{9.79}$$

At high z it gives $c' = 2.60$, the Flory value. The above two formulae are illustrated in Figure 9.16. Computer simulation results in the region of z below the asymptotic regime have been fitted to an interpolation formula as well (Domb and Barrett, 1976; Lax, Barrett and Domb, 1978)

$$\alpha_R^2 = (1 + 20z + 155.54z^2 + 591.86z^3 + 325z^4 + 1670z^6)^{1/15}. \tag{9.80}$$

Notice that the asymptotic relation, equation (9.75), value of $c = 1.64$ referred to above is consistent with the high z limit of this equation.

Renormalization group theory has been applied to this aspect as well although approximations are necessary. Douglas and Freed (1984) treated both the poor and good solvent region. In the good solvent regime, these authors found, in common with the results discussed in Section 6.4, that the exponent, v, is slightly less than $\frac{3}{5}$. Their high z relation, based on

$v = 0.5918$, is

$$\alpha_R^2 = 1.63 z^{0.367}. \tag{9.81}$$

This relation is shown in Figure 9.16.

The expansion factor for the radius of gyration, α_s, has also been of interest. The series expansion at low z is known to two terms,

$$\alpha_s^2 = \langle s^2 \rangle / \langle s^2 \rangle_0 = 1 + S_1 z + S_2 z^2 + \cdots, \tag{9.82}$$

where $S_1 = 134/105$ (Zimm, Stockmayer and Fixman, 1953), $S_2 = -2.082$ (Yamakawa, Aoki and Tanaka, 1966).

At the high z limit of good solvents it is thought that the molecular weight exponent for $\langle s^2 \rangle^{1/2}$ is the same as for the end-to-end distance, $\langle R^2 \rangle^{1/2}$. That is, v is $\frac{3}{5}$ or a little less. Only the c coefficient in

$$\alpha_s^2 = c z^{2(2v-1)}; \quad z \to \text{large} \tag{9.83}$$

is different. Computer simulations (Domb and Barrett, 1976) indicate (i.e., $v = \frac{3}{5}$)

$$\alpha_s^2 = 1.53 z^{2/5}; \quad z \to \text{large}, \tag{9.84}$$

where the factor 1.53 is to be compared for the same simulations to 1.64 for α_R^2. The Douglas–Freed (1984) theory gives

$$\alpha_s^2 = 1.60 z^{0.367}; \quad z \to \text{large}, \tag{9.85}$$

where the factor 1.60 is to be compared with 1.63 in equation (9.80).

Interpolation formulae have been proposed; the most commonly used one is (Yamakawa and Tanaka, 1967)

$$\alpha_s^2 = 0.541 + 0.459(1 + 6.04z)^{0.46}. \tag{9.86}$$

Computer simulations below the asymptotic region embraced by equation (9.84) have been fit to a relation for the *ratio* of the $\langle R^2 \rangle$, $\langle s^2 \rangle$ expansion factors giving (Domb and Barrett, 1976; Suzuki, 1982)

$$\alpha_s^2 / \alpha_R^2 = 0.933 + 0.067 \, e^{(0.85z - 0.67z^2)}. \tag{9.87}$$

It is possible to measure α_s^2 experimentally. That is, the radius of gyration can be measured by light scattering (Section 3.4). For example, the dependence of α_s^2 on molecular weight may be determined. Such measurements in the good solvent region are consistent with the $z^{\sim 2/5}$ relations above. The measurements also probe variations in the excluded volume parameter β over various solvents.

9.6.3 Second virial coefficient

The second virial coefficient, see equation (3.6), was formulated in terms of the statistical mechanical theory of liquids by Zimm (1946). Since it represents two-body chain–chain interactions, it involves the product of the chain distribution functions for a pair of chains. When applied to the two-parameter model the following expression results,

$$A_2 = -(N_A/2VM^2) \int \int p(\mathbf{r}_1)p(\mathbf{r}_2)(e^{-\beta \sum_i \sum_j \delta(\mathbf{r}_j - \mathbf{r}_i)} - 1)\, d\mathbf{r}_1\, d\mathbf{r}_2, \quad (9.88)$$

where $p(\mathbf{r}_1)$, $p(\mathbf{r}_2)$ are the chain distribution functions, equation (9.71), of a pair of chains, M is the molecular weight and V the volume. As in the case of the expansion factor, a cluster expansion is used to develop A_2 in a series in the excluded volume strength, β. The results is of the form

$$A_2 = (N_A \beta N_K^2/2M^2)h(z), \quad (9.89)$$

where N_K is the number of Kuhn segments,

$$h(z) = 1 + D_1 z + D_2 z^2 + \cdots + \quad (9.90)$$

and z is the excluded volume parameter of equation (9.74). Evaluation of the coefficients in the series for $h(z)$ is even more difficult than for the expansion factor case. Only the first two are known (see Tanaka and Solc (1982) for currently accepted values); $D_1 = -2.865$, $D_2 = 13.928$. Barrett (1985) has given an interpolation formula based computer simulations for the high z region and the above low z series expansion,

$$h(z) = (1 + 14.3z + 57.3z^2)^{-0.2}. \quad (9.91)$$

This function reduces at high z to the asymptotic relation,

$$h(z) = 0.45z^{-0.4}; \quad z \to \text{large}. \quad (9.92)$$

It has become common practice to define another dimensionless parameter, Ψ, called the *interpenetration function*, that is based on both A_2 and the radius of gyration, as

$$\Psi = zh(z)/\alpha_s^3. \quad (9.93)$$

Substituting the relation between A_2 and $h(z)$ in equation (9.89) and the definition of z, equation (9.74), into the above gives,

$$\Psi = A_2 M^2/4\pi^{3/2}N_A\langle s^2 \rangle^{3/2}. \quad (9.93)$$

The interpenetration function has the advantage of being completely experimentally determinable, i.e., through measurements of A_2 and $\langle s^2 \rangle^{3/2}$. It is apparent from its being a function of z, equation (9.74), that Ψ depends on molecular weight through N_K and $\langle s^2 \rangle$. It also depends on the excluded volume cluster integral, β. The latter is a strong function of temperature since the same polymer–solvent system may be at the theta temperature, where $\beta = 0$, or be at some higher temperature where the medium becomes a good solvent and a large positive value for β results. Thus both temperature and molecular weight are variables for exploring the behaviour of the interpenetration function.

Since asymptotic and interpolation functions have been given as discussed above for both α_s^2 and $h(z)$ as a function of z, it follows from equation (9.93) that such relations are available for Ψ. The asymptotic form of $h(z)$, equation (9.92), and that of α_s^2, for $v = 3/5$, equation (9.84), indicate that, asymptotically, Ψ should become independent of z at high z. Therefore it should also become independent of α_s at large values of the latter. Thus the high z α_s, or high molecular weight asymptotic value of Ψ is of considerable interest. This applies both to establishing it experimentally and to its theoretical prediction.

If the computer simulation results for α_R^2, equation (9.80), for α_s^2/α_R^2, equation (9.87), and for $h(z)$, equation (9.91) are used to calculate Ψ via $\Psi = zh(z)/\alpha_s^3$, i.e., equation (9.93), the results plotted against α_s^3 appear as in Figure 9.17. The asymptotic value of Ψ is 0.235.

There are renormalization group theoretical predictions of Ψ and the asymptotic high z value. The Douglas–Freed theory gives the relation

$$\Psi = \frac{z}{1 + 4.83z} + \frac{1.45z^2}{(1 + 4.83z)^2}. \tag{9.94}$$

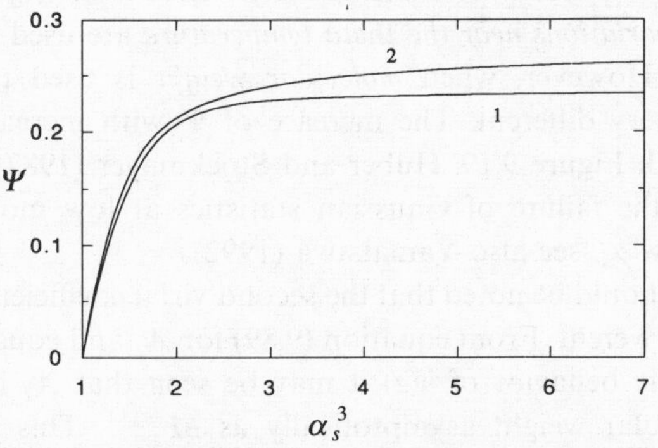

Fig. 9.17 Interpenetration function. Curve 1 from Domb–Barrett computer simulation interpolation formulae. Curve 2 from Douglas–Freed theory.

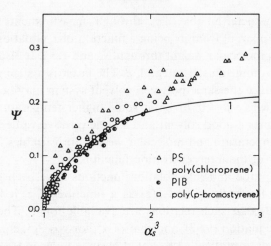

Fig. 9.18 Interpenetration function vs expansion factor (cubed) for the polymers indicated. Data is from light scattering. The curve labeled 1 is curve 1 from Figure 9.17. The changes in Ψ are accomplished through *temperature* variation near the θ point for high molecular weight samples. From Fujita (1990) where the data sources are listed.

This gives the asymptotic value of 0.269. The function is plotted in Figure 9.17. In other theoretical calculations the same values of the limiting Ψ have been found (Witten and Schaefer, 1978; des Cloizeaux, 1981). Thus a limiting value of ~ 0.25 seems established from theory and simulations.

Experiments using both molecular weight and temperature as variables seem to confirm the above expectations with respect to the limiting value of Ψ. The general shape of the dependence of Ψ on α_s^3 is experimentally in approximate accord with the relations shown in Figure 9.17 when *temperature variations near the theta temperature* are used to vary α_s, see Figure 9.18. However, when *molecular weight* is used to vary α_s the situation is very different. The increase of Ψ with increasing α_s^3 is not even observed, Figure 9.19. Huber and Stockmayer (1987) attribute this behavior to the failure of Gaussian statistics at low molecular weight and hence low α_s, see also Yamakawa (1992).

Finally, it should be noted that the second virial coefficient, A_2, depends on molecular weight. From equation (9.89) for A_2 and equation (9.92) for the asymptotic behavior of $h(z)$ it may be seen that A_2 is predicted to follow molecular weight asymptotically as $M^{-1/4}$. This appears to be consistent with experiment. However, consonant with the slow approach

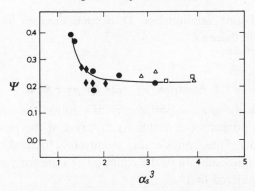

Fig. 9.19 Interpenetration function vs expansion factor (cubed) for polystyrene in toluene. The variation in Ψ is accomplished through *molecular weight*, cf. Figure 9.18. From Huber and Stockmayer (1987, with permission *Macromolecules*, © 1987, American Chemical Society).

of Ψ experimentally to its asymptotic limit, A_2 appears to approach this limit slowly (Fujita, 1990; Yamakawa, 1992).

9.7 Equation-of-state theories

The mean-field Flory–Huggins lattice model was found wanting even in concentrated solutions as evidenced by the fact that the apparent χ parameter deduced from experiment is very often not a constant but χRT depends on both concentration and temperature. These shortcomings are, of course, to be associated with the simplicity of the lattice that is assumed to represent the structure of the solution. The most serious shortcoming is the assumption of rigidity of the lattice, there was no allowance for the system to be able to change volume with temperature, except to incorporate the result on an *ad hoc* indirect basis by invoking enthalpic and entropic components for χ. Thus there have been considerable efforts made to formulate mixing theories that incorporate, at the beginning on a molecular basis, the ability for the mixture to change volume with temperature. One way to accomplish this is to formulate the free energy as an explicit function of *both* volume and temperature. (Note that the Flory–Huggins free energy, equation (9.27), is explicitly a function of temperature only through $\chi = B/RT$.) That is, the Helmholtz free energy is to be expressed as $A = A(T, V)$. The pressure then follows as $p = -(\partial A/\partial V)_T$. Thus an *equation of state*, $p = p(T, V)$, is obtained. The volume as a function of temperature at fixed pressure is found by inverting the equation-of-state relationship. Thus such theories are often called equation-of-state theories.

9.7.1 Equations of state for pure fluids

In keeping with the general philosophy of solution theories, the goal is to describe the properties of solutions in terms of the properties of the pure components. Thus considerable attention is focused on developing a satisfactory representation of the equation of state for pure fluids. This question is considered first.

Cell models for chain molecules

In Appendix 9.3 it is shown that the configurational partition function for a monatomic fluid in the cell model approximation can be written as the product of two terms, or

$$Z = [e^{-\phi(0)/k_B T}]^{N/2} v_f^N. \tag{9.95}$$

The first term involves $\phi(0)$, the energy of interaction of one of the N particles, held at its cell center or rest position, with its neighbors and represents the partition function of the static lattice. The second term v_f^N is the contribution of the vibrational energy, where v_f is an integration over the cell, equation (A9.25), that has the dimensions of volume. The configurational Helmholtz free energy, $A = -k_B T \ln Z$, is correspondingly a sum of two terms, one involving the static lattice energy and the other the vibrational free energy,

$$A = N\phi(0)/2 - Nk_B T \ln v_f. \tag{9.96}$$

For a Lennard–Jones interaction potential between the particles, the lattice energy may be written as a function of the cell size and therefore of the volume per particle, v, as

$$\phi(0) = z\varepsilon[A(v^*/v)^4 - 2B(v^*/v)^2], \tag{9.97}$$

where v^* is a constant that represents the repulsive core volume of the particles, ε is the well depth of the interaction potential, z is the coordination number and A and B are constants characteristic of the lattice assumed to represent the fluid. The free volume integral, v_f, is often approximated by representing the vibrational potential as a square well. In this approximation,

$$v_f = C(v^{1/3} - \gamma v^{*1/3})^3, \tag{9.98}$$

where C and γ are constants characteristic of the geometry of the lattice and v^* is the same parameter representing the repulsive core volume of the particles. Differentiation of equation (9.96), to find $p = -(\partial A/\partial V)_T$ using equations (9.97), (9.98) gives the equation of state. In terms of the dimensionless *reduced variables*, $\tilde{p}$, $\tilde{v}$, $\tilde{T}$,

$$\frac{\tilde{p}\tilde{v}}{\tilde{T}} = \frac{\tilde{v}^{1/3}}{(\tilde{v}^{1/3} - \gamma)} + \frac{2}{\tilde{T}}(A\tilde{v}^{-4} - B\tilde{v}^{-2}), \qquad (9.99)$$

where

$$\tilde{v} = v/v^*,$$

$$\tilde{T} = T/T^*; \qquad T^* = z\varepsilon/k_B,$$

$$\tilde{p} = p/p^*; \qquad p^* = z\varepsilon/v^* = k_B T^*/v^*.$$

In the treatment for a monatomic fluid, v_f represents an integration over the three spatial coordinates of a particle in its cell, equation (A9.25). In the more complex case of a chain molecule it would be presumed that this integration would involve a chain bead or mer rather than the complete chain. On the other hand it is clear that each bead should not be allowed full three-dimensional freedom in the integration since it is covalently bonded to its neighbors. These latter stiff forces would not be expected to respond to volume change to nearly the degree that is associated with the much softer intermolecular non-bonded forces (and represented by the Lennard–Jones function). This situation has been handled empirically (Prigogine *et al.*, 1953a, b; Prigogine, 1957) by the device of replacing the exponent of value 3 in equation (9.98) with a factor $3c$, where $3c$ is the number of intermolecular or external degrees of freedom per mer. Thus the term v_f^N in the partition function, equation (9.95), becomes

$$v_f^N = [C(v^{1/3} - \gamma v^{*1/3})^3]^{rNc}, \qquad (9.100)$$

where r is the number of mers per molecule, N is the number of molecules and c expresses the external degrees of freedom per mer. The volume v is defined now as the volume per mer, $v = V/rN$. (The notation r is used here for the number of mers instead of X as in Sections 9.3 and 9.4, in order to preserve the specialized connotation of equation (9.24) that X is the ratio of molar volumes in a two-component system. Thus, if in a two-component system r_1, r_2 are defined as the number of mers for each molecule type, $r_2/r_1 = X$.)

The lattice energy term involving $\phi(0)$ in equation (9.96) will be unchanged except that the effective coordination number in equation

(9.97) might be expected to be reduced in magnitude. Therefore equation (9.96) is now

$$A = Nr\phi(0)/2 - rNck_B T \ln[C(v^{1/3} - \gamma v^{*1/3})^3]. \qquad (9.101)$$

Thus equation (9.99) for the equation of state still holds but the reduced variables are now defined as

$$\left.\begin{array}{ll} \tilde{v} = v/v^*; & v^* = \Gamma r^*, \\[4pt] \tilde{T} = T/T^*; & T^* = z\varepsilon/(ck_B), \\[4pt] \tilde{p} = p/p^*; & p^* = z\varepsilon/v^* = ck_B T^*/v^*. \end{array}\right\} \qquad (9.102)$$

Notice that, with the introduction of the factor c expressing the external degrees of freedom, there are now *three* independent parameters: the energy parameter, $z\varepsilon$, the hard core volume, v^*, and c, or, as alternatively expressed, v^*, p^* and T^*.

The Flory–Orwoll–Vrij–Eichinger (FOVE) model

The cell model in its imposition of crystalline-like order about the central molecule results in the lattice energy being strongly dependent on the volume, i.e., as expressed by equation (9.97). At the larger volumes associated with a liquid, the numerical value of the energy tends to be dominated by the attractive term indicating a v^{-2} dependence of energy on volume. There is some reason to think that in a liquid this v^{-2} dependence is too strong and that a smaller power is in order. The Flory–Orwoll–Vrij–Eichinger model (Flory, Orwoll and Vrij, 1964; Eichinger and Flory, 1968) replaces equation (9.97) with an expression that is v^{-1} in the volume dependence, or,

$$\phi(0) = -s\eta/v, \qquad (9.103)$$

where η is an energy parameter associated with mer–mer contacts (analogous to ε), s is the number of such contacts per bead or mer (analogous to z), and v is the volume per mer. The Prigogine form for v_f, equation (9.100), is adopted, with γ chosen as $\gamma = 1$, thus leading to the free energy as

$$A = -rNs\eta/2v - 3rNck_B T \ln[C(v^{1/3} - v^{*1/3})^3]. \qquad (9.104)$$

In reduced form the Gibbs free energy, $G = A + pV$, becomes

$$\tilde{G} \equiv G/rNp^*v^* = -\tilde{v}^{-1} - 3\tilde{T} \ln[Cv^{*1/3}(\tilde{v}^{1/3} - 1)] + \tilde{p}\tilde{V}, \quad (9.105)$$

where V = system volume = rNv^* and where the reduced variables are defined by

$$\left.\begin{array}{ll}
\tilde{v} = v/v^*; & \tilde{V} = V/V^* = V/rNv^*, \\
\tilde{T} = T/T^*; & T^* = s\eta/2v^*ck_B, \\
\tilde{p} = p/p^*; & p^* = s\eta/2v^{*2} = ck_B T^*/v^*.
\end{array}\right\} \tag{9.106}$$

Differentiation of equation (9.104) or (9.105) leads to the reduced equation of state

$$\frac{\tilde{p}\tilde{v}}{\tilde{T}} = \frac{\tilde{v}^{1/3}}{\tilde{v}^{1/3} - 1} - \frac{1}{\tilde{T}\tilde{v}}. \tag{9.107}$$

The reduced variables in equation (9.106) are essentially the same as in the cell model, equation (9.102), and differ only in the fact that η in equation (9.103) has the dimensions of energy times volume whereas the well depth, ε, is purely energetic and in the absorption of the factor of 2 in the first term in equation (9.104) into T^*. Again there are three parameters, the energy parameter, η, the hard core volume, v^*, and c, the external degrees of freedom parameter.

Hole models and the lattice fluid model

As seen above, the cell model introduces a volume-dependent free energy through the effect of the average particle separation in changing the energy via the distance-dependent intermolecular potential function. Another way to accomplish this is to assume that the lattice underlying the cell approximation is not fully occupied. The unoccupied sites, or 'holes', thus result in a lower density for the system. They also increase the energy of the system since the attractive particle–particle interactions are reduced in number. The concentration of holes is found by minimizing the free energy with respect to their number. Thus the number of holes will be temperature-dependent and will increase with the latter. The lattice fluid model is the simplest implementation of this concept. The temperature dependence of the density arises entirely from hole effect, the underlying lattice is assumed to be rigid.

The mean-field lattice fluid model of a pure fluid (Sanchez and Lacombe, 1976, 1977) is analogous to or *isomorphous* with the lattice theory of regular solutions if one of the components in the latter is identified with the fluid molecules and the other with the holes. In the case of polymer liquids, the lattice fluid theory is isomorphous with the Flory–Huggins model. The holes play the role of the solvent molecules. Thus the equations of the earlier sections of the chapter (Sections 9.1 and

9.3) are applicable. For example in the case of a polymeric fluid, the free energy of mixing of polymer and holes is given by equation (9.27). Although expressed as Gibbs free energy this equation was based on a constant volume rigid lattice where the distinction between Gibbs and Helmholtz free energy disappears.

In the case now treated, the free energy is an explicit function of the volume so that the equation actually refers to the Helmholtz free energy. The pressure may be found by differentiating equation (9.27) with respect to volume. This is accomplished by assigning a *fixed* volume v^* to each of the lattice sites and then writing the volume, V, as

$$V = (N_0 + rN)v^*, \tag{9.108}$$

where N_0 is the number of holes, N is the number of polymer molecules and each occupies r sites. The notation N, r used here is to be identified with N_2, X in Section 9.3. Let $f =$ *the fraction of sites that are occupied* $= N/(N_0 + rN)$. The energy of the lattice is, according to Appendix A9.2, equal to $rNfz\varepsilon_{22}/2$ if $\varepsilon_{11}, \varepsilon_{12} = 0$ as is appropriate for the interaction of holes when the latter are denoted as 1 and the filled sites as 2. If the energy ε_{22} is denoted simply as ε, then equation (9.27) may be recast as

$$A/k_B T = (N_0 + rN)[f^2 z\varepsilon/2k_B T + (1 - f)\ln(1 - f) + (f/r)\ln f]. \tag{9.109}$$

The pressure may be found by differentiating equation (9.109) with respect to N_0 since in the present case, as may be seen from equation (9.108), this is equivalent to differentiating with respect to V/v^* and the change in N_0 is the source of volume change. By analogy with the differentiation leading to equation (9.29), this results in the equation of state,

$$-pv^* = k_B T[\ln(1 - f) + (1 - 1/r)f] - f^2 z\varepsilon/2.$$

The above two equations can be put into reduced form. For the Gibbs free energy, $G = A + PV$, this gives

$$\tilde{G} \equiv G/(Nr\varepsilon^*) = -\tilde{\rho} + \tilde{T}\tilde{v}[(1 - \tilde{\rho})\ln(1 - \tilde{\rho}) + (\tilde{\rho}/r)\ln \tilde{\rho}] + \tilde{p}\tilde{V} \tag{9.110}$$

and for the equation of state there results

$$\tilde{p}/\tilde{\rho}\tilde{T} = -\tilde{\rho}^{-1}[\ln(1 - \tilde{\rho}) + (1 - 1/r)\tilde{\rho}] - \tilde{\rho}/\tilde{T}, \tag{9.111}$$

where the definitions

$$v = \text{volume per mer} = V/Nr = (N_0 + rN)/Nr$$

ρ = number density = $1/v = f$ = the fraction of occupied sites

$$= Nr/(N_0 + rN)$$

$$\varepsilon^* = -z\varepsilon/2$$

are used and the reduced variables are defined as

$\tilde{v} = v/v^*$; v^* = volume per lattice site or closed packed polymer
volume per mer

$\tilde{\rho} = 1/\tilde{v}$

$\tilde{T} = T/T^*$; $T^* = \varepsilon^*/k_B$

$\tilde{p} = p/p^*$; $p^* = \varepsilon^*/v^* = k_B T^*/v^*$.

$$(9.112)$$

There are thus two independent reducing parameters, the close packed mer volume, v^*, and the mer contact energy ε^*, or, v^*, T^*. The third parameter, r, the number of mers per molecule does not reduce. For high molecular weight polymers, the equation of state reduces to a two-parameter equation. Some parameters for the lattice fluid model for simple liquids are given in Table 9.2 and some for polymers in Table 9.3.

Since the parameters in the equations of state are determined by fitting experimental data, it may be expected that any of the above equations of state, equation (9.99) for the cell model, equation (9.107) for the FOVE model, and equation (9.111) for the lattice fluid model should give good representations of volume over reasonable ranges of temperature and pressure. This is found to be the case. Figure 9.20 shows fits for the FOVE and lattice gas models for polystyrene. In a more critical comparison (Dee and Walsh, 1988; Walsh and Dee, 1989) it has been found that, in the temperature range associated with polymer liquids, cell models and cell-hole models fit better over wide temperature and pressure ranges than the lattice fluid or FOVE model. However, at higher temperatures and lower volumes, that are close to critical conditions, the fit associated with the latter two models improves but the cell-based models are degraded in performance.

9.7.2 *Application to solutions*

The extension of equation-of-state models to solutions or mixtures is accomplished by making the assumption *that the form of the free energy equation and therefore the equation of state that holds for the separate pure*

Table 9.2 *Simple liquids: equation of state parameters for the lattice fluid model.*[a]

	T^* K	v^* cm³/mol	r	p^* MPa	ρ^* g/cm³
methane	224	7.52	4.26	248	0.500
ethane	315	8.00	5.87	327	0.640
propane	371	9.84	6.5	314	0.690
n-butane	403	10.4	7.59	322	0.736
isobutane	398	11.49	7.03	288	0.720
n-pentane	441	11.82	8.09	310	0.755
isopentane	424	11.45	8.24	308	0.765
neopentane	415	12.97	7.47	266	0.744
cyclopentane	491	10.53	7.68	388	0.867
n-hexane	476	13.28	8.37	298	0.775
cyclohexane	497	10.79	8.65	383	0.902
n-heptane	487	13.09	9.57	309	0.800
n-octane	502	13.55	10.34	308	0.815
n-nonane	517	14.00	11.06	307	0.828
n-decane	530	14.47	11.75	305	0.837
n-undecane	542	14.89	12.40	303	0.846
n-dodecane	552	15.28	13.06	300	0.854
n-tridecane	560	15.58	13.79	299	0.858
n-tetradecane	570	15.99	14.36	296	0.864
n-nonadecane	596	17.26	15.83	287	0.880
benzene	523	9.80	8.02	444	0.994
fluorobenzene	527	10.39	8.05	422	1.150
chlorobenzene	585	11.14	8.38	437	1.210
bromobenzene	608	11.13	8.73	454	1.620
toluene	543	11.22	8.50	402	0.966
p-xylene	561	12.24	9.14	381	0.949
m-xylene	560	12.11	9.21	384	0.952
o-xylene	571	12.03	9.14	395	0.965
carbon tetrachloride	535	11.69	7.36	381	1.790
chloroform	512	9.33	7.58	456	1.690
methylene chloride	487	7.23	7.64	560	1.540

[a] From Sanchez and Lacombe (1978).

fluids is valid also for the mixture. The constants or parameters for the latter are to be derived by combining rules from the parameters of the pure fluids. In addition, the partition function associated with the equation of state is modified by multiplying by a combinatorial factor that represents the configurational contribution of the different ways of mixing the component molecules. This gives rise to a purely entropic contribution

Table 9.3 *Some common polymers, $r = \infty$; equation of state parameters for the lattice fluid model.*[a]

	T^* K	v^* cm^3/mol	p^* MPa	ρ^* g/cm^3	Temp. range K	max p MPa
poly(dimethyl siloxane)	476	13.1	302	1.104	298–343	100
poly(vinyl acetate)	590	9.64	509	1.283	308–373	80
poly(n-butyl methacrylate)	627	12.1	431	1.125	307–473	200
polyisobutylene	643	15.1	354	0.974	326–383	100
polyethylene (linear)	649	12.7	425	0.904	426–473	100
polyethylene (branched)	673	15.6	359	0.887	408–471	100
poly(methyl methacrylate)	696	11.5	503	1.269	397–432	200
poly(cyclohexyl methacrylate)	697	13.6	426	1.178	396–472	200
polystyrene	735	17.1	357	1.105	388–468	200
poly(o-methyl styrene)	768	16.9	378	1.079	412–471	160

[a] From Sanchez and Lacombe (1978).

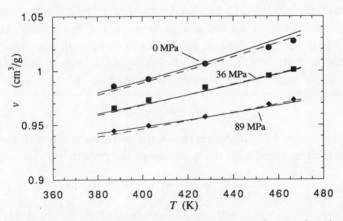

Fig. 9.20 Specific volume vs temperature at three pressures for atactic polystyrene. The full curve is the lattice fluid model and the dashed one is for the FOVE model. Points are experimental. Parameters for the lattice fluid model are $T^* = 735$ K, $\rho^* = v^{*-1} = 1.105$ g/cm^3 (Sanchez and Lacombe, 1978). For the FOVE model they are $T^* = 8104$ K, $v^* = 0.823$ cm^3/g, $p^* = 474$ MPa, and are from Dee and Walsh (1988) except for an adjustment to v^*.

to the mixing free energy. This modification thus simply requires adding a Flory–Huggins entropic term to the free energy.

Although the goal of solution theory is to represent the solutions in terms of pure component properties alone, in practice it is found that the geometric mean rule, equation (A9.10), is not sufficiently accurate to represent many of the properties adequately. Thus the pair contact exchange energy is usually left as a single adjustable parameter for the mixture. It is experimentally determined, usually from heat of mixing data.

FOVE model

It is recalled that the relevant parameters, equation (9.106), can be taken as v^*, T^* and the degree of freedom parameter, c, and that $p^* = ck_B T^*/v^*$. Thus three combining rules are needed. The local segment core volume, v^*, can be defined as being the *same for both components* by the expedient of defining the number of mers per molecule, r_1, r_2 for both components to make this consistent. Thus the equation-of-state data for pure 1 and 2 can be reduced with *molar volumes* to find V_1^* and V_2^*. Therefore $V_1^* = r_1 v^*$, $V_2^* = r_2 v^*$ and $r_1/r_2 = V_1^*/V_2^*$. However, since the number of mers, r, appears in the free energy expression to be used for the mixture, a combining rule will be necessary for this parameter. Thus the combining rules can be considered to be based on either T^* or P^* and c and r.

Arguments concerning the energy term in the free energy, equation (9.104) or (9.105), are used to establish the combining rule for P^*. This proceeds as follows (Flory, 1965). According to equation (9.104) the energy of the pure fluid is given by the first term, or

$$-E = rNs\eta/2v. \tag{9.113}$$

From Appendix A9.1, if the energy of the solution is taken in the lattice approximation, then, with the notation of the present section,

$$-E = (N_{11}\eta_{11} + N_{22}\eta_{22} + N_{12}\eta_{12})/v, \tag{9.114}$$

where N_{11}, N_{22}, N_{12} are the total numbers of $(1, 1)$, $(2, 2)$, $(1, 2)$ pair contacts. The number of contacts per local segment (analogous to z in Appendix A9.1), s_1, s_2, are *not* assumed to be the same. The total numbers of contacts at type 1 mers and type 2 mers are given by,

$$\left.\begin{array}{l} 2N_{11} + N_{12} = s_1 r_1 N_1, \\ 2N_{22} + N_{12} = s_2 r_2 N_2. \end{array}\right\} \tag{9.115}$$

The number of unlike pairs is assumed to be approximated, with $N = N_1 + N_2$, by

$$N_{12} = (s_1 r_1 N_1)(s_2 r_2 N_2)/srN \tag{9.116}$$

where

$$r = (r_1 N_1 + r_2 N_2)/N, \qquad (9.117)$$

$$s = (s_1 r_1 N_1 + s_2 r_2 N_2)/rN. \qquad (9.118)$$

These relations give

$$-E = \left[s_1 r_1 N_1 \eta_{11} + s_2 r_2 N_2 \eta_{22} - \left(\frac{s_1 r_1 N_1 s_2 r_2 N_2}{srN} \right) \Delta\eta \right] \Big/ 2v, \quad (9.119)$$

where $\Delta\eta = \eta_{11} + \eta_{22} - 2\eta_{12}$. Segment fractions are defined as

$$\phi_1 = r_1 N_1/rN; \qquad \phi_2 = r_2 N_2/rN. \qquad (9.120)$$

With these, the energy becomes

$$-E = rN[s_1\phi_1\eta_{11} + s_2\phi_2\eta_{22} - (s_1 s_2/s)\phi_1\phi_2 \,\Delta\eta]/2v, \quad (9.121)$$

which can be rewritten as

$$-E = rN[\phi_1 p_1^* + \phi_2 p_2^* - \phi_1\phi_2(s_1 s_2/s) \,\Delta\eta/(2v^{*2})]v^{*2}/v, \quad (9.122)$$

where the relations

$$p_1^* = s_1\eta_{11}/2v^{*2}, \qquad p_2^* = s_2\eta_{22}/2v^{*2} \qquad (9.123)$$

for the pure fluids, equation (9.106), have been invoked. According to equation (9.106), the energy, equation (9.113), can be written as,

$$-E = rNp^*v^{*2}/v. \qquad (9.124)$$

Comparing with equation (9.122) it is seen that p^* is to be defined as

$$p^* = \phi_1 p_1^* + \phi_2 p_2^* - \Delta p_{12}^* \phi_1\phi_2, \qquad (9.125)$$

where

$$\Delta p_{12}^* = (s_1 s_2/2sv^{*2}) \,\Delta\eta. \qquad (9.126)$$

If the external degrees of freedom variable, c, is written as

$$c = (c_1 r_1 N_1 + c_2 r_2 N_2)/rN$$

$$= \phi_1 N_1 + \phi_1 N_2, \qquad (9.127)$$

then from $T^* = p^*v^*/ck_B$, $T_1^* = p_1^*v^*/c_1 k_B$, $T_2^* = p_2^*v^*/c_2 k_B$, equation (9.106), the mixing rule for T^* is,

$$1/T^* = (\phi_1 p_1^*/T_1^* + \phi_2 p_2^*/T_2^*)/p^*. \qquad (9.128)$$

The free energy for the mixture is then given by equation (9.105) but with a Flory–Huggins combinatorial entropic contribution, equation

(9.25) with X replaced by r_2/r_1, added, or

$$\tilde{G} \equiv G/rNp^*v^* = -\tilde{v}^{-1} - 3\tilde{T}\ln[Cv^{*1/3}(\tilde{v}^{1/3} - 1)] + \tilde{p}\tilde{V}$$
$$+ \tilde{T}c^{-1}(\phi_1 r_1^{-1} \ln \phi_1 + \phi_2 r_2^{-1} \ln \phi_2) \qquad (9.129)$$

and where it is understood that the reduced variables are determined using the c, r, p^*, T^* mixing rules given above. The equation of state is identical to equation (9.107).

The enthalpy of mixing, on ignoring the small $p \, \Delta V$ contribution, results from the inverse reduced volume first term in equation (9.129), i.e., the term without the reduced temperature multiplier, or, on subtracting the enthalpies of the pure components,

$$\Delta H^m = E - E_1 - E_2$$

and

$$\Delta H^m = v^* r N(-p^*/\tilde{v} + \phi_1 p_1^*/\tilde{v}_1 + \phi_2 p_2^*/\tilde{v}_2), \qquad (9.130)$$

where p^* is given by equation (9.125). The entropy of mixing arises from the terms in equation (9.129) with the reduced temperature multiplier,

$$\Delta S^m = 3rNv^*\left[\phi_1\left(\frac{p_1^*}{T_1^*}\right)\ln\left(\frac{\tilde{v}^{1/3} - 1}{\tilde{v}_1^{1/3} - 1}\right) + \phi_2\left(\frac{p_2^*}{T_2^*}\right)\ln\left(\frac{\tilde{v}^{1/3} - 1}{\tilde{v}_2^{1/3} - 1}\right)\right]$$
$$+ rNk_B(\phi_1 r_1^{-1} \ln \phi_1 + \phi_2 r_2^{-1} \ln \phi_2). \qquad (9.131)$$

The chemical potential of component 1 is found by differentiating equation (9.129) with respect to N_1,

$$\mu_1 - \mu_1^0 = \left(\frac{v_1^*}{\tilde{v}}\right)\Delta p_{12}^*\left(\frac{s_2}{s}\right)\phi_2^2 + p_1^* v_1^*\left[3\tilde{T}_1 \ln\left(\frac{\tilde{v}_1^{1/3} - 1}{\tilde{v}^{1/3} - 1}\right) + (\tilde{v}_1^{-1} - \tilde{v}^{-1})\right]$$
$$+ k_B T[\ln \phi_1 + (1 - X^{-1})\phi_2], \qquad (9.132)$$

where $X = r_2/r_1$. The first term represents the contact exchange contribution, the second term is from the equation of state and the final term, with $k_B T$ multiplier, is from the combinatorial entropy of mixing. It is to be noticed that the factor (s_2/s) now occurs in the contact exchange term. This is due to the dependence of s, equation (9.118), on composition. In the FOVE model s_1, s_2 are taken to be proportional to the surface areas of the respective segments and are estimated from measurements on molecular models. The ratio s_2/s_1 suffices to determine s_2/s.

Lattice fluid model

The relevant parameters, equation (9.112), are v^* and $T^* = \varepsilon^*/k_B$, or alternatively, v^*, $p^* = \varepsilon^*/v^*$. In addition, the number of mers per molecule,

r, appears. The volume per lattice site, v^*, in the mixture is uniform over the lattice but this volume, in general, is not the same as this quantity in either pure liquid. The combining rules for v^* and r are found from the following assumptions (Sanchez and Lacombe, 1978). Let r_1^0, r_2^0 be the number of mers per respective molecule in the pure liquids and r_1, r_2 be the corresponding quantities in the mixed state. The first assumption is that the close packed volume of the mixture is additive in that of the pure liquids. This gives

$$V^* = r_1^0 N_1 v_1^* + r_2^0 N_2 v_2^* = (r_1 N_1 + r_2 N_2)v^*, \qquad (9.133)$$

which implies $V_1^* = r_1^0 v_1^* = r_1 v^*$, $V_2^* = r_2^0 v_2^* = r_2 v^*$. Thus $r_1/r_2 = V_1^*/V_2^*$, the same as in the FOVE theory.

The second assumption is that the number of pair interactions in the solution is the sum of the number of pair interactions in the pure states. Thus, with $N = N_1 + N_2$,

$$(r_1^0 N_1 + r_2^0 N_2)z/2 = (r_1 N_1 + r_2 N_2)z/2 \equiv rNz/2. \qquad (9.134)$$

These relations lead to the mixing rules

$$r = x_1 r_1^0 + x_2 r_2^0 \qquad (9.135)$$

and

$$v^* = (r_1^0 x_1 v_1^* + r_2^0 x_2 v_2^*)/r, \qquad (9.136)$$

where x_1, x_2 are *mole fractions*, N_1/N, N_2/N.

The occupied site fractions, ϕ_1, ϕ_2, are given by

$$\phi_1 = r_1 N_1/rN = x_1 r_1/r, \qquad \phi_2 = 1 - \phi_1. \qquad (9.137)$$

The remaining mixing rule is to be set for either ε^* or p^* in equation (9.112). In earlier applications to small molecule fluids, Sanchez and Lacombe (1976) chose ε^* but in later application to polymers Sanchez and Lacombe (1978) chose p^*. The p^* version is given here. It follows the FOVE version except that the contacts per segment, s_1, s_2, are assumed to be equal. Thus

$$p^* = \phi_1 p_1^* + \phi_2 p_2^* - \phi_1 \phi_2 \, \Delta p_{12}^*, \qquad (9.138)$$

where Δp_{12}^* can be directly taken as a composition-independent parameter since $s_1 = s_2$, cf. equations (9.125) and (9.126). The parameters ε^* and T^* follow directly from $\varepsilon^* = k_B T^* = p^* v^*$, equation (9.112).

The free energy is obtained by adding the term representing the entropy of mixing of the molecules to equation (9.110), or

$$G \equiv (Nr\varepsilon^*)\tilde{G},$$

where

$$\tilde{G} = -\tilde{\rho} + \tilde{T}\tilde{v}[(1 - \tilde{\rho})\ln(1 - \tilde{\rho}) + (\tilde{\rho}/r)\ln\tilde{\rho}] + \tilde{p}\tilde{v}$$
$$+ \tilde{T}[(\phi_1/r_1)\ln\phi_1 + (\phi_2/r_2)\ln\phi_2]. \tag{9.139}$$

The reduced variables are defined as in equation (9.112) and with r and the reducing parameters, v^*, ε^*, defined in terms of those of the pure components as above. Since the entropy of mixing of the molecules is written in terms of the occupied site fractions, ϕ_1, ϕ_2 it does not depend on volume. Thus this term does not contribute to the equation of state itself and the latter is given by equation (9.111).

The enthalpy of mixing, in the approximation of neglecting $p\,\Delta V$, arises from the $-\tilde{\rho}$ leading term in equation (9.139). On subtracting the enthalpy of the pure components there results

$$\Delta H^m/N = -\tilde{\rho}r\varepsilon^* - (-\tilde{\rho}_1 x_1 r_1^0 \varepsilon_1^* - \tilde{\rho}_2 x_2 r_2^0 \varepsilon_2^*),$$

which on introducing $p^* = \varepsilon^*/v^*$ from equation (9.138) becomes

$$\Delta H^m/N = rv^* \tilde{\rho}\phi_1\phi_2 \,\Delta p_{12}^* + rv^*(\tilde{\rho}_1 - \tilde{\rho})\phi_1 p_1^* + rv^*(\tilde{\rho}_2 - \tilde{\rho})\phi_2 p_2^*. \tag{9.140}$$

The entropy contribution to the free energy arises from the terms in equation (9.139) with the multiplier of reduced temperature. On subtracting the entropies of the pure components, the mixing entropy is found as

$$\Delta S^m/Nk_B = -r\left\{\frac{\phi_1}{r_1}\ln\phi_1 + \frac{\phi_2}{r_2} + \tilde{v}\left[(1 - \tilde{\rho})\ln(1 - \tilde{\rho}) + \frac{\tilde{\rho}}{r}\ln\tilde{\rho}\right]\right\}$$
$$+ r\phi_1^0 \tilde{v}_1\left[(1 - \tilde{\rho}_1)\ln(1 - \tilde{\rho}_1) + \frac{\tilde{\rho}_1}{r_1^0}\ln\tilde{\rho}_1\right]$$
$$+ r\phi_2^0 \tilde{v}_2\left[(1 - \tilde{\rho}_2)\ln(1 - \tilde{\rho}_2) + \frac{\tilde{\rho}_2}{r_2^0}\ln\tilde{\rho}_2\right], \tag{9.141}$$

where ϕ_1^0, ϕ_2^0 are occupied site fractions based on $\phi_1^0 = r_1^0 N_1/rN$, $\phi_2^0 = r_2^0 N_2/rN$.

The chemical potential of component 1 is

$$\mu_1 - \mu_1^0 = \tilde{\rho}r_1 v^* \,\Delta p_{12}^* \phi_2^2$$
$$+ r_1^0 k_B T[-\tilde{\rho}/\tilde{T}_1 + \tilde{p}_1 \tilde{v}/\tilde{T}_1 + \tilde{v}(1 - \tilde{\rho})\ln(1 - \tilde{\rho}) + \tilde{v}(\tilde{\rho}/r_1^0)\ln\tilde{\rho}]$$
$$+ k_B T[\ln\phi_1 + (1 - X^{-1})\phi_2], \tag{9.142}$$

where $X = r_2/r_1$.

9.7.3 Comparison with experiment

It is recalled, Section 9.4.1, that a major failing of the Flory–Huggins model was that the χ parameter does not experimentally commonly follow the implied $1/T$ temperature dependence, a circumstance necessitating creating both enthalpic and entropic components to χ. In fact, the entropic component is often large and positive and can outweigh the enthalpic component which can even be negative. The latter circumstance is, of course, at odds with the simple lattice mixing picture when the geometric mean rule is invoked. There the enthalpy is expected to be positive. Further, it was found that the apparent χ, when computed from activity data at various concentrations was often concentration-dependent. Quite importantly, it was seen, Section 9.4.3, that polymer solutions commonly phase separate again as temperature continues to increase above the first critical solution temperature (UCST) and thus exhibit the LCST phenomenon. It was seen that this behavior implies that the enthalpic and entropic components of the χ parameter are further temperature-dependent through a ΔC_P contribution of negative sign. It is of considerable interest to see to what degree these shortcomings are addressed by the equation-of-state theory approach.

The resolution of χ into χ_H and χ_S and the concentration dependence of χ is taken up first. As was seen, Section 9.4.1, χ is often dominated by the *entropic* term, χ_S. The latter term is usually positive indicating a large negative contribution of the entropy itself. For the FOVE model, it is apparent from equation (9.125) that the enthalpy of mixing contains terms, in addition to the conventional contact exchange term represented by $\phi_1\phi_2\,\Delta p_{12}^*$, that arise through the equation of state and the fact that the reduced volumes of the solution and pure components are not, in general, equal. It is also evident that the entropy of mixing has equation-of-state terms. Similar conclusions are apparent with respect to the lattice fluid model, equations (9.140) and (9.141).

In both models, the contact interaction parameter, Δp_{12}^*, is determined by fitting experimental enthalpy of mixing data. With this accomplished, a completely determined free energy function is available, provided, of course, that the characteristic parameters of the pure components are known. The chemical potential of the solvent, equation (9.132) or (9.142) resulting from this function can be used to compute an apparent χ parameter or *reduced residual chemical potential* as it is sometimes called

$$\chi \equiv [(\mu_1 - \mu_1^0)/RT - \ln \phi_1 - (1 - 1/X)\phi_2]/\phi_2^2$$

or

$$\chi \equiv [\ln a_1 - \ln \phi_1 - (1 - 1/X)\phi_2]/\phi_2^2. \qquad (9.143)$$

This can then be compared with the apparent χ determined from experimental activity data using the same definition. The temperature dependence of χ can be used to establish χ_H and χ_S, Section 9.4.1. Although the results are somewhat mixed in quantitative aspects, the equation-of-state approach is by and large successful in producing χ parameters that are characterized by a large positive χ_S component.

Results from application of the FOVE model are shown in Figures 9.21 and 9.22. It may be seen that the agreement although not totally quantitative is nevertheless impressive.

Results for the lattice fluid theory (Sanchez and Lacombe, 1978) are given in Table 9.4. The χ, χ_H and χ_S parameters are expressed at the extremes of composition range. That is, χ_1 is the leading term in an expansion of χ in concentration about $\phi_2 = 0$ and is the infinite dilute value; $\chi_{H;1}$ and $\chi_{S;1}$ are the accompanying enthalpic and entropic components. The parameter χ_∞ is the corresponding value at $\phi_2 = 1$. It may be seen that the agreement for the polymer dilute side is generally quite

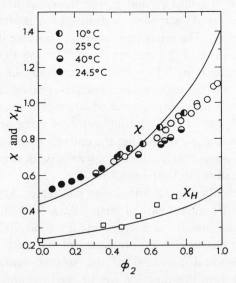

Fig. 9.21 Apparent χ and χ_H parameters vs segment fraction, ϕ_2, for polyiso-butylene in benzene. The χ_S parameter is the difference between χ and χ_H. Points are experimental. The curves are calculated from the FOVE equation-of-state theory (Flory, 1970).

Table 9.4 *Lattice fluid model. Comparison of calculated and experimental*
χ *parameters for polyisobutylene in various solvents, 298 K.*[a]

	χ_1		$\chi_{H;1}$		$\chi_{S;1}$		χ_∞	
	exp	calc	exp	calc	exp	calc	exp	calc
n-pentane	0.49	0.76	−0.42	−0.33	0.91	1.09	0.93	0.69
n-hexane		0.56		−0.27		0.83		0.47
n-heptane		0.49		−0.18		0.67		0.43
n-octane	0.46	0.43	−0.17	−0.13	0.63	0.57	>0.5	0.38
n-decane		0.32		−0.09		0.41		0.28
cyclohexane	0.47	0.34	0.00	−0.02	0.47	0.36	0.5	0.36
benzene	0.50	0.63	0.26	0.67	0.24	−0.04	1.15	0.74

[a] The subscript 1 refers to the infinitely dilute in polymer state, $\phi_2 \to 0$; ∞ refers to $\phi_2 \to 1$.

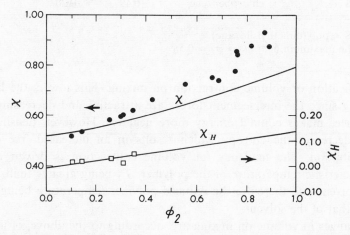

Fig. 9.22 Apparent χ and χ_H parameters vs segment fraction, ϕ_2, for polystyrene in methyl ethyl ketone at 25 °C. Points are experimental. The curves are calculated from the FOVE equation of state theory (Flory, 1970).

good. Negative enthalpic and large positive entropic components to χ_1 are predicted for the cases where they are observed experimentally. The principal failing is seen to be connected with the polymer-rich side. The calculated values for χ_∞ are low. Thus the commonly experimentally observed substantial increase in χ in proceeding from dilute to concentrated polymer solutions is underestimated.

The lattice fluid model ascribes the tendency toward small or negative χ_H to the fact that, at constant temperature, there is often a slight

Table 9.5 *Changes in volume on mixing.[a] Results from the FOVE and lattice fluid theories compared with experiment, 298 K.*

polymer	solvent	$(\Delta v^m/v^0) \times 10^2$		
		exp	FOVE	lattice fluid
polyisobutylene	n-pentane	-1.27	-1.54	-1.83
polyisobutylene	n-hexane	-0.86		-1.25
polyisobutylene	n-heptane	-0.62		-0.92
polyisobutylene	n-octane	-0.48	-0.54	-0.73
polyisobutylene	n-decane	-0.29		-0.48
polyisobutylene	cyclohexane	-0.14	-0.07	-0.44
polyisobutylene	benzene	0.34	0.75	0.20
polystyrene	methyl ethyl ketone	-0.82	-0.42	
polystyrene	ethyl benzene	-0.31	-0.27	
polystyrene	cyclohexane	-0.14	-0.14	
PDMS	cyclohexane	0.06	0.55	
PDMS	chlorobenzene	-0.49	0.20	

PDMS = poly(dimethylsiloxane).
[a] At the maximum near $\phi_1 = \phi_2 = 0.5$.

densification or volume contraction on mixing. This lowers the heat of mixing since the intersegment forces are attractive and decreasing their distances makes contact energy more negative. However, densification also decreases the entropy which results in an increased free energy contribution. The tendency for volume contraction is driven by the characteristic temperature of the polymer, T_2^* being greater than T_1^* for the solvent and therefore the reduced polymer temperature being lower than that of the solvent.

Changes in volume on mixing are, according to the above, of importance in the context of interpretation of the solution thermodynamics. Since they can be directly measured experimentally it is of interest to compare such values with the theoretical predictions. From the theories the relative change in volume on mixing is given by

$$\Delta \tilde{v}/\tilde{v}^0 = \tilde{v}/\tilde{v}^0 - 1, \tag{9.144}$$

where

$$\tilde{v}^0 = \phi_1 \tilde{v}_1 + \phi_2 \tilde{v}_2.$$

Table 9.5 displays results for both the FOVE (Flory, 1970) and lattice fluid models (Sanchez and La Combe, 1978). It may be seen that both approaches predict volume changes that approximately mimic the experiments.

Turning to the behavior of the apparent χ over a wide range of temperature, it is found that the equation-of-state theories (Flory *et al.*, 1964; Sanchez and Lacombe, 1978) predict the occurrence of phase separation at high temperatures, i.e., the presence of an LCST. The lattice fluid model predicts this for all solutions of high molecular weight polymers showing a UCST provided the temperature is high enough. The enthalpy of mixing at constant composition decreases with increasing temperature, often strongly. Thus a basis is provided for the occurrence of negative ΔC_P as discussed in Section 9.4.3. As developed there negative ΔC_P implies the eventual observation of a LCST at high temperature. Although the equation-of-state theories provide a natural and satisfactory explanation for the phenomenon the quantitative representation is not necessarily highly accurate (Sanchez and Lacombe, 1978).

As a further observation on UCST/LCST behavior, it is found (Sanchez and Lacombe, 1978) that the position, i.e., the predicted numerical value, of the *UCST* is exceedingly sensitive to the value of the enthalpic mixing parameter, Δp^*_{12}, and to its deviation from the geometric mean rule. From the standpoint of *predicting* solubility behavior of polymer/solvent pairs this is unfortunate. It means that it is unlikely, without measurements on the solutions themselves, that the values of the parameter can be predicted with sufficient accuracy to be useful in establishing the temperature region of miscibility.

On a more optimistic note than the above, it is very interesting to observe that gas solubilities in polymers apparently *can* be predicted with impressive accuracy. Henry's Law constants predicted with the lattice fluid theory from the properties of the pure substances alone, using the geometric mean rule, are in good agreement with experiment (Sanchez and Rodgers, 1990, 1992). This results from the Henry's Law constants being much less sensitive to the gas–polymer interaction than is the case with general liquid–liquid equilibria.

Some further elaborations on the theories are worth noting. The regular solution approach to computing the energy of mixing, with its assumption of local randomness to the contacts implies the lack of strong specific interactions. Such interactions are often important in the solubility behaviour of real systems. The FOVE model has been generalized on the basis of the quasi-chemical approximation to include non-random specific interactions (Renuncio and Prausnitz, 1976; Panayiotou and Vera, 1980). The lattice fluid model has also been extended to include the effects of specific interactions (Sanchez and Balazs, 1989). Elaboration of the mean-field lattice fluid approach has been carried out by Koningsveld and Beckman and their coworkers (Koningsveld, Kleintjens and Leblans-Vinckl, 1987; Beckman, Porter and Koningsveld, 1987; Beckman, Koningsveld and Porter, 1990). The requirements of the free energy function necessary to be useful in representing phase behavior have been discussed on an empirical basis by Qian, Mumby and Eichinger (1991). Empirical representations are also discussed by Fujita (1990).

9.8 More rigorous extensions of theory

The Flory–Huggins model, Section 9.3, and the equation of state generalizations of it, Section 9.7, are motivated by the desire to find the *simplest* way or at least a *simple* way to describe the behavior of solutions. Both the Flory–Huggins model and the lattice fluid theory for pure fluids are based on a well-defined model. That is, a lattice that can be occupied by chain molecules and by solvent molecules and/or holes results in a formulated problem to be addressed by statistical mechanics. This can be called the treatment of the *standard lattice*. In the Flory–Huggins model the statistical treatment is subjected to the mean-field approximation. It obviously is important to separate the failings of the standard lattice model from those of the approximations, such as mean-field, that are used to make it tractable. In the case of the lattice fluid equation-of-state theory, although the important feature of variable volume is introduced, it is subject to the same uncertainties with respect to realistic statistics as the original theory. In proceeding to mixtures or solutions it makes the further important *assumption* that the fluid is to be described in essentially the same way as the pure component fluids. This is accomplished by construction of combining rules for the characteristic parameters. In effecting this step the treatment is no longer a well-defined statistical mechanical problem, but is now to some degree an empirical model. The FOVE model is an empirical treatment from the beginning. There is no well-defined statistical mechanical model that leads, in approximation, to the starting point assumption that the energy in a relatively dense fluid is inversely proportional to volume.

Based on the above comments, it is seen that there is every reason to pursue approaches that improve the statistical treatment. This is true even if the exercise is inherently more complicated. There has been significant activity and very promising progress, mainly by Freed and coworkers, in advancing the statistical mechanics of the standard lattice beyond the Flory–Huggins mean-field approximation (Freed, 1985; Bawendi, Freed and Mohanty, 1986, 1987; Bawendi and Freed, 1988; Pesci and Freed,

1989; Nemirovsky, Bawendi and Freed, 1987; Dudowicz, Freed and Madden, 1990; Dudowicz and Freed, 1991). Not only has this included the basic Flory–Huggins lattice itself but has been extended to include the lattice fluid model, or as alternately put, a compressible lattice, by incorporation of holes (Dudowicz and Freed, 1991). To some degree the structure of the participant molecules can be accommodated through assigning lattice sites to parts of the molecules (Dudowicz *et al.*, 1990).

In Appendix A9.2 the combinatorial factor that represents the partition function of a set of chains occupying the lattice is written in the mean-field approximation, equation (A9.15). It is possible to write *formal exact expressions* for this factor. This is most conveniently done in terms of constraints that express the chemical connectivity of the constituent molecules. These constraints, which take the form of Kronecker delta functions, can be expressed in terms of, or transformed to, crystallographic reciprocal space. When this is carried out, the combinatorial partition function can be arranged so that a sum over the lattice vectors takes the form of

$$z/N\left[1 + z^{-1}\sum_{q \neq 0} f(\mathbf{q})\,e^{i\mathbf{q}\cdot(\mathbf{r}_i - \mathbf{r}_j)}\right],$$

where z is the lattice coordination number ($z = 6$ for the three-dimensional cubic lattices treated) and the summation involves a wave vector, $\mathbf{q}$, and the structure factor, $f(\mathbf{q})$, of the nearest neighbors. When lattice sums of this type are inserted into the formal expression for the partition function, it is found *the leading term, the factor of unity, in the above gives rise to the Flory–Huggins result*. Thus terms of the remaining type, those in z^{-1}, generate the exact correction to the Flory–Huggins expression. Most importantly such terms occur as products such that they generate *systematic corrections in powers of* z^{-1}. Thus a cluster expansion theory results that bears much resemblance to the Mayer theory of imperfect gases (Dudowicz *et al.*, 1990).

The evaluation of the partition function as outlined above is in the same spirit as the Flory–Huggins model. It is appropriate for a system without energetic interactions, i.e., for an *athermal* system. In Flory–Huggins, the energy of mixing is calculated separately on a random mixing basis and the entropic combinatorial partition function is assumed to be unperturbed by the energetics. However, again on a formal basis, it is possible to write the complete partition function, including the energetics, by inserting a Boltzmann factor term involving the contact energies, ε, for each configuration. It is found that another cluster

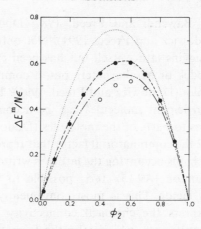

Fig. 9.23 Dimensionless energy of mixing for a polymer of DP $= 100$ with a single bead solvent. Comparison of the Flory–Huggins model and the lattice cluster theory with essentially exact Monte Carlo simulations. The dotted curve is the Flory–Huggins model. The filled points are for Monte Carlo simulation for $\varepsilon/k_B T = 0$. The dashed curve is the lattice cluster theory for the same condition. The open circles are Monte Carlo for $\varepsilon/k_B T = \frac{1}{3}$. The dashed-dot curve is the lattice cluster theory for this condition. The energy ε is twice the contact exchange energy, ω, of equation (A9.8), the coordination number, $z = 6$, is for a cubic lattice (from Dudowicz et al., 1990, with permission Macromolecules, © 1990, American Chemical Society).

expansion, this time in powers of $\varepsilon/k_B T$, can be effected. Thus by means of a double cluster expansion, one in z^{-1} and the other in $\varepsilon/k_B T$, a free energy expression is obtained that contains corrections for both the mean-field packing entropy and the random contacts assumed for the energy of mixing in the Flory–Huggins theory.

Some results obtained with the *lattice cluster* theory are shown in Figures 9.23 and 9.24. These results involve comparison with simulations. The latter are Monte Carlo calculations on the standard lattice and may be regarded as exact results for the model. Comparison with simulations thus has the advantage of precisely elucidating the accuracy of the theoretical treatment of the lattice problem. Figure 9.23 shows the accuracy of the theory for the energy of mixing. In the Flory–Huggins model, the energy, or enthalpy, of mixing, equation (9.26), is independent of temperature. With correlation corrections introduced, the non-random nature of the energy contacts appears and the mixing energy becomes temperature-dependent. In Figure 9.23 lattice cluster theory and Monte Carlo results are compared for two temperatures, i.e., $\varepsilon/k_B T$ ratios. Also shown is the Flory–Huggins result. The theory and simulations show

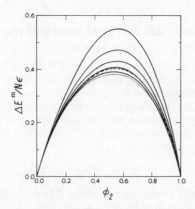

Fig. 9.24 Dimensionless energy of mixing showing the effect of solvent structure as calculated from the lattice cluster theory. Polymer has DP = 100. The solid curves are for linear solvents of from 1 to 5 beads, top to bottom. Dashed curve is for a branched 4-mer and the dotted curve is for a neopentane-like 5-mer (from Dudowicz *et al.*, 1990, with permission *Macromolecules*, © 1990, American Chemical Society).

significant reduction in the magnitude of the energy of mixing over the Flory–Huggins theory. At $\varepsilon/k_B T = 0$, the theory is essentially in exact agreement with the simulation. Under this condition the effect of the expansion in $\varepsilon/k_B T$ is suppressed and the effectiveness of the z^{-1} expansion is demonstrated. At $\varepsilon/k_B T = \frac{1}{3}$, where the effect of the $\varepsilon/k_B T$ expansion is more prominent, the theoretical result noticeably deviated from the simulation but nevertheless captures much of the effect of the non-randomness of mixing on the energy.

An illustration of the promise of the lattice cluster theory in representing structural effects is shown in Figure 9.24. The dimensionless energy of mixing is shown for a variety of solvent structures. These include linear molecules of size spanning 1–5 lattice sites and two branched solvent molecules. Substantial effects that would be totally ignored in the Flory–Huggins formulation are to be observed.

9.9 Further reading

General books on polymer solutions are those by Fujita (1990), Morawetz (1975) and Yamakawa (1971). For matters concerning scaling the book by de Gennes (1979) is standard. A comprehensive work on scaling and its relation to experiments is available (des Cloizeaux and Jannik, 1991).

Appendix A9.1 Entropy and energy of mixing of equal-sized molecules on a lattice

Figure 9.25 shows a lattice of $N = N_1 + N_2$ sites occupied by two kinds of molecules of number N_1, N_2 respectively. The molecules are assumed to occupy the sites at random. The total number of ways of mixing the molecules on this basis is $\Omega = N!/(N_1!\, N_2!)$. The resulting entropy, S, of the mixture is $S = k_B \ln \Omega$, which is also the entropy of mixing as Ω for the pure components is 1. On application of Stirling's approximation, $\ln N! \cong N \ln N - N$, the entropy of mixing in terms of mole fractions x_1, x_2 is

$$\Delta S^m = -k_B N(x_1 \ln x_1 + x_2 \ln x_2). \tag{A9.1}$$

The energy of mixing is evaluated as follows. The molecules are assumed to interact through pair interaction energies, ε. The total energy is the sum of the pair energies,

$$E = N_{11}\varepsilon_{11} + N_{22}\varepsilon_{22} + N_{12}\varepsilon_{12}, \tag{A9.2}$$

where 11, 22 refer to like contacts, 12 to unlike and the Ns to the number of pairs of each kind. Under the assumption of *random* mixing or the *mean-field approximation*, the chance that a site is occupied by a type 1 molecule is x_1, its mole fraction. The chance that a given type 1 molecule has a type 1 on one of its neighboring sites is x_1 and the chance that it has 1 as a neighbor on any of its z neighboring sites is zx_1. The total number of N_{11} pairs in terms of the number of type 1 molecules, N_1, then is

$$N_{11} = zN_1 x_1/2, \tag{A9.3}$$

where division by 2 is included to avoid counting twice. Also

$$N_{22} = zN_2 x_2/2 \tag{A9.4}$$

and

$$N_{12} = zN_1 x_2. \tag{A9.5}$$

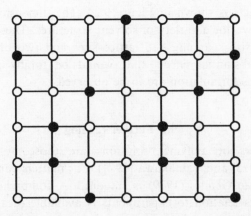

Fig. 9.25 Lattice model for a 'strictly' regular solution of equal sized molecules.

The change in energy on mixing N_1, N_2 molecules from pure states on separate lattices is

$$\Delta E^m = (zN_1x_1/2)\varepsilon_{11} + (zN_2x_2/2)\varepsilon_{22} + (zN_1x_2)\varepsilon_{12} - (zN_1/2)\varepsilon_{11} - (zN_2/2)\varepsilon_{22}$$

$$= (N_1 + N_2)Bx_1x_2, \tag{A9.6}$$

where $B = z(\varepsilon_{12} - \varepsilon_{11}/2 - \varepsilon_{22}/2)$. If N_1, N_2 are taken as being in units of moles rather than total number of molecules,

$$B = zN_A\omega, \tag{A9.7}$$

where ω defines the *contact exchange energy*,

$$\omega = (\varepsilon_{12} - \varepsilon_{11}/2 - \varepsilon_{22}/2), \tag{A9.8}$$

and N_A is Avogadro's number. There are thus three parameters, necessary to quantify the model, $\varepsilon_{12}, \varepsilon_{11}, \varepsilon_{22}$. The energies for the like contacts, $\varepsilon_{11}, \varepsilon_{22}$ can be determined from the properties of the pure components. The energies of vaporization of one mole of pure components are

$$\Delta E_1^{vap} = -zN_A\varepsilon_{11}/2, \qquad \Delta E_2^{vap} = -zN_A\varepsilon_{22}/2. \tag{A9.9}$$

It is usually assumed that the unlike pair contact energy can be expressed in terms of the like ones via the *geometric mean rule* as

$$-\varepsilon_{12} = (\varepsilon_{11}\varepsilon_{22})^{1/2}. \tag{A9.10}$$

In this case equation (A9.7) becomes

$$B = [(\Delta E_1^{vap})^{1/2} - (\Delta E_2^{vap})^{1/2}]^2. \tag{A9.11}$$

Appendix A9.2 Entropy of polymer chains on a lattice

There are N_2 polymer molecules each with X monomeric units. The latter are considered to be beads of the same size as the solvent molecules. Either a bead or one of the N_1 solvent molecules can occupy a lattice site, Figure 9.26. If

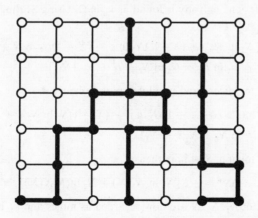

Fig. 9.26 Lattice model for a solution of flexible polymer molecules and solvent.

polymer chains are added to the lattice of $N = N_2 X + N_1$ sites, one-by-one, the number of sites available to the first monomeric unit of the first chain is N. The number available to the second unit is equal to the coordination number, z, times the chance that the neighboring sites of the first unit are unoccupied. It is assumed that this latter chance is given by the *average* fraction of sites that are unoccupied, i.e., in the *mean-field approximation*. The number of sites available to the third unit is $z' = z - 1$ times the chance that the sites are unoccupied. Again the latter chance is assumed to be given by an average fraction or random basis. Computing the chance that a site is unoccupied on the basis of average or random occupancy is obviously only an approximation, albeit a necessary one to make the counting process tractable. Proceeding in this fashion the number of configurations of the first chain is,

$$N \cdot z \left(\frac{N-1}{N} \right) \cdot z' \left(\frac{N-2}{N} \right) \cdot \cdots \cdot z' \left(\frac{N-x+1}{N} \right), \tag{A9.12}$$

for the second chain,

$$(N-X) \cdot z \left(\frac{N-X-1}{N} \right) \cdot z' \left(\frac{N-X-2}{N} \right) \cdot \cdots \cdot z' \left(\frac{N-2X+1}{N} \right) \tag{A9.13}$$

and for the last

$$(N - (N_2 - 1)X) \cdot z \left(\frac{N - (N_2 - 1)X - 1}{N} \right) \cdot$$
$$z' \left(\frac{N - (N_2 - 1)X - 2}{N} \right) \cdot \cdots \cdot z' \left(\frac{N - N_2 X + 1}{N} \right). \tag{A9.14}$$

The total number of configurations, Ω, is the product of these divided by $N_2!$, or

$$\Omega = \left(\frac{z}{z'} \right)^{N_2} \left(\frac{z'}{N} \right)^{(X-1)N_2} \frac{N!}{N_1! \, N_2!}. \tag{A.9.15}$$

The division by $N_2!$ is necessary because the chains once on the lattice are indistinguishable. The entropy is found as $k_B \ln \Omega$. Using Stirling's approximation gives

$$S = k_B [N_2 \ln(z/z') + (X-1)N_2 \ln z' - (X-1)N_2 \ln N + N \ln N$$
$$- N_1 \ln N_1 - N_2 \ln N_2 - N + N_1 + N_2]. \tag{A9.16}$$

The entropy of the disordered undiluted pure polymer is

$$S_0 = k_B [N_2 \ln(z/z') + (X-1)N_2 \ln z' - (X-1)N_2 \ln N_2 X + N_2 X \ln N_2 X$$
$$- N_2 \ln N_2 - N_2 X + N_2]. \tag{A9.17}$$

The entropy of mixing, $S - S_0$ is then

$$\Delta S^m = -R[N_1 \ln(N_1/N) + N_2 \ln(N_2 X/N)], \tag{A9.18}$$

where the units of N_1, N_2, N are now regarded as moles and k_B is replaced by the gas constant.

Appendix A9.3 Cell models for monatomic fluids

Cell models for simple fluids have a long history (Lennard-Jones and Devonshire, 1937, 1938) and were motivated by the desire to find the simplest model that would be explicitly molecular in nature and incorporate temperature and volume. Atomistically, the volume becomes a function of temperature at constant pressure, i.e., thermal expansion occurs, because the potential energy of interaction between molecules is *anharmonic*. That is, the potential function increases less steeply at intermolecular distances larger than the minimum in the energy curve than it does at smaller distances. As temperature increases, thermal vibrational amplitudes increase and, because of the anharmonic effect, average displacements of nearby molecules increase. The cell model incorporates this effect in the simplest possible way. Vibrations are allowed but the approximation is made that each atom in a monatomic fluid (or internally rigid molecule in a more complex fluid) vibrates against its neighbors *fixed at their average positions*. With this assumption, it is a relatively simple matter to evaluate the free energy for an assumed fixed average interatomic (or intermolecular) distance. The central atom or molecule being focused on along with its neighbors constitutes the *cell*, Figure 9.27.

The classical partition function for a set of N indistinguishable particles is given by

$$Q = \Lambda^{-3N} Z(T, V)/N!, \tag{A9.19}$$

where $\Lambda = (h/2\pi m k_{B} T)^{1/2}$, the configurational partition function, Z, is an integration over the system volume, V,

$$Z(T, V) = \int \cdots \int_{V} e^{-U(\mathbf{r})/k_{B}T} \, d\mathbf{r} \tag{A9.20}$$

and the potential energy of the system, $U(\mathbf{r})$, is a function of the position vectors, $\mathbf{r}_1, \mathbf{r}_2, \ldots, \mathbf{r}_i, \ldots, \mathbf{r}_N$. The system energy $U(\mathbf{r})$ is considered to be a sum

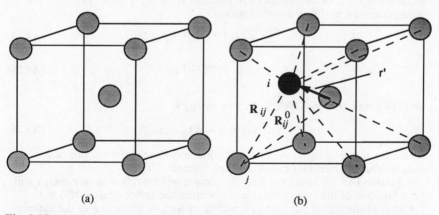

(a)   (b)

Fig. 9.27 (a) In cell theories of liquids, the local structure is considered to have crystalline order. (b) The energy is computed in the approximation that the central particle (dark shading), i, can move about the cell with displacement, $\mathbf{r}'$, away from its rest point while its neighbors remain fixed at their rest points.

of pair interaction potentials, $u(R_{ij})$ that are functions of the interparticle separations, or, $U(\mathbf{r}) = \sum_{i<j} u(R_{ij})$, $R_{ij} = |\mathbf{r}_j - \mathbf{r}_i|$.

In a crystal, the constituent atoms or molecules can be considered to be confined to spaces near their equilibrium positions. In that case, equation (A9.19) becomes

$$Q = \Lambda^{-3N} Z(T, V) \tag{A9.21}$$

with

$$Z(T, V) = \int \cdots \int_{\Delta} e^{-U(\mathbf{r})/k_B T} \, d\mathbf{r}, \tag{A9.22}$$

where the $N!$ is dropped because the localized particles can be considered to be distinguishable and the integration of each $\mathbf{r}_i$ is over a volume, Δ, associated with the cells comprising the crystal. Thus Δ could be taken as over a volume $\sim a^3$ if a is a lattice parameter. Since $U(\mathbf{r}) = \sum_{i<j} u(R_{ij})$ is a coupled function of the particle coordinates, the integration cannot be carried further without simplifying assumptions. In the cell model, the potential experienced by a given particle, i, $\phi(\mathbf{r}_i) = \sum_j u(R_{ij})$ is approximated by limiting the sum over j to near neighbors (or perhaps also including next near neighbors), and, more importantly, by assuming that each neighbor is to be taken at its fixed lattice position. Thus $\phi(\mathbf{r}_i)$ is a function only of the position of particle i and the integrations in equation (A9.22) are decoupled. It is convenient to write $\phi(\mathbf{r}_i)$ as the sum of two terms, one expressing the energy of the static lattice, with the particles at the lattice positions, and the other representing the vibrational energy. If a position vector $\mathbf{r}_i'$ of each particle is now considered to be centered on its static lattice position, i.e., $\mathbf{r}_i' = 0$ is the static lattice position, then

$$\left.\begin{aligned} \phi(\mathbf{r}_i') &= \phi(0) + \psi(\mathbf{r}_i'), \\ \psi(\mathbf{r}_i') &= [\phi(\mathbf{r}_i') - \phi(0)], \end{aligned}\right\} \tag{A9.23}$$

where $\phi(0)$ is the static energy and $\Psi(\mathbf{r}_i')$ is the vibrational energy. The total static energy, U_0, of the system of N particles is $U_0 = N\phi(0)/2$. The configurational partition function is now

$$Z = [e^{-\phi(0)/k_B T}]^{N/2} v_f^N \tag{A9.24}$$

$$v_f = \int_{\Delta} e^{-\Psi(\mathbf{r}')/k_B T} \, d\mathbf{r}' \tag{A9.25}$$

and the configurational Helmholtz free energy is

$$A = N\phi(0)/2 - Nk_B T \ln v_f, \tag{A9.26}$$

where the integration, v_f, over the cell volume Δ has the dimensions of a volume and is sometimes called the 'free volume'.

Equation (A9.26) forms the basis for simple cell models. The free energy will be a function of the system volume, V, because the lattice energy $\phi(0)$ is a function of the lattice parameter, a, which, in turn, is a function of the volume and because the potential $\psi(\mathbf{r}')$, equation (A9.23), in the v_f integration also involves the lattice parameter. When the potentials $\phi(0)$, $\psi(\mathbf{r}')$ are expressed in terms of V through the lattice parameter then the pressure, p, can be found from $p = -(\partial A/\partial V)_T$. An equation of state is the result.

The expression of the potentials in terms of lattice parameter is often based on the Lennard–Jones '6–12' potential,

$$u(R_{ij}) = \varepsilon[(r^*/R_{ij})^{12} - 2(r^*/R_{ij})^6], \tag{A9.27}$$

where ε is the well depth and r^* is the distance at the minimum. The lattice energy is found from $\phi(r'_i = 0) = \sum_j u(R^0_{ij})$, where the sum is over neighbors with values of R^0_{ij} fixed by the lattice parameter. As an example, for the BCC lattice of Figure 9.27, if only near neighbors were included, $R^0_{ij} = 3^{1/2}a/2$. Therefore $\phi(0) = z\varepsilon[(3^{1/2}/2)^{12} - 2(3^{1/2}/2)^6(r^*/a)^6]$ where z is the coordination number and is equal to 8 for BCC. For this example, since there are two particles per cell, the volume per particle, v, is related to a as $v = V/N = a^3/2$. Thus replacing $(r^*/a)^3$ by v^*/v, it results that $\phi(0) = z\varepsilon[(3^{1/2}/2)^{12}(v^*/v)^4 - 2(3^{1/2}/2)^6(v^*/v)^2]$ where $v^* = r^{*3}/2$. In general, for any type of lattice assumed, the lattice energy may be written as

$$\phi(0) = z\varepsilon[A(v^*/v)^4 - 2B(v^*/v)^2], \tag{A9.28}$$

where $v^* = \Gamma r^{*3}$ and Γ, A and B are constants characteristic of the lattice.

The evaluation of the vibrational contribution to Z, i.e., the v_f term in equation (A9.26) requires performing the integration involving the vibrational potential, $\psi(r')$ in equation (A9.25). This can be done directly from $\psi(r') = [\phi(r') - \phi(0)]$, equation (A9.23), by using the Lennard–Jones potential, equation (A9.26), for $u(R_{ij})$ in $\phi(r') = \sum_j u(R_{ij})$ and equation (A9.27) for $\phi(0)$. This requires expressing R_{ij}, the interparticle distance with i off the static position but j on the rest position, in terms of r', the displacement vector of i relative to the static position, and $\mathbf{R}^0_{ij}$, the interparticle vector for the rest positions. This integration has to be carried out numerically (Hirschfelder, Curtiss and Bird, 1954).

It is also possible to approximate this integration. Two methods have been popular. One is to expand equation (A9.23) in a series about $r' = 0$ through harmonic terms. This results in the vibrating center being a harmonic oscillator whose force constant, K, is related to the Lennard–Jones potential constants, ε, r^*. The integration can then be carried out giving v_f as the familiar classical three-dimensional harmonic oscillator partition function $(k_B T/hv)^3$ where $v = (2\pi)^{-1}(K/m)^{1/2}$ and m is the particle mass. The most widely used approximation is based on representing the Lennard–Jones function as a square-well potential (Prigogine, 1957). The depth is given by ε and the repulsive hard-core distance is set at $\sigma \ (= 2^{-1/6}r^*)$, the point where the Lennard–Jones potential crosses zero. On forming the vibration energy, $\phi(\mathbf{r}) - \phi(0)$, $\psi(\mathbf{r})$ becomes a hard-sphere potential depending only on σ. The integration over the cell leads to V_f having the form $v_f = C'(a - \gamma'p)^3$ where C' and γ' are constants characteristic of the lattice. This can be re-written in analogy with equation (A9.28) as

$$v_f = C(v^{1/3} - \gamma v^{*1/3})^3, \tag{A9.29}$$

where C and γ are again constants characteristic of the lattice. Introduction of equations (A9.28) and (A9.29) into equation (A9.26) and finding the pressure from $p = -(\partial A/\partial V)_T$ results in

$$\frac{pv}{k_B T} = \frac{v^{1/3}}{(v^{1/3} - \gamma v^{*1/3})} + \frac{2z\varepsilon}{k_B T}\left[A\left(\frac{v^*}{v}\right)^4 - B\left(\frac{v^*}{v}\right)^2\right].$$

For a hexagonal close packed lattice $\gamma = 2^{-1/6}$, $A = 1.2045$, $B = 1.011$. In terms of the dimensionless *reduced variables*, $\tilde{p}$, $\tilde{v}$, $\tilde{T}$,

$$\frac{\tilde{p}\tilde{v}}{\tilde{T}} = \frac{\tilde{v}^{1/3}}{(\tilde{v}^{1/3} - \gamma)} + \frac{2}{\tilde{T}}[A\tilde{v}^{-4} - B\tilde{v}^{-2}], \qquad (A9.30)$$

where

$$\tilde{v} = v/v^*; \qquad v^* = \Gamma r^*,$$

$$\tilde{T} = T/T^*; \qquad T^* = z\varepsilon/k_B,$$

$$\tilde{p} = p/p^*; \qquad p^* = z\varepsilon/v^* = k_B T^*/v^*.$$

Thus there are two parameters, beyond the constants characteristic of the lattice (Γ, γ, A, B) and k_B, involved. They may be regarded as the potential function constants $z\varepsilon$ and r^* or, alternatively, any two of T^*, v^*, p^*. From the latter point of view, this pair (T^*, v^*, for example) would be adjustable constants determined empirically to fit experimental data.

Nomenclature

a_1, a_2 = activity of component 1 (solvent), 2 (polymer)

A_2, A_3, ... = second, third, ... virial coefficients

B = enthalpy of mixing interaction parameter in a strictly regular solution

C = weight concentration of polymer

G, ΔG, ΔG^m = Gibbs free energy, Gibbs free energy change, change in Gibbs free energy on mixing

H, ΔH, ΔH^m = enthalpy, enthalpy change, change in enthalpy on mixing

N = total number of lattice sites in a lattice model, or,

$\quad = N_1 + N_2 X$ total moles of solvent and polymer bead; $= N_1 + N_2$ for a regular solution where $X = 1$; also number of bonds in a polymer molecule, Sections 9.5, 9.6.

N_1, N_2 = moles of component 1 (solvent), 2 (polymer)

p = pressure

$\mathbf{r}$, $\mathbf{r}_i$ = position vector, position vector of center, i

$\mathbf{r}'$ = position vector relative to the static position in a cell model

r_1, r_2 = the number of lattice sites occupied by a solvent molecule, polymer molecule respectively in equation of state theories

r^* = position of the minimum in the Lennard–Jones potential

R = gas constant, J/(mole K)

S, ΔS, ΔS^m = entropy, entropy change, change in entropy on mixing

v_2^0 = specific volume of pure polymer

V_1^0, V_2^0 = molar volume of pure component 1 (solvent), 2 (polymer)

x_1, x_2 = mole fraction of component 1, 2

X = degree of polymerization; taken as ratio of molar volumes in Flory–Huggins theory, equation (9.24); $= r_2/r_1$ in equation-of-state theories, the ratio of the number of lattice sites occupied by a polymer molecule and a solvent molecule

z = excluded volume parameter; also
= lattice coordinate number in Flory–Huggins and other solution models

$\alpha_{\mathbf{R}}, \alpha_s$ = expansion factor for end-to-end vector, $\mathbf{R}$; radius of gyration, s

β = binary cluster integral for strength of excluded volume interaction

δ_1, δ_2 = solubility parameter of component 1 (solvent), 2 (polymer), equation (9.4)

ε = interaction energy, well-depth in Lennard–Jones potential

$\varepsilon_{11}, \varepsilon_{22}, \varepsilon_{12}$ = interaction energies of 11, 22, 12 type pairs in lattice theory

$\phi(\mathbf{r})$ = potential experienced by a particle in a cell model, a function of position, $\mathbf{r}$

ϕ_1, ϕ_2 = volume fraction of component 1 (solvent), 2 (polymer)

μ_1, μ_2 = chemical potential of component 1 (solvent), 2 (polymer); superscript 0 denotes standard state

Π = osmotic pressure

χ = enthalpy (or, usually, free energy) of mixing interaction parameter divided by RT in a Flory–Huggins solution

$\psi(\mathbf{r}')$ = vibrational energy in a cell model, a function of the displacement from the static position, $\mathbf{r}'$

Ψ = interpenetration parameter expressing the interaction between polymer molecule pairs in dilute solution

ω = exchange interaction energy in lattice theory, $= \varepsilon_{12} - \varepsilon_{11}/2 - \varepsilon_{22}/2$

Problems

9.1 (a) Find an equation for the binodals of a symmetric strictly regular solution by applying equation (9.21) to equation (9.11).

 (b) Verify the relation in Section 9.2 for the CST of a strictly regular solution, $T_c = B/2R$. Apply equation (9.16) to equation (9.11) and notice where the roots on solving for x_1 can no longer be satisfied.

9.2 (a) Calculate the critical solution temperature of n-butane and aniline. Use equation (9.2) for ΔH^m but calculate B from $B \cong V(\delta_1 - \delta_2)^2$ where V is the average of V_1^0 and V_2^0 (101,

92 cm^3/mole respectively), δ_1, δ_2 are 6.70 and 10.70 (cal/cm^3)$^{1/2}$ respectively. Under these assumptions the T_c-relation for a strictly regular solution in (*b*) above applies.

    (*b*) Calculate the composition of the two phases of the above at 0 °C. Use the results of 9.1(*a*). Solve the equation numerically.

9.3 (*a*) From the plots in Figure 9.3 determine the value of the exchange energy parameter, *B*, used in computing the curves.

    (*b*) Do the same thing for the plots in Figure 9.4.

9.4 (*a*) Use the UCST data in Figure 9.8 to generate a plot of T_c^{-1} vs $1/2X + X^{-1/2}$. Calculate X from equation (9.24) using the specific volumes of pure polystyrene (0.95 cm^3/g) and cyclohexane (1.29 cm^3/g) at room temperature. (*b*) From the plot in (*a*) find values for H_χ/R, S_χ/R, equations (9.43) and (9.44). From these find χ_c for each X. Put all of the numbers in a table.

    (*c*) Calculate the Flory–Huggins ϕ_2^c for each X. Compare with the experimental values in a table. Notice that the plots in Figure 9.8 are in weight fraction. Volume fraction can be defined from weight fractions w_1, w_2 and pure component densities, ρ_1, ρ_2 as $\phi_1 = w_1\rho_1^{-1}/(w_1\rho_1^{-1} + w_2\rho_2^{-1})$.

    (*d*) Calculate the Flory–Huggins binodals for one of the X values. Use two temperatures, one 5 °C and one 10 °C, below T_c. Suggestions: (1) First set up the Flory–Huggins free energy function, equation (9.27). At each of the two temperatures, calculate χ for equation (9.27) from equation (9.37) and the H_χ/R, S_χ/R values from part (*b*). (2) The slope of the free energy curve, Figure 9.2, is dG^m/dx_1, an intercept is $\mu_1 - \mu_1^0$. Find an equation for the slope from equation (9.27) and use equation (9.29) for the intercept. Plot both of these on the same graph, different scales, vs ϕ_2. There will be many places where the slope separately has the same value at two compositions, ϕ_2. The same is true for the intercept. Search for two compositions where *both* the slope *and* the intercept have equal values at the *same two values* of ϕ_2. Closely spaced composition points will be necessary in the plots.

9.5 Show mathematically that the term in square brackets multiplying ΔC_P in equation (9.47) is always negative for $T > \theta$ and that if ΔC_P is negative in equation (9.47), then regardless of its magnitude there will according to this equation always be a LCST.

9.6 Find a scaling relation between the osmotic pressure and the correlation length valid in the semidilute concentration region.

References

Abe, A., Jernigan, R. L. & Flory, P. J. (1966) *J. Amer. Chem. Soc.*, **88**, 631.
Adam, M. (1991) *Makromol. Chem. Macromol. Symp.*, **45**, 1.
Ainberg, L. O. (1964) in *Vulcanization of Elastomers* (G. Alliger and I. J. Sjothu, eds.). New York: Reinhold.
Aklonis, J. J., MacKnight, W. J. & Shen, M. (1972) *Introduction to Polymer Viscoelasticity*. New York: Wiley-Interscience.
Alexandrowicz, Z. (1969) *J. Chem. Phys.*, **51**, 561.
Alexandrowicz, Z. (1983) *Phys. Rev. Lett.*, **50**, 736.
Alfrey, Jr., T & Goldfinger, G. (1944) *J. Chem. Phys.*, **12**, 205.
Alfrey, Jr., T. & Price, C. C. (1947) *J. Polym. Sci.*, **2**, 101.
Alfrey, Jr., T. & Young, L. J. (1964) in *Copolymerization* (G. E. Ham, ed.). New York: Wiley.
Allegra, G., Calagaris, M., Randaccio, L. & Moraglio, G. (1973) *Macromolecules*, **6**, 397.
Allen, G. & Bevington, J. C., eds. (1989) *Comprehensive Polymer Science*. 7 Volume Series. Oxford: Pergamon.
Arlman, E. J. & Cossee, P. (1964) *J. Catalysis*, **3**, 99.
Badgley, W. G. & Mark, H. (1949) *High Molecular Weight Compounds (Frontiers in Chemistry)*, Vol. 6 (R. E. Burke and O. Grummitt, eds.). New York: Interscience.
Baher, I., Zuniga, I., Dodge, R. & Mattice, W. L. (1991) *Macromolecules*, **24**, 2986.
Ball, R. C., Doi, M., Edwards, S. F. & Warner, M. (1981) *Polymer*, **22**, 1010.
Bamford, C. H. (1985) in *Alternating Copolymerization* (J. M. G. Cowie, ed.). New York: Plenum.
Barrett, A. J. (1985) *Macromolecules*, **18**, 196.
Bartell, L. S. & Kohl, D. A. (1963) *J. Chem. Phys.*, **39**, 3097.
Barton, J. (1968) *J. Polym. Sci.*, *A-1*, **6**, 1315.
Bawendi, M. G., Freed, K. F. & Mohanty, U. (1986) *J. Chem. Phys.*, **84**, 7036.
Bawendi, M. G., Freed, K. F. & Mohanty, U. (1987) *J. Chem. Phys.*, **87**, 5534.
Bawendi, M. G. & Freed, K. F. (1988) *J. Chem. Phys.*, **88**, 2741.
Beckman, E. J., Porter, R. & Koningsveld, R. (1987) *J. Phys. Chem.*, **91**, 6429.
Beckman, E. J., Koningsveld, R. & Porter, R. (1990) *Macromolecules*, **23**, 2321.
Benoit, H. (1947) *J. Chim. Phys.*, **44**, 18.
Benoit, H., Rempp, P. & Grubisc, Z. (1967) *J. Polym. Sci., Polymn. Lett. Ed.*, **5**, 753.

Benson, S. W. & North, A. M. (1962) *J. Am. Chem. Soc.*, **84**, 935.

Berne, B. J. & Pecora, R. (1976) *Dynamic Light Scattering*. New York: Wiley.

Birshtein, T. M. (1959) *Vyosokomolekul. Soedin.*, **1**, 798.

Birshtein, T. M. & Ptitsyn, O. B. (1959) *Zh. Tekhn. Fiz.*, **29**, 523.

Birshtein, T. M. & Ptitsyn, O. B. (1966) *Conformations of Macromolecules* (Vol. 22 in *High Polymers*). New York: Interscience.

Bishop, M., Clarke, J. H. R., Rey, A. & Freire, J. J. (1991) *J. Chem. Phys.*, **95**, 4589.

Boor, J. (1979) *Ziegler–Natta Catalysts and Polymerizations*. New York: Academic.

Bovey, F. S. (1969) *Polymer Conformation and Configuration*. New York: Academic.

Bovey, F. S. (1982) *Chain Structure and Conformation of Macromolecules*. New York: Academic.

Boyd, R. H. (1968) *J. Chem. Phys.*, **49**, 2574.

Boyd, R. H. & Breitling, S. M. (1972a) *Macromolecules*, **5**, 1.

Boyd, R. H. & Breitling, S. M. (1972b) *Macromolecules*, **5**, 279.

Boyd, R. H., Breitling, S. M. & Mansfield, M. (1973) *AICHE Journal*, **19**, 1016.

Boyd, R. H. & Kesner, L. (1981a) *J. Polym. Sci., Polym. Phys. Ed.*, **19**, 375.

Boyd, R. H. & Kesner, L. (1981b) *J. Polym. Sci., Polym. Phys. Ed.*, **19**, 393.

Boyd, R. H. (1989) *Macromolecules*, **22**, 2477.

Brandrup, J. and Immergut, E. H. (1966) *Polymer Handbook*, 2nd ed. New York: Interscience (pp. II-141 and II-291).

Brandrup, J. & Immergut, E. H. (1989) *Polymer Handbook*, 3rd ed. New York: Wiley-Interscience.

Broadhurst, M. G., Davis, G. T., McKinney, J. E. & Collins, R. E. (1978) *J. Appl. Phys.*, **49**, 4992.

Brotzman, R. W. & Eichinger, B. E. (1982) *Macromolecules*, **15**, 531.

Brotzman, R. W. & Mark, J. E. (1986) *Macromolecules*, **19**, 667.

Burchard, W. (1981) *Pure Appl. Chem.*, **63**, 1519.

Burchard, W. (1982) *Adv. Polym. Sci.*, **44**, 1.

Carazzolo, G. A. (1963) *J. Polymer Sci., Part A*, **1**, 1573.

Carazzolo, G. A. & Mammi, M. (1963) *J. Polymer Sci., Part A*, **1**, 965.

Casassa, E. F. (1976) *J. Polym. Sci. Polym. Symp.*, **54**, 53.

Chang, S. J., McNally, D., Shary-Tehrany, S., Hickey, M. J. & Boyd, R. H. (1970) *J. Amer. Chem. Soc.*, **92**, 3109.

Charlesby, A. & Pinner, S. H. (1959) *Proc. Roy. Soc. (Lond.) A*, **249**, 367.

Chu, B. (1974) *Laser Light Scattering*. New York: Academic.

Ciferri, A. (1961) *Makromol. Chem.*, **43**, 152.

Daoud, M., Cotton, J. P., Famoux, B., Jannink, G., Sema, G., Benoit, H., Duplessix, R., Picot, C. & de Gennes, P.-G. (1975) *Macromolecules*, **8**, 804.

Davis, G. T., Broadhurst, M. G., McKinney, J. E. & Roth, S. C. (1978) *J. Appl. Phys.*, **49**, 4998.

Dayantis, J. & Palierne, J.-F. (1991) *J. Chem. Phys.*, **95**, 6088.

De Boer, A. P. & Pennings, A. J. (1976) *J. Polym. Sci. Polym. Phys. Ed.*, **14**, 187.

de Gennes, P.-G. (1979) *Scaling Concepts in Polymer Physics*. Ithaca, NY: Cornell.

de Pyun, C. W. & Fixman, M. (1966) *J. Chem. Phys.*, **44**, 2107.

Debye, P. (1947) *J. Phys. Chem.*, **51**, 18.

Dee, G. T. & Walsh, D. J. (1988) *Macromolecules*, **21**, 811.

des Cloizeaux, J. (1974) *Phys. Rev.*, **A10**, 1665.

des Cloizeaux, J. (1975) *J. Phys. (Paris)*, **36**, 281.

des Cloizeaux, J. (1980) *J. Phys. (Paris)*, **41**, 223.
des Cloizeaux, J. (1981) *J. Phys. (Paris)*, **42**, 635.
des Cloizeaux, J. & Jannik, G. (1991) *Polymers in Solution*. New York: Oxford/Hanser.
Domb, C. (1963) *J. Chem. Phys.*, **38**, 2957.
Domb, C. & Barrett, A. J. (1976) *Polymer*, **17**, 179.
Dossin, L. M. & Graessley, W. W. (1979) *Macromolecules*, **12**, 123.
Douglas, J. F. & Freed, K. F. (1984) *Macromolecules*, **17**, 1854, 2344.
Douglas, J. F. & McKenna, G. B. (1992) in *Elastomeric Polymer Networks* (J. E. Mark and B. Erman, eds.). Englewood Cliffs: Prentice-Hall.
Dudowicz, J., Freed, K. F. & Madden, W. G. (1990) *Macromolecules*, **23**, 4803.
Dudowicz, J. & Freed, K. F. (1991) *Macromolecules*, **24**, 5076.
Duiser, J. A. & Staverman, J. A. (1965) *The Physics of Non-crystalline Solids* (J. A. Prins, ed.). Amsterdam: North-Holland.
Ebdon, J. R., Towns, C. R. & Dodson, K. J. (1986) *Macromol. Sci., Rev.*, **C26**, 523.
Edwards, S. F. & Stockmayer, W. H. (1973) *Proc. Roy. Soc. (Lond.) A*, **332**, 439.
Edwards, S. F. & Vilgis, T. (1986) *Polymer*, **27**, 483.
Edwards, S. F. & Vilgis, T. (1988) *Rep. Prog. Phys.*, **51**, 243.
Eichinger, B. E. & Flory, P. J. (1968) *Trans. Faraday Soc.*, **64**, 2035, 2053, 2061, 2066.
Eichinger, B. E. (1970) *J. Chem. Phys.*, **53**, 561.
Eichinger, B. E. (1983) *Ann. Rev. Phys. Chem.*, **34**, 359.
Erman, B. & Flory, P. J. (1978) *J. Chem. Phys.*, **68**, 5363.
Erman, B. & Flory, P. J. (1982) *Macromolecules*, **15**, 806.
Erman, B. & Mark, J. E. (1989) *Ann. Rev. Phys. Chem.*, **40**, 351.
Erman, B. & Monnerie, L. (1989) *Macromolecules*, **22**, 3342.
Erman, B. & Mark, J. E. (1992) *Macromolecules*, **25**, 1917.
Ewen, J. A. (1984) *J. Am. Chem. Soc.*, **106**, 6355.
Eyring, H. (1932) *Phys. Rev.*, **39**, 746
Feller, W. (1968) *Introduction to Probability Theory and Its Applications, Vol I*, 3rd edn. New York: Wiley.
Ferry, J. D. (1981) *Viscoelastic Properties of Polymers*, 3rd edn. New York: J. Wiley.
Fisher, M. E. & Sykes, M. F. (1959) *Phys. Rev.*, **114**, 45.
Fisher, M. E. & Hiley, B. J. (1961) *J. Chem. Phys.*, **34**, 1253.
Fisher, M. E. (1966) *J. Chem. Phys.*, **44**, 616.
Fixman, M. (1955) *J. Chem. Phys.*, **23**, 1656.
Flory, P. J. (1936) *J. Am. Chem. Soc.*, **58**, 1877.
Flory, P. J. (1939) *J. Am. Chem. Soc.*, **61**, 3334.
Flory, P. J. (1940a) *J. Am. Chem. Soc.*, **62**, 1561.
Flory, P. J. (1940b) *J. Am. Chem. Soc.*, **62**, 2261.
Flory, P. J. (1942) *J. Chem. Phys.*, **10**, 51.
Flory, P. J. (1943) *J. Am. Chem. Soc.*, **65**, 372.
Flory, P. J. & Rehner, J. (1943a) *J. Chem. Phys.*, **11**, 512.
Flory, P. J. & Rehner, J. (1943b) *J. Chem. Phys.*, **11**, 521.
Flory, P. J. (1946) *Chem. Rev.*, **39**, 137.
Flory, P. J. (1950) *J. Chem. Phys.*, **18**, 108.
Flory, P. J. (1953) *Principles of Polymer Chemistry*. Ithaca, NY: Cornell.
Flory, P. J., Hoeve, C. A. J. & Ciferri, A. (1959) *J. Polym. Sci.*, **34**, 337.
Flory, P. J. (1964) *Proc. Nat. Acad. Sci. (USA)*, **51**, 1060.
Flory, P. J., Orwoll, R. A. & Vrij, A. (1964) *J. Am. Chem. Soc.*, **86**, 3507, 3515.

Flory, P. J. (1965) *J. Am. Chem. Soc.*, **87**, 1833.
Flory, P. J. & Jernigan, R. L. (1965) *J. Chem. Phys.*, **42**, 3509.
Flory, P. J., Mark, J. E. & Abe, A. (1966) *J. Am. Chem. Soc.*, **88**, 639.
Flory, P. J. (1969) *Statistical Mechanics of Chain Molecules*. New York: Interscience.
Flory, P. J. (1970) *Disc. Farad. Soc.*, **49**, 7.
Flory, P. J. (1976) *Proc. Roy. Soc. (Lond.) A*, **351**, 351.
Flory, P. J. (1977) *J. Chem. Phys.*, **66**, 5720.
Flory, P. J. & Pak, H. (1979) *J. Polym. Sci. Polym. Phys. Ed.*, **17**, 1845.
Flory, P. J. & Erman, B. (1982) *Macromolecules*, **15**, 800.
Freed, K. F. (1985) *J. Phys. A*, **18**, 871.
Freed, K. F. (1989) *Renormalization Group Theory of Macromolecules*. New York: Wiley.
Fujita, H. (1990) *Polymer Solutions*. Amsterdam: Elsevier.
Gaylord, R. J. (1982) *Polym. Bull.*, **8**, 325.
Gaylord, R. J. & Douglas, J. F. (1987) *Polym. Bull.*, **18**, 347.
Gaylord, R. J. & Douglas, J. F. (1990) *Polym. Bull.*, **23**, 529.
Gee, G. (1944) *Trans. Farad. Soc.*, **40**, 261.
Gordon, M. & Ross-Murphy, S. B. (1975) *Pure Appl. Chem.*, **43**, 1.
Gordon, M. & Ross-Murphy, S. B. (1979) *J. Phys.*, **A12**, L155.
Gordon, M. & Torkington, J. A. (1981) *Pure Appl. Chem.*, **53**, 1461.
Gottlieb, M., Macosko, C. W., Benjamin, G. S., Meyers, K. O. & Merrill, E. W. (1981) *Macromolecules*, **14**, 1039.
Gottlieb, M. & Gaylord, R. J. (1983) *Polymer*, **24**, 1644.
Gottlieb, M. & Gaylord, R. J. (1984) *Macromolecules*, **17**, 2024.
Gottlieb, M. & Gaylord, R. J. (1987) *Macromolecules*, **20**, 130.
Gough, J. (1805) *Mem. Lit. & Phil. Soc. (Manchester)*, **6**, 288.
Graessley, W. W. (1974) *Adv. Polym. Sci.*, **16**, 1.
Graessley, W. W. (1975a) *Macromolecules*, **8**, 186.
Graessley, W. W. (1975b) *Macromolecules*, **8**, 865.
Guth, E. & Mark, H. (1934) *Monats. Chem.*, **65**, 93.
Guth, E. & James, H. M. (1941) *Ind. Eng. Chem.*, **33**, 625.
Guttman, J. Y. & Guillet, J. E. (1970) *ACS Org. Coat. Prepr.*, **30**, 177.
Ham, G. E., ed. (1967) *Vinyl Polymerization*. New York: Dekker.
Harpell, G. & Walrod, D. H. (1973) *Rubber Chem. Tech.*, **46**, 1007.
Heinrich, G., Straube, E. & Helmis, G. (1988) *Adv. Polym. Sci.*, **85**, 33.
Higgs, P. G. & Gaylord, R. J. (1990) *Polymer*, **31**, 70.
Hildebrand, J. H. & Scott, R. L. (1950) *Solubility of Non-Electrolytes*, 3rd edn. New York: Reinhold.
Hirschfelder, J. O., Curtiss, C. F. & Bird, R. B. (1954) *Molecular Theory of Gases and Liquids*. New York: Wiley.
Hoeve, C. A. J. (1960) *J. Chem. Phys.*, **32**, 888.
Hogan, J. P. & Banks, R. L. US Patent 2825721.
Hopfinger, A. J. (1973) *Conformational Properties Of Macromolecules*. New York: Academic.
Huber, K. and Stockmayer, W. H. (1987) *Macromolecules*, **20**, 1400.
Huggins, M. L. (1942a) *Ann NY Acad. Sci.*, **41**, 1.
Huggins, M. L. (1942b) *J. Phys. Chem.*, **46**, 151.
Huggins, M. L. (1942c) *J. Am. Chem. Soc.*, **64**, 1712.
Huggins, M. L. (1945) *J. Chem. Phys.*, **13**, 37.
Imamura, Y. (1955) *J. Chem. Soc. (Japan)*, **76**, 217.
Irvine, P. & Gordon, M. (1980) *Macromolecules*, **13**, 761.

James, H. M. & Guth, E. (1942) *Ind. Eng. Chem.*, **34**, 1365.

James, H. M. & Guth, E. (1943a) *J. Chem. Phys.*, **11**, 455.

James, H. M. & Guth, E. (1943b) *J. Chem. Phys.*, **11**, 470.

James, H. M. (1947) *J. Chem. Phys.*, **15**, 651.

James, H. M. & Guth, E. (1947) *J. Chem. Phys.*, **15**, 669.

Kajiwara, K. & Gordon, M. (1973) *J. Chem. Phys.*, **59**, 3623.

Kaminsky, W., Kulper, K., Brintsinger, H. H. & Wild, F. R. W. P. (1985) *Angew. Chem.*, **97**, 507.

Kaminsky, W. & Sinn, H., eds. (1988) *Transition Metals and Organometallics as Catalysts*. Berlin: Springer-Verlag.

Karakelle, M. (1986) PhD dissertation. University of Utah.

Keii, T. (1972) *Kinetics of Ziegler–Natta Polymerization*. London: Chapman & Hall.

Keller, A. & Patel, G. N. (1973) *J. Polym. Sci., Polym. Lett. Ed.*, **11**, 737.

Kemp, J. D. & Pitzer, K. S. (1936) *J. Chem. Phys.*, **4**, 749.

Kemp, J. D. & Pitzer, K. S. (1937) *J. Am. Chem. Soc.*, **59**, 276.

Kennedy, J. P. (1975) *Cationic Polymerization of Polyolefins: A Critical Inventory*. New York: Wiley.

Kennedy, J. P. & Chou, R. T. (1982) *J. Macromol. Sci., Chem.*, **A18**, 17.

Kennedy, J. P. & Marechal, E. (1982) *Carbocationic Polymerization*. New York: Wiley.

King, J. S., Boyer, W., Wignall, G. D. & Ullman, R. (1985) *Macromolecules*, **18**, 709.

Kittel, C. (1956) *Introduction to Solid State Physics*, 2nd edn. New York: Wiley.

Koningsveld, R., Kleintjens, L. A. & Leblans-Vinckl, A. (1987) *J. Phys. Chem.*, **91**, 6423.

Kotera, A., Shima, M., Fujisaki, N., Kobayahsi, T. (1962) *Bull. Chem. Soc. Japan*, **35**, 1117.

Kramers, H. A. & Wannier, G. (1941) *Phys. Rev.*, **60**, 252.

Kranbuehl, D. E. & Verdier, P. H. (1992) *Macromolecules*, **25**, 2557.

Kuhn, H. (1947) *J. Chem. Phys.*, **15**, 843.

Kuhn, W. (1934) *Kolloid Z.*, **68**, 2.

Kuhn, W. (1936) *Kolloid Z.*, **76**, 258.

Kuhn, W. & Gruen, F. (1942) *Kolloid Z.*, **101**, 248.

Langley, N. R. (1968) *Macromolecules*, **1**, 348.

Lax, M., Barrett, A. J. & Domb, C. (1978) *J. Phys. A*, **11**, 361.

le Guillou, J. C., Zinn-Justin, J. (1977) *Phys. Rev. Lett.*, **39**, 95.

LeFevre, R. J. W. & Sundarum, K. M. S. (1962) *J. Chem. Soc.*, 1494.

LeFevre, C. G., LeFevre, R. J. W. & Parkins, G. M. (1960) *J. Chem. Soc.*, 1814.

Lennard–Jones, J. E. & Devonshire, A. F. (1937) *Proc. Roy. Soc. (Lond.) A*, **163**, 53.

Lennard–Jones, J. E. & Devonshire, A. F. (1938) *Proc. Roy. Soc. (Lond.) A*, **165**, 1.

Lenz, R. W. (1967) *Organic Chemistry of Synthetic High Polymers*. New York: Interscience.

Lifson, S. (1959) *J. Chem. Phys.*, **30**, 964.

Lovinger, A. (1982) in *Developments in Crystalline Polymers* (D. C. Bassett, ed.). London: Applied Science.

Margerison, D. & East, G. C. (1967) *Introduction to Polymer Chemistry*. London: Pergamon.

Mark, H. & Whitby, G. S., eds. (1940) *Collected Papers of Wallace Hume*

Carothers on High Polymeric Substances. H. (Vol. I in *High Polymers*, H. Mark, ed.) New York: Interscience.

Mark, J. E. (1972a) *J. Chem. Phys.*, **56**, 451.

Mark, J. E. (1972b) *J. Chem. Phys.*, **56**, 458.

Mark, J. E. (1973) *Rubber. Chem. Tech.*, **46**, 593.

Mark, J. E. (1985) *Accts. Chem. Res.*, **18**, 202.

Mark, J. E. & Erman, B. (1988) *Rubber-like Elasticity. A Molecular Primer.* New York: Wiley-Interscience.

Mark, J. E. & Erman, N., eds. (1992) *Elastomeric Polymer Networks.* Englewood Cliffs: Prentice-Hall.

Matsumoto, M. & Ohyanagi, Y. (1961) *J. Polym. Sci.*, **50**, S1.

Matsuo, K. & Stockmayer, W. H. (1975) *Macromolecules*, **8**, 660.

Mayo, F. R. & Lewis, F. M. (1944) *J. Am. Chem. Soc.*, **66**, 1594.

McKenna, G. B., Flynn, K. M. & Chen, Y.-H. (1988) *Polym. Commun.*, **29**, 272.

McKenna, G. B., Flynn, K. M. & Chen, Y.-H. (1989) *Macromolecules*, **22**, 4507.

McKenna, G. B., Douglas, J. F., Flynn, K. M. & Chen, Y.-H. (1991) *Polymer*, **32**, 2128.

McKenzie, D. S. (1976) *Phys. Rep.*, **27C**, 35.

McKenzie, D. S. & Moore, M. A. (1971) *J. Phys.*, A4, L82.

Memmler, K. L. (1934) *The Science of Rubber*, translated by Dunbrook, R. F. and Morris, V. N. New York: Reinhold.

Meyer, K. H., von Susich, G. & Valko, E. (1932) *Kolloid Z.*, **59**, 208.

Miller, R. L. (1989) in Brandrup, J. & Immergut, E. H. *Polymer Handbook*, 3rd edn. New York: Wiley-Interscience.

Mooney, M. (1940) *J. Appl. Phys.*, **11**, 582.

Mooney, M. (1948) *J. Appl. Phys.*, **19**, 434.

Morawetz, H. (1975) *Macromolecules in Solution* (Volume 21 in *High Polymers*), 2nd edn. New York: Wiley-Interscience.

Morawetz, H. (1965) *Polymers – The Origins and Growth of a Science.* New York: Wiley-Interscience.

Morgan, P. W. (1979) *Chem. Tech.*, **9**, 316.

Morgan, P. W. (1985) *J. Polym. Sci., Polym. Symp.*, **72**, 27.

Morrison, R. T. & Boyd, R. N. (1966) *Organic Chemistry*, 2nd edn. Boston: Allyn and Bacon.

Muthukumar, M. & Nickel, B. G. (1984) *J. Chem. Phys.*, **80**, 5839.

Nagai, K. J. (1959) *J. Chem. Phys.*, **31**, 1169.

Nakajima, A. & Kato, K. (1966) *Makromol. Chem.*, **95**, 52.

Nairn, J. A. (1990) *Lattice 4.0^{TM}, A Macintosh Application.* Salt Lake City: Dept. Materials Sci. & Eng. U. Utah.

Natta, G. (1960) *J. Polym. Sci.*, **48**, 219.

Natta, G. & Corradini, P. (1960) *Suppl. Nuovo Cimento*, **15**, 40.

Natta, G., Corradini, P. & Ganis, P. (1960) *Makromol. Chem.*, **39**, 238.

Natta, G., Corradini, P. & Ganis, P. (1962) *J. Polym. Sci.*, **58**, 1191.

Natta, G. & Pasquon, I. (1959) in *Advances in Catalysis and Related Subjects* (D. D. Eley, P. W. Sellwood and B. Weisz, eds.), Vol. II, p. 2. New York: Academic Press.

Natta, G., Pasquon, I., Corradini, P., Peraldo, M., Pegoraro, M. & Zambelli, A. (1960) *Rend. Acc. Naz. Lincei*, **28**, 539.

Natta, G., Peraldo, M. & Allegra, G. (1964) *Makromol. Chem.*, **75**, 215.

Natta, G. & Danusso, F., eds. (1967) *Stereoregular Polymers and Stereospecific Polymerizations.* New York: Pergamon.

Nemirovsky, A. M., Bawendi, M. G. & Freed, K. F. (1987) *J. Chem. Phys.*, **87**, 7272.

Noda, I., Kato, N., Kitano, T. & Nagasawa, M. (1981) *Macromolecules*, **14**, 668.

North, A. M. & Phillips, P. J. (1967) *Trans. Farad. Soc.*, **63**, 1537.

Odian, G. (1970) *Principles of Polymerization*. New York: McGraw-Hill.

Odian, G. (1991) *Principles of Polymerization*, 3rd edn. New York: McGraw-Hill.

Oka, S. (1942) *Proc. Phys. Math. Soc. Japan*, **24**, 657.

Overberger, C. G. & Bonsignore, P. V. (1958) *J. Am. Chem. Soc.*, **80**, 5427.

Panayiotou, C. & Vera, J. H. (1980) *Fluid Phase Equil.*, **5**, 55.

Patel, G. N. & Keller, A. (1975) *J. Polym. Sci., Polym. Phys. Ed.*, **13**, 323.

Pearson, D. S. & Graessley, W. W. (1978) *Macromolecules*, **11**, 528.

Pearson, D. S. & Graessley, W. W. (1980) *Macromolecules*, **13**, 1001.

Pepper, D. C. (1954) *Quart. Rev.*, **8**, 88.

Person, W. B. & Pimentel, G. C. (1953) *J. Am. Chem. Soc.*, **75**, 532.

Pesci, A. I. & Freed, K. F. (1989) *J. Chem. Phys.*, **90**, 2003.

Pitzer, K. S. (1940) *J. Chem. Phys.*, **8**, 711.

Plesch, P. H., ed. (1963) *The Chemistry of Cationic Polymerization*. New York: Macmillan.

Porod, G. (1949) *Monatsh. Chem.*, **80**, 251.

Porter, C. H. & Boyd, R. H. (1971) *Macromolecules*, **54**, 589.

Prigogine, I., Trappeniers, N. & Mathot, V. (1953a) *Disc. Farad. Soc.*, **15**, 93.

Prigogine, I., Trappeniers, N. & Mathot, V. (1953b) *J. Chem. Phys.*, **21**, 559.

Prigogine, I. (1957) *The Molecular Theory of Solutions*. Amsterdam: North-Holland.

Pryor, W. A. (1966) *Free Radicals*. New York: McGraw-Hill.

Qian, C., Mumby, S. J. & Eichinger, B. E. (1991) *Macromolecules*, **24**, 1655.

Queslel, J. P. & Mark, J. E. (1985) *Adv. Polym. Sci.*, **71**, 229.

Rabek, J. (1980) *Experimental Methods in Polymer Chemistry*. New York: Wiley-Interscience.

Rayleigh (Lord) (J. W. Strutt) (1871) *Phil. Mag.*, **41**, 107, 274, 447.

Rayleigh (Lord) (J. W. Strutt) (1899) *Phil. Mag.*, **47**, 375.

Rayleigh (Lord) (J. W. Strutt) (1919) *Phil. Mag.*, **37**(6), 321.

Renuncio, J. A. R. & Prausnitz, J. M. (1976) *Macromolecules*, **9**, 895.

Rivlin, R. S. (1948) *Phil. Trans. Roy. Soc. London Ser. A*, **241**, 379.

Rivlin, R. S. & Saunders, D. W. (1951) *Proc. Roy. Soc. (Lond.) A*, **243**, 251.

Rodgers, P. A. & Sanchez, I. C. (1990) *Pure Appl. Chem.*, **62**, 2107.

Rodgers, P. A. & Sanchez, I. C. (1993) *J. Polym. Sci., Part B, Polym. Phys.*, **31**, 273.

Rodriguez, L. A. M. & Gabant, J. A. (1963) *J. Polym. Sci.*, **C4**, 125.

Rodriguez, L. A. M. & Gabant, J. A. (1966) *J. Polym. Sci. A1*, **4**, 1971.

Ronca, G. & Allegra, G. (1975) *J. Chem. Phys.*, **63**, 4990.

Saeki, S., Kuwahara, S., Konno, S. & Kaneko, M. (1975) *Macromolecules*, **6**, 246.

Sanchez, I. C. & Lacombe, R. H. (1976) *J. Phys. Chem.*, **80**, 2352, 2568.

Sanchez, I. C. & Lacombe, R. H. (1977) *J. Polym. Sci. Polym. Lett. Ed.*, **15**, 71.

Sanchez, I. C. & Lacombe, R. H. (1978) *Macromolecules*, **11**, 1145.

Sanchez, I. C. & Balazs, A. C. (1989) *Macromolecules*, **22**, 2325.

Sato, M., Koshiishi, Y. & Asahina, M. (1963) *J. Polym. Sci., Polym. Lett. Ed.*, **1**, 233.

Saville, B. & Watson, A. A. (1976) *Rubber Chem. Tech.*, **40**, 100.

Schleyer, P. V. R., Williams, J. E. & Blanchard, K. R. (1970) *J. Amer. Chem. Soc.*, **92**, 2377.

Schmitz, K. S. (1990) *An Introduction to Dynamic Light Scattering by Macromolecules*. New York: Academic.

Schroder, E., Muller, G., and Arndt, K.-F. (1989) *Polymer Characterization* (rev. edition). Oxford: Oxford University Press.

Schultz, A. R. & Flory, P. J. (1952) *J. Am. Chem. Soc.*, **74**, 4760.

Schulz, G. V. (1939) *Z. Physik. Chem.*, **B44**, 227.

Schulz, G. V. & Doll, H. (1952) *Z. Elektrochem.*, **56**, 248.

Schulz, G. V. (1956) *Z. Physik. Chem.*, **8**, 290.

Simha, R. & Branson, H. (1944) *J. Chem. Phys.*, **12**, 253.

Smith, G. D. & Boyd, R. H. (1991) *Macromolecules*, **24**, 2731.

Smith, W. V. & Ewart, R. H. (1948) *J. Chem. Phys.*, **16**, 592.

Sheppard, N. & Szasz, G. J. (1949) *J. Chem. Phys.*, **17**, 86.

Sorensen, R. A., Liau, W. B., Kesner, L. & Boyd, R. H. (1988) *Macromolecules*, **21**, 200.

Starkweather, Jr., H. W. & Boyd, R. H. (1960) *J. Phys. Chem.*, **64**, 410.

Stauffer, D., Coniglio, A. & Adam, M. (1982) *Adv. Polym. Sci.*, **44**, 104.

Stauffer, D. & Aharony, A. (1991) *Introduction to Percolation Theory*. Bristol: Taylor and Francis.

Staverman, A. J. (1982) *Adv. Polym. Sci.*, **44**, 74.

Stegen, G. E. & Boyd, R. H. (1978) *Polym. Prepr. Am. Chem. Soc. Div. Polym. Chem.*, **19**(1), 595.

Stevens, M. P. (1990) *Polymer Chemistry*, 2nd edn. New York, Oxford: Oxford.

Stockmayer, W. H. (1943) *J. Chem. Phys.*, **11**, 45.

Stockmayer, W. H. (1944) *J. Chem. Phys.*, **12**, 125.

Stockmayer, W. H. (1952) *J. Polym. Sci.*, **9**, 69; see also (1954) *J. Polym. Sci.*, **11**, 424.

Stockmayer, W. H. (1977) *Brit. Polym. J.*, **9**, 89.

Strong, K. A. & Brugger, R. M. (1967) *J. Chem. Phys.*, **47**, 421.

Suter, U. W. & Flory, P. J. (1975) *Macromolecules*, **8**, 765.

Suzuki, H. (1982) *Brit. Polym. J.*, **14**, 137.

Szasz, G. F., Sheppard, N. & Rank, D. H. (1948) *J. Chem. Phys.*, **16**, 704.

Tadokoro, H., Yasumoto, T., Murahashi, S. & Nitta, I. (1960) *J. Polymer Sci.*, **44**, 266.

Tadokoro, H. (1979) *Structure of Crystalline Polymers*. New York: Wiley-Interscience.

Takeda, M., Imamura, Y., Okamura, S. & Higashimura, T. (1960) *J. Chem. Phys.*, **33**, 631.

Takehashi, Y. & Tadokoro, H. (1979) *J. Polym. Sci., Polym. Phys. Ed.*, **17**, 123.

Tanaka, G. & Solc, K. (1982) *Macromolecules*, **15**, 791.

Tarazona, M. P. & Saiz, E. (1983) *Macromolecules*, **16**, 1128.

Taylor, W. J. (1947) *J. Chem. Phys.*, **15**, 412.

Taylor, W. J. (1948) *J. Chem. Phys.*, **16**, 257.

Thirion, P. & Weil, T. (1984) *Polymer*, **25**, 609.

Treloar, L. R. G. (1943) *Trans. Faraday Soc.*, **34**, 36.

Treloar, L. R. G. (1944) *Trans. Faraday Soc.*, **40**, 59.

Treloar, L. R. G. (1946) *Trans. Faraday Soc.*, **42**, 77.

Treloar, L. R. G. (1975) *The Physics of Rubber Elasticity*, 3rd edn. Oxford: Clarendon.

Trommsdorf, E., Kohle, H. & Lagally, P. (1948) *Makromol. Chem.*, **1**, 169.

Uchida, T., Kurita, Y. & Kubo, M. (1956) *J. Polymer Sci.*, **19**, 365.

Usami, T., Gotoh, Y. & Takayama, S. (1986) *Macromolecules*, **19**, 2722.

Valanis, K. C. & Landel, R. F. (1967) *J. Appl. Phys.*, **38**, 2997.

Verma, A. L., Murphy, W. R. & Berstein, H. J. (1974) *J. Chem. Phys.*, **60**, 1540.
Volkenstein, M. V. (1963) *Configuration Statistics of Polymeric Chains*. New York: Interscience.
Wall, F. T. (1942) *J. Chem. Phys.*, **10**, 132.
Wall, F. T. (1943) *J. Chem. Phys.*, **11**, 527.
Wall, F. T. (1944) *J. Am. Chem. Soc.*, **66**, 2050.
Wall, F. T. & Flory, P. J. (1951) *J. Chem. Phys.*, **19**, 1435.
Wall, F. T., Hiller, L. A. & Wheeler, D. J. (1954) *J. Chem. Phys.*, **22**, 1036.
Wall, F. T., Widmer, S. & Gans, P. J. (1963) in *Methods of Computational Physics*, Vol. 1. New York: Academic.
Walsh, D. J. & Dee, G. T. (1989) *J. Supercrit. Fluids*, **2**, 57.
Weber, L. E. (1926) *The Chemistry of Rubber Manufacture*. London: Griffin.
Weiss, S. & Leroi, G. E. (1968) *J. Chem. Phys.*, **48**, 962.
Wilson, E. B. (1962) *Pure Appl. Chem.*, **41**, 1.
Witt, D. R. (1974) in *Reactivity, Mechanism and Structure in Polymer Chemistry* (A. D. Jenkins and A. Ledwith, eds.). New York: Wiley-Interscience.
Witten, T. & Schaefer, L. (1978) *J. Phys. A*, **11**, 1843.
Yamakawa, H., Aoki, A. & Tanaka, G. (1966) *J. Chem. Phys.*, **45**, 1938.
Yamakawa, H. & Tanaka, G. (1967) *J. Chem. Phys.*, **47**, 3991.
Yamakawa, H. (1971) *Modern Theory of Polymer Solutions*. New York: Harper & Row.
Yamakawa, H. (1992) *Macromolecules*, **25**, 1912.
Yoon, D. Y., Suter, U., Sundararajan, P. R. & Flory, P. J. (1975) *Macromolecules*, **8**, 784.
Ziegler, K., Holzkamp, E., Breil, H. & Martin, H. (1955) *Angew. Chem.*, **67**, 541.
Zimm, B. H. (1946) *J. Chem. Phys.*, **14**, 164.
Zimm, B. H. & Myerson, I. (1946) *J. Am. Chem. Soc.*, **68**, 911.
Zimm, B. H. (1948) *J. Chem. Phys.*, **16**, 1099.
Zimm, B. H., Stockmayer, W. H. & Fixman, M. (1953) *J. Chem. Phys.*, **21**, 1716.

Index